Wave F

WAVE FORMS

A Natural Syntax for Rhythmic Language

JAMES H. BUNN

STANFORD UNIVERSITY PRESS

Stanford, California

2002

Stanford University Press
Stanford, California
© 2002 by the Board of Trustees of the
Leland Stanford Junior University

Printed in the United States of America
on acid-free, archival-quality paper

Library of Congress Cataloging-in-Publication Data

Bunn, James H.
    Wave forms : a natural syntax for rhythmic language / James H. Bunn.
    p. cm.
    Includes bibliographical references and index.
    ISBN 0-8047-4178-6 (alk. paper) -- ISBN 0-8047-4507-2 (pbk : alk paper)
    1. Grammar, Comparative and general--Syntax. 2. Language and
    languages--Rhythm. I. Title.

P291 .B86 2002
415 21

                                                            2001057653

Typeset by Publication Services in 10.5/13 Minion

Original Printing 2002
Last figure below indicates year of this printing:
11   10   09   08   07   06   05   04   03   02

CONTENTS

ILLUSTRATIONS

PREFACE

The physical trope most often used in this century is perhaps one representing a continuum between space and time as a series of curvilinear waves oscillating in nature. We are spatial creatures, with bilaterally symmetrical arms and legs that rhythmically wave us over the ground. The rhythm of walking is truly the counterrhythm of arms and legs alternating in natural arcs of S-shaped curves. But in space we are haunted by time, by what is past, passing, and proleptic. Time, like space, is an abstraction from our feelings of fleeting impressions, for which we invent grasping metaphors that have been transferred from feelings of matter in forceful motion. We grasp for staying power, and we clutch at straws in the wind. We seek a continuum. We desire equilibrium in space and time.

Our minds seek bodily homeostasis, and we express that need in moving proportions of change and order. A bodily need for balanced energy is, I think, the reason why we use these S-curved tropes as compensating metaphors that stand for a desire to abide rhythmically in motion. If metabolic exchange, in which we draw energy from the environment in exchange for waste, is the basic sign of life, then it may be that other forms of symbolic exchange derive from this lifelong urge. We imaginatively transfer material from the physical world to use as bolstering metaphors for living in the time of our transit from life through generation into decline. However that may be, we surely use S-curved frequencies of sound waves and light waves—the main transmitters of our saying and seeing—as constant proportions for our experiences of living rhythmically in scale. Just as sunlight is the main energy source for metabolic exchange among living creatures, so its constant speed of about 186,000 miles per second is the constant measure for all other frequencies along the electromagnetic spectrum. The gravitational power of the sun, as we shall see, is the prime shaper of bilateral symmetry among animals. So the periodic intervals of the sun, its metaphorical steps and stations and other forms of orienting by symmetrical directions in space and time, will be the subject of Chapter 3.

In this book then, I pursue a curvilinear wavelike form that serves as a primary archetype for generating patterns of writing and speaking across the arts and sciences. I think of it as a helical model for describing the shapes of in*form*ation. I shall define this relatively unfashionable term *archetype* in the Introduction. For now, I mean by *archetype* not a Jungian form coming out of a collective unconscious mind but a consciously understood model of a way in which the world is seen to be shaped. In this sense an archetype is a world model. For ages the archetype of a wavelike form has been used by artists, poets, natural philosophers, and scientists to measure the moving proportions of wind and water, light, and sound. Thus, for my purposes, the early use of a wavy form is an archetype of prescience, of foresight about the structures of the world, not just a prescientific model of ill-understood significance.

What is urgent about my thesis? I seek to reshape a theory of natural language and help restore it for literary studies. As always when elemental issues are at stake, poetry is featured. Natural-language theory is common among linguists and scientists untroubled by the hermeneutic circle of Continental theory. I do not intend to review that debate; instead I start with the physical theory of wave frequency. But because I have studied Continental linguistics and hermeneutics with respect and appreciation, I want to bridge some of the differences and try a different point of departure.

A serviceable starting point for any theory of natural language lies in the physical carrier of a message. The carrier is always in the form of a wave, such as a light or sound wave. There is nothing scientifically controversial about the significance of wavelike forms. Nowadays, concepts of energy, mass, and matter, as well as the elements of the periodic table, are all gathered into discourse on electromagnetic waves. Like fish in water or birds in air, we live within a total environment of electromagnetic waves, though we are often unaware of them. Thus the archetype of a moving wave-form can help to resolve one of the more contested issues in language theory. In my approach, a continuously recurring wave sequence is the helical carrier for generating and measuring the syntactic forms of a language. I would like to explore the idea of patterned recurrence as the basis for a kind of primal thinking in frequencied rhymes. A moment's thought will suggest that a helix, however small, is a beginning for any model of recursion, that is, for any model that appears to work backward and forward in a twist.

In the Introduction I discuss the idea that waves carry all messages to the brain in certain significant shapes. Though these waves may seem to be formless or inchoate, they already have characteristic form. If these waves have significant shapes as they are transmitted to the brain, then those wave shapes constitute a significant form that the brain processes and transforms into sound images or light images. Curvilinear wave forms can be seen to embody transformations that compose an a priori syntax common to languages. Later, I shall define *syntax* as a pattern of physical transformations, derived from the form of wave, that inheres in all speech acts. But I want to reserve for *grammar* the compositional orderings that are specific to different cultures.

This approach retains the generative elements of Universal Grammar that are so controversial in current linguistic theory but not so much discussed in most literary circles. For instance, the origin of languages, and the origin of species, and the origin and generation of computational language have been brought together in Daniel C. Dennett's engagingly polemical book, *Darwin's Dangerous Idea: Evolution and the Meanings of Life.*[1] In the Introduction I shall discuss this book further. I mention it here not only because it argues against the much contested issue of Universal Grammar but also because Dennett uses a graphic pair of metaphors—cranes versus sky hooks—by which to illustrate the difference between his approach and that of Noam Chomsky and Stephen Jay Gould. Dennett compares his evolutionary approach to the workaday crane, an honest lever of pulleys that he enjoys watching at construction sites. The crane's sequence of lifts represents Dennett's algorithmic approach to evolution and the constructive development of language processes that occur one step at a time. But to his way of thinking, Chomsky and Gould are hopeful believers who think that the origin of language is attached to a nonexistent sky hook, a "mind-first" power from which hangs a preexistent grammar as an ideal form (*DDI*, 76). Their more controversial idea is that language, because it is a set of innate patterning rules, could not have evolved in a typical Darwinian process but instead followed certain other adaptations of growth and form (*DDI*, 391). *Skyhook* was originally aeronautical slang for the hopeful impossibility of keeping an airplane aloft beyond its capabilities. Although this excursus into evolution may seem far from the topic of a natural syntax of poetic form, it is not, for I want to substitute a different kind of hypothetical model, one that artists, poets and philosophers have employed and studied for ages. For Dennett, skyhooks and cranes are mere metaphors, but in this book the curvilinear wave serves as an archetype for a primary forming idea of patterns in nature and in culture, in art and science.

In the Introduction I shall try to show where this kind of model came from, so that I shall not be accused of privileging a Platonic form as archetype. There I introduce the form of a wave that carries a message, and I describe a biologist's "old sequence" of coded instructions that propels animal bodies through water and then over land in an alternating pattern of strokes in the form of a periodic wave action. I argue that the old sequence of coded neural instructions will suffice for a biological theory of natural syntax. My thesis then is essentially a body-based theory of linguistic cognition.

In Chapter 1, "Art / Frequency," I set forth the idea that a carrier wave is usually the suppressed part of any sign in a communication signal, and I introduce an aesthetic theory of the sublimated carrier wave. In Chapter 2 I study some archaic examples of natural languages and wave theories. But since my thesis depends upon the anatomy of bodies and their relations to the forms of waves that propel them, I continuously cite examples from primal cultures that interrelate archetypal body symmetries and wave forms. This symmetrical body plan is the main burden of Chapter 3. There I review some symmetry theory in order to describe the basic transformations of wave forms. I show that all one needs to begin to generate the elements of a common natural syntax are a few transformations—translation, rotation, and twist—and that these operations can be derived from wave forms. In Chapter 4, I study the idea of natural syntax as a set of physical transformations that affect grammatical constructions. Natural syntax is primarily a coded set of a few transformations that coordinate mental perceptions with body movements in action. In Chapter 5, I suggest that in poetry the physical nature of language, its rhymes and alliterations, specifically governs grammatical composition.

Why should literary and artistic people interest themselves in the sometimes recondite theory of symmetry? In every art form one finds a rhythmic pattern as a base. These patterns, though formal, are everywhere evidence of material in action. Principles of symmetry provide a way of explaining how aesthetic patterns are enactments of the very principles that structure the universe in rhythmic patterns. Every artwork, whatever its nature, is constructed of materials that make the patterns develop at the same deep level as the laws of physics and biology. Perhaps the most important thesis is that the principles of symmetry can help explain the ways that nature distributes patterns as *stablizing* structures. If symmetry conserves structures in rhythmic patterns of material, works of art also should enact those same kinds of harmonic principles but in wonderfully strange and sometimes discordant harmonies of form. So a fair answer to the question is, I believe, that symmetry theory can explain why the arts are not just an "add-on," but that they demonstrate in different media and by different enactments the ways that the world works, moves, and stabilizes itself in rhythms. What I have called *natural syntax* is a way of describing these physical transformations of pattern.

How is symmetry theory inherently related to syntax? In the Conclusion I review the hypothesis that symmetry is inherently part of the conservation of energy. Symmetry theory explores the invariance of natural forms in terms of their geometrical and mathematical groups. So I try to distinguish some plain-language uses of conservation and symmetry by physicists. If symmetrical transformations may be seen as the turns within carrier waves themselves, then the syntax of any composed sentence will conform to the same kind of stabilizing shapes as in other physical transformations that retain their essential forms. The physical elements of syntax in speech acts are the same kinds of physical elements that conserve all other forms of nature in symmetries. And the curvilinear form of light waves can be described by an aesthetic whose point of view includes some formalities from the sciences but is not swallowed by the sciences. Thus each chapter takes a different point of view regarding curvilinear waves as archetypal forms that transform physical shapes into aesthetic acts. At the beginning of each chapter, I have provided a Synopsis; readers who want a more complete description of the development of the thesis are welcome to skip ahead.

ACKNOWLEDGMENTS

I want to thank the people who have helped and encouraged me to improve parts of this manuscript: Albert Cook, Franklin R. Rogers, Charles Fourtner, Robert Innis, Irving Massey, Jim Swan, Bill Rapaport, Carol Jacobs, Emanuele Licastro, and Fred See. At Stanford University Press I am especially grateful to Helen Tartar for her diplomacy and to Anna Friedlander for her care. My colleague Henry Sussman provided many leads and much enthusiasm along the way. Some years ago I was a visiting fellow at the Yale Center for British Art. Some results of my study of William Blake at the Yale Center appear in the Introduction and Conclusion. Also, I wish to thank Barbara Evans, of the Art and Photographic Services office at the University at Buffalo, for her renderings of some of my sketchy diagrams. Several years ago I gave an early version of this thesis in a lecture at the Center for Cognitive Science, University at Buffalo. I am grateful to the participants for their shrewd comments.

This book is truly dedicated to our family: Jude, Libby, Scott, and Susannah.

INTRODUCTION

Synopsis

In this book I study variations upon a theme of what has been called "generative grammar." I find generating principles not in the grammars of languages but in a basic form of nature, the helical wave. When I use the controversial term *natural syntax,* I mean that physical nature enters human communication literally by way of a transmitting wave frequency. This premise addresses one of the twentieth century's principal questions concerning symbolism: how are our ideas symbolically related to physical reality? I describe a theory of symbolic communication in which nature is not reached by reference to an object. My goal is not a theory of referentiality. For me, nature is the part of a message that is known only tacitly as the wavy carrier of a sign or signal. One does not refer to nature, even though one might intend to; one refers with nature as carrier vehicle.

A natural language of transmission has an inherent physical syntax of naturally patterned wave forms, which can also be described as certain "laws of form," a phrase used by D'Arcy Thompson, L. L. Whyte, Noam Chomsky, Stephen Jay Gould, and others. I describe a syntax inherent in natural languages that derives from the rhythmic form of a propelling wave. Instead of the "laws" of a wave's form, however, I prefer to speak of its elements of rhythmic composition, first because *rhythmos* means "wave" in Greek, and second because *composition* is a term used across the arts—the composition of a poem, of a sentence, of music, of a painting. So I am pursuing a philosophy of rhythmic composition.

1

How is it that we see wave patterns everywhere in nature? Wherever in the world you toss a pebble into a still pond, waves will radiate outward in widening ringlets of moving form. Whenever you strike a gong or clash a cymbal or pluck a taut string, acoustic waves will radiate from the metal and cross over into the air in characteristic wave forms. Notice that the wave form itself reverberates across the metal into another physical medium, the less dense air. In solid metal, in less solid air, and then upon the reverberating ear drum, the wave form transfers and carries a message. Wave forms are the main agencies of transmission from one natural state to another.

These examples illustrate an important idea for my thesis, to be discussed later, called "symmetry sharing": wherever a symmetrical pattern is disrupted in one place, a symmetrical pattern will be repeated at another place. The forms of carrier waves in nature are sufficient pattern generators for the making of patterns in other places. It is the form of the wave that is the pattern generator and, I hope to show, the language generator. The form of a wave and the format of speech acts are archetypally related. If Universal Grammar is thought of as a brain function that inherently generates linguistic patterns, then symmetry sharing is a principle that nicely describes the patterned transformations in the physical parts of syntactical transformation. Seen as a set of patterned transformations throughout the physical world, symmetry sharing is a principle that, though perhaps not a Universal Grammar, may provide a good way to describe a common natural syntax that physically carries a message from one natural state to another.

Although the general study of symmetry is ancient, contemporary symmetry theory specifically encompasses a series of orderly transformations that may help describe the formal structures of all things in nature. Therefore I have used a few transformations from symmetry theory in order to describe changes that inhere in the helical form of a wave: *Translate, Rotate,* and *Twist* are the three main transformations, while *Order, Position,* and *Shape* are three ways of describing the location of figures. These three kinds of transforming strokes may suffice to describe the compositions in the syntactic orders of natural language. Like any alphabet, symmetry theory is a way of composing or assembling an indefinite number of constructions from just a few elements of composition. To the extent that syntax is a principle of connecting the physical elements of the sounds of speech acts, syntax should perform its connections in accord with certain symmetrical principles of transformation. In that sense I prefer to speak of a common syntax and not of a Universal Grammar.

These symmetrical elements of transformation allow me to describe a physical syntax that is common to the several arts, including language itself.

Throughout the book I have used examples from the plastic arts because I have found that the study of formal modeling in different dimensions helps to set off and to overcome the limitations of describing the primarily linear order of syntax in sentences.

The book attempts to carry over some of the findings about natural languages from cognitive science to poetics, and it attempts to reposition poetics as a central study of natural language by way of the physical rhymes and rhythms that constitute the natural syntax of poems. Rhymes are my way of describing what some cognitive scientists call *algorithms*. A way of celebrating syntactic rhyme is described in the course of the book.

———————

In the late eighteenth century the poet J. W. Goethe sought a primal form, an *Urform*, that could generate other living forms.[1] During the late Enlightenment and early Romantic movement, some natural philosophers had also thought that several known kinds of energy—gravity, lightning, electricity, magnetism, and chemical reactions—might be unified under one *Ur*-force. Because Goethe was one of a few poets who celebrated an organic form of energy he also searched eagerly in his natural history excursions for an *Urpflanze*, an original plant, whose shape could give clues about the metamorphoses of leaf arrangements on the stems of all plants. As a natural historian engaged in what is now called *phyllotaxis*, Goethe eventually described a spiral form of leaf arrangement. (I describe this form in later chapters.) Secretly, he also used a helical shape from the "depths of Nature" as his prime model for pursuing an original metaphysics :

> By means of levers and rollers it is possible to transport loads of considerable weight: to move the pieces of the obelisk it was necessary to use winches, pulleys, etc. The heavier the load or the greater the precision required of a thing—take a clock, for example—the more complicated or ingenious the mechanism has to be and, at the same time, the more perfect the unity of its internal structure. The same is true of all hypotheses, or rather, all *general principles.* The person who has nothing much to move grabs the lever and scorns my pulley; what can the stonemason do with an endless screw?
>
> During the metaphysical discussions, I have often noticed with silent amusement that "they" did not take me seriously. Being an artist, I didn't care. It might suit me much better if the principle upon which I work remains a secret. By all means, let them stick to their lever; I have been using my endless screw for a long time now and shall go on using it with ever greater ease and delight.[2]

What exactly did Goethe mean by this oddly tantalizing entry in his journal? A clue seems to be the analogy of moving heavy things by means of mechanical contrivances as carrier vehicles. But what possible association could Goethe have made between winches and pulleys and an endless screw? How could he think of a screw as a mechanical contrivance? Was he misled by his metaphysical pursuit of a helical form? Is there a false analogy in the transfer of formal motion from mechanics to metaphysics? The answer to the first two questions is that an endless or continuous screw had been used, at least since the Roman era, as an essential form of motion in mechanical devices for moving loads. In his pursuit of a helical form of motion, Goethe would have found in Vitruvius's *Ten Books on Architecture* an entire book on machines and implements, including machines for raising heavy loads.[3] The tenth book includes chapters on water wheels and water mills; in addition, a separate chapter describes continuous water screws. Their principles of motion are studied, and their principles of construction are discussed and illustrated. Another section called "The Elements of Motion" considers vertical and horizontal motion, rotary motion, and helical motion.

Because Vitruvius's book was the first great systematic study of the basic symmetry theories of building with material forms, I shall discuss it in the chapter on symmetry theory. Here I simply want to mention that Daniel Dennett's metaphorical crane, serving in his case as a sufficient algorithm for constructing the evolution of language, may be considered as a useful example of the elements of formal motion, even though it was designed to move heavy loads. For instance, Dennett illustrates four such cranes from Diderot and D'Alembert's *Encyclopedie* (*DDI*, 219). Studying the illustrations, one sees that the basic forms of motion—vertical and horizontal, rotary, and helical—are integrated in their composition. A crane is essentially an angular arm that is hoisted off angle from a vertical post or strut. The armature can rotate around this vertical axis. A rope runs from a spindle, spool, drum, or windlass up the arm of the crane to its tip, where the rope drops vertically to the load that is to be lifted and turned. In each of the four illustrations, the rope is wrapped around the spool in a continuous helix. Each crane is built upon the principle of a continuous screw, which helps to transform vertical into horizontal motion. Wherever in real space one wants to transform one kind of motion into another—from straight motion to rotary motion for instance—the most useful contrivance is some kind of implement that twists or torques or screws the motion into another dimension.[4] I use these three elemental transforming motions to describe the changes of syntax in the wave forms of natural languages.

Why is the continuous helix an archetype for the generation of language and not merely a metaphor? Since there are only a few regular forms of motion for the movement of things in the physical world, these formal motions are basic forms of transformation as well as transportation. In this study I show that there is a basic syntax of moving and re-composing physical things in the world of space and time. For me, there is an existential syntax of acts. The forms of motion, though here represented by the English language, are not representations; they are actual enacted movements through three-dimensional space. And because the speech acts of language are physical elements that puff and billow through waves of breath, they too may be seen to follow the elementary forms of actual motion. These few formal motions compose a coded syntax of elementary principles of composition that also describe the transformations within natural language. Just as the alphabet or the periodic table of elements comprise only a few components, which can combine in an indefinite number of compositions or compounds, so there are only a few elementary forms of actual motion that which can be used to describe the moving proportions of the rest of the world. So when Dennett and Gould quarrel over the analogies between building cathedrals (specifically, say, the supporting arcs and columns of Gothic cathedrals) and building languages, they are both right in different ways. Both depend tacitly upon the essentially *physical* act of generating natural languages, and both depend upon the underlying forms of physical motion to make their cases. Thus Goethe's idea of an endless screw as transforming principle is less eccentric than it is oblique, like the angle of the water screw that lifts water from one horizontal level to another level by means of an helical form constructed on a strut that lifts up and carries the water at an angle.

So instead of featuring either Dennett's crane or his skyhook (mentioned in the Preface), I want to employ the natural form that twists periodically into a curvilinear helix. In Chapter 3 we will see that the continuous screw is the most basic symmetry transformation in nature, for the twisted helix is the model for polarized light, protein chains, and the DNA molecule. It serves then as an essential model for the physical transmission of information. Its form satisfies Goethe's tacit guide for a leading principle of inquiry that has a more perfect unity. The representation of an S-curved series is the common form of these physical transformations. In this study, I want to follow the lead of some cognitive scientists who think of the brain and its extensions as transformers of wave frequencies into images. If the brain transforms helical wave frequencies into images, light and sound waves into pictures and speeches, then the brain is always following the moving form of a wave in its compositions.

If one seriously follows Chomsky's suggestion about Goethe's *Urform*, one learns that Goethe used the study of form itself to unify his studies of biology and art. Goethe saw not only the static form of an object but also a temporal aspect combined with the spatial aspects of form. Form was for Goethe a sign of metamorphosis, of transformation in space and time. Ernst Cassirer, the philosopher of symbolic form, stated the centrality of trans*form*ation for Goethe: "Form belongs not only to space but to time as well, and it must assert itself in the temporal. . . . It is remarkable how everything developed logically and consistently from this one original and basic concept of Goethe."[5] And it was another exemplary philosophical scientist, Hermann von Helmholz, who noted Goethe's use of a trans*forma*tion as discovery: "A fortunate glance at a broken sheep's skull, which Goethe found by accident at the Lido in Venice, suggested to him that the skull consisted of a series of very much altered vertebrae."[6] The idea that form itself may be the medium for bridging nature and culture, between poetry and biology, is not a very surprising notion, but the notion that form is the transformation principle between spatial things and temporal actions, as a principle of discovery and change, is a very rare and useful idea, one that I will apply later in breaking down the form of a wave into a few symmetry operations. To link space with time in a rhythmic proportion is the crux of form, I believe, and the helical form of a wave will be seen as the essential model for a curvilinear model of space-time itself.

As late as the 1960s the philosopher and historian of science L. L. Whyte could write that nobody had as yet discovered the "primary secret" of the "genesis of spatial forms."[7] Writing before symmetry theory had become widely known, Whyte observed that there must be one "one general law under which spatial forms are generated" (620). If that kind of premise were tested, there might be seen to emerge a symbiotic rhyme between the patterned transformations in nature and the patterns of poetic language. When, for instance, Herbert Read wrote about the morphology of art forms, he featured this kind of symbiotic patterning of forms: "The whole world is 'patterned,' and there can be little wonder, therefore, in the fact that the psychic element in life *conforms* to the all-envisioning physical world."[8] Later in the same paragraph Read says, "The freedom which we inevitably associate with the creative activity, can perhaps be explained as an apparently infinite series of variations on a relatively few fixed forms." A few patterned forms of wave transformations—which include straight motion, rotary motion, and helical motion—are the elements of a physical syntax of forms to be studied here as generators of language and art. A few fixed forms of natural transformations suffice for the compositional elements of most arts. More recently than Read and Whyte, the linguist Chomsky and the paleontologist Stephen Jay Gould have furthered the appeal for a "science of form," claiming that there must be a special kind of innate pattern recognition in the human brain that generates linguistic forms and furthers creativity.[9]

My study contributes to this quest by highlighting an essential form for the "transport," as Goethe called it, or transmission of messages in any medium. This essential form is a natural wave, and is most commonly represented as an S-curved series. Information of all kinds is sent in natural wave frequencies. Speech is sent and received in sound waves. Print is sent and received by light waves. The sounds we hear on television are transmitted by radio waves. A radiological image is sent by X-rays. Some television images are sent by lasers. Information is now being transmitted over most of the electromagnetic spectrum. "Information" connotes that meaning is chunked "informs." Syntactic units are coded and sent over wave frequencies that have their own characteristic wave-lengths, which are visualized and diagrammed by their own S-curved forms. Later I will show that, across the electromagnetic spectrum, waves are the *Urforms* for packing information into meaningful shapes in series. I think that the *Urform* of a continuous wave, seen as an S-curved series, is sufficient to generate a number of patterned art forms, poetic forms, scientific forms, and, primally, the linguistic forms of syntax.

S-Curved Waves

Figure 1 represents a static image or still frame for what in most cases should be imagined as a moving proportion. It suggests a rhythmic motion. The serried half circles can be seen as S-shaped curves and as moving forms of waves. This kind of illustration is a conventional visual diagram used in physics texts to accompany most formulas of frequency for

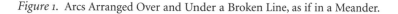

Figure 1. Arcs Arranged Over and Under a Broken Line, as if in a Meander.

the electromagnetic spectrum. This is also the visual image I have chosen to represent wave action because it is one of the most widespread images of moving proportion in decoration and in conceptual art. Notice that if one looks downward at the image from above, the diagram can represent a top-down image of an undulating series, so that, for instance, it can stand for the left and right undulation of a sidewinder. But exactly the same image can also represent a sideways perspective on a wave series undulating up and down, as if a whip were being snapped. The image is a two-dimensional, bilaterally symmetrical slice of an action in three dimensions of space and in one dimension of time. In addition, the same illustration could also represent, with some shading of light and dark, strands of any objects twisting over and under each other. As I show later, it could also be a continuous helix of tube-like waves continuously moving in space-time.

My conjecture about an archetypal form for the transmission of messages revisits an issue of natural languages. Across the electromagnetic spectrum, there is only one universal form perceivable as such, and that is a continuous wave; it is represented most often as a series of S-shaped curves. By giving examples from poetry and art, I propose a hypothesis that ventures an answer for the "fundamental question," posed by Jerry Fodor for any theory of language, namely, "How is it possible for speakers and hearers to communicate by the production of acoustic wave forms?"[10] My hypothesis begins with a study of the form of wave motion. I suggest that the forms of wave motion are ready to be decoded and transformed in every person's brain.

Wave Patterns in the Brain

If the energies of light waves that glance off objects are themselves patterned in serial packets from those objects, then brains in organisms have evolved to sort and to transform those patterned waves into structured images of the physical world. The brain transforms some of the physical frequencies, such as light and sound patterns, into picture and word images. As we shall see in later chapters, this kind of patterned transformation from brain to bodily control of movement may be involved in what some cognitive scientists have called *metaphor*.[11] George Lakoff and Mark Johnson have extensively explained how the brain is neurally integrated into the environment by a *bodily* set of spatial and kinesthetic relationships, which would later come to be incorporated in the language of *deixis*. I discuss this body-based theory of languages in later chapters. When we perceive and think, Lakoff and Johnson say, we characteristically think in spatial tropes such as metaphor and metonymy. Literary metaphor, for some cognitive scientists, is part of this fundamental processing of the brain through bodily orientation.

To introduce the idea here, I suggest that the radical act of metaphor is the transformation of some patterned wave frequency of a carrier signal into a meaningful image. If the carrier signal is a sound wave for speech, then the sound waves are shaped as phonemes into morphemes and words. Through the senses and their magnified instruments of reception, the brain transforms light waves into visual images of objects, sound waves into speech or shrieks or songs, odors and rubbings into coded systems of perfumes and hand shakes. Since the brain transforms patterned forms and carrier waves into meaning, one of its essential activities is metaphor. The essential metaphor is the act of joining together, into a compound sign the wave frequency and the image, so that we see-as. I take up these issues in the next chapters. Speaking too reductively, one may think of the brain as a multi-channeled pattern-transformer that can process and re-compose light, sound, strokes, and smells almost at once. Consider, for example, that we may sit at a festive table listening to the conversation, with music in the background, looking at, and smelling and tasting, the food, talking to a neighbor, perhaps with an occasional nudge of the shoulder. All the physical frequencies transmitting these messages to and from bodies are effortlessly and happily transformed into gusto and pleasure. Sometimes we focus on the taste of the food, sometimes on the music. Sometimes we are arrested by a neighbor's wit, so that the rest of the impressions become background noises to one's aroused interest or desire or aversion. Around the dinner table everyone transforms all these different physical frequencies into bodily locations of serial orders, positions, and shapes. Our brains, as transformers, orient all of these impressions into spatial and temporal coordinates around our bodies. The metaphorical transformation occurs when we see the light waves not as such, but as visual images of a neighbor whom we can identify by name. We do not hear the sound waves as such; instead we convert them into speech acts of American idioms, or perhaps Spanish words, or maybe even French expressions. Because we do not attend to these wave frequencies in themselves, we often forget that they are the fundamental carriers of all messages in all languages.

Poets, however, often heed the primary ways in which waves carry our life-sustaining messages. In Robert Frost's poem "West-Running Brook," for example, two lovers imagine their love to be like a brook that they have just discovered to run contrary to the eastward slope of most brooks that empty toward the Atlantic Ocean. When they also discover in the brook a standing wave surging from a sunken rock, the speaker imagines that the wave has been standing there ever since rivers began:

> "Speaking of contraries, see how the brook
> In that white wave runs counter to itself.
> It is from that in water we were from
> Long, long before we were from any creature." [12]

The speaker says that the wave could represent existence, that it flows over and between us and with us. Like the standing wave, the lover and we live and speak in contrary dialogues; all the while the river reminds of its origin and its destination:

> "The brook runs down in sending up our life.
> The sun runs down in sending up the brook.
> And there is something sending up the sun.
> It is this backward motion toward the source,
> Against the stream that we see ourselves most in,
> The tribute of the current to the source.
> It is from this in nature we are from.
> It is most us." (65–72)

This S-curved form of a wave, which is defined by the wave's running back against its source as it flows towards its destination, is the *Urform* or archetype of the formal motion that runs through and around our bodies in space-time. This forward-backward wave motion defines us. "It is most us." Long before we produced wave forms in contrary conversations, we were, as the poet says, from current. This form of the standing wave, with its reflexive pattern of retrospection and anticipation, its swaying backward and forward, defines the helical physical carrier or vehicle that I pursue in this study.

The Form of a Motion

There is a strange quality to the forms of waves. This strangeness prompts my conjecture that it is the formality of the wave itself that appears in so many archetypal art forms and that generates syntactical articulations in one's brain. The wave is the source of Goethe's *Urform*, and I wish to suggest that the universal form of waves is a sufficient topology for a poetics of forms, if not the science of common forms sought by Whyte, Chomsky, and Gould. Like so many other latent paradigms hitherto unstressed by science, this common syntax can be found substantially in the tropes of poetry and the arts. Moreover, in a recent book, *The Matter Myth,* Paul Davies and John Cribben use the history of physics to argue that contemporary science is in the midst of a vast paradigm shift away from "matter," as the very basis of physical reality, to a wavelike model.[13]

The wavelike form of natural frequency, for instance, can be found deep in A. R. Ammons's "Swells":

> The very longest swell in the ocean, I suspect,
> carries the deepest memory, the information of actions
> summarized (surface peaks and dibbles and local sharp
>
> slopes of windstorms) with a summary of summaries
> and under other summaries a deeper summary: well, maybe
> deeper, longer for length here is the same as deep
>
> time: so that the longest swell swells least; that
> is, its effects in immediate events are least perceptible,
> a pitch to white water rising say a millimeter more
>
> because of an old invisible presence: and on the ocean
> floor an average so vast occurs it moves in a noticeability
> of a thousand years, every blip, though, of surface and
>
> intermediacy moderated into account: I like to go
> to old places where the effect dwells, summits or seas
> so hard to summon into mind, even with the natural
>
> ones hard to climb or weigh: I go there in my mind
> (which is, after all, where these things negotiably are)
> and tune in to the wave nearly beyond rise or fall in its
>
> staying and hum the constant, universal assimilation: the
> information, so packed, nearly silenced with majesty
> and communicating hardly any action: go there and
>
> rest from the ragged and rapid pulse, the immediate threat
> shot up in a disintegrating spray, the many thoughts and
> sights unmanageable, the deaths of so many, hungry or mad.[14]

Presumably, many poets, artists, visionaries, and plain folks wish to reach those elemental places. Deep in the speaker's memory is the form of a natural swell, which repeats those other long swells of space and time. The form of a wave is in fact an old invisible presence. Ammons does not mean that the form of a swell is literally in one's mind. As I note later in the book, however, neurobiologists often speak of neurons generating feedback patterns whose signals oscillate in patterned waves through the body. But before describing the natural code for the wave, which is coded in the nervous system, I must look more closely at the form of the swelling motion itself.

Since I am pursuing a poetics and aesthetics of natural frequency, an elementary syntax of forms and not a science of form, let me introduce the strangeness of the dynamics of waves via "Swells." What is the formal description of Ammons's wave? Harold Bloom and others, for comparison, have studied Ammons's long poem, "Sphere: the Form of a Motion" (*AA*, 151–53). What does Ammons mean by the "form" of a motion? Bloom was the first to notice that this formal preoccupation follows after Ralph Waldo Emerson's transcendental meditation in "Circles," his observation that circles are universal designs of fluid volatility (*AA*, 81). I examine forms of fluid volatility throughout this study.

What internal dynamics characterize formal motion or moving proportion? The oddity is that, though physical frequencies carry all messages in a most pedestrian way, there is still something very *unthingly* about a wave— or there is something the matter with traditional models of matter. Listen to another poet's meditation. Emerson used the formal oddity of a wave as an archetype for society: "Society is a wave. The wave moves onward, but the water of which it is composed does not. The same particle does not rise from the valley to the ridge. Its unity is only phenomenal. The persons who make up a nation to-day, next year die, and their experience dies with them."[15] To understand this watery archetype, one must have observed the formal quality of water waves, which seems separate from their content. Any fisherman watching a bobber knows that when the swell from a passing boat jiggles the bobber, the bobber does not surf along before the crest of a wave like a surfboard but instead just bounces up and down as the wave moves onward. It seems as though the form of the wave moves onward whereas the particles of the water molecules just jostle together. The wave appears to be an independent form, regardless of the contents of kinds of physical elements involved, whether air, water, light, or any frequency on the electromagnetic spectrum. In Chapter 1 I study this paradox of formal wave motion more carefully, by describing three forms of energy propagation by wave motion.

The independence of the wave form is the reason why the S-shaped curves in Figure 1 can stand for different physical media. Formal autonomy is the reason why the wave can serve Emerson as a trope for a democratic society whose tacit measure is a phenomenal quality of unity in variety. Society moves onward like a wave in a pleasing aesthetic that has no particular regard for individuals, or for particular matters, even as its propulsion depends upon the crowding of particulars. Ammons's poem attends to the idea that the longest swell is least perceptible, whereas the smallest swell or ripple has a memory so deep that few know where it came from.

Karl Pearson thought about the occult form of waves in a different way. He challenged the simple empiricism of the assertion that it is matter that moves: "The wave consists of a particular form of motion in the substratum which for the time constitutes a wave. The form of motion itself moves along the surface of the water."[16] Considering that "matter in motion" was the empirical object of scrutiny characterizing the Scientific Revolution, Thomas Hobbes's political commonwealth, and John Locke's psychology of mind, this shift of scrutiny toward the form of a wave motion has significant consequences for inquiry in disciplines other than physics. But what an elegant way for Pearson to express the idea that there are moving forms of motion. It is this formal wave motion that I use as the generator of a common syntax that moves, positions, orders, and redistributes sounds in the sentences of poems.

As Ammons implies, it is the forming swell of the wave that oscillates the particulars. It may be that the particular atoms or molecules, which do the jostling, are not matter at all, but rather tightly folded waves within waves. A deep memory contains that formal torrent which impels all things and thinking things. Like a Wordsworthian form of nature romanticism, this old presence resided in the unconsciousness of some sciences: form-motion is so simply an archetypal drive that it is rarely understood as such. But poets think and feel the effect of its character. In "The Poet," for instance, Emerson pursued these rare ideal forms because he Platonically believed Spenser's lines from "An Hymn in Honour of Beautie":

> For, of the soul, the body form doth take,
> For soul is form, and doth the body make (*SE*, 66).

Echoing the long tradition of defenses of poetry, Emerson says in "Experience," that "human life is made up of the two elements, power and form, and the proportion must be invariably kept if we would have it sweet and sound" (*SE*, 298). My thesis about the power of wave forms is that the moving proportion of carrier waves in frequencies is the implicit measure of other generative forms in nature. But the wave form is so rarely understood that most of us, even some scientists and humanists, see and feel the effects of its powerful force rather than the rare proportion of the form itself.

The Propelling Wave

Chomsky has asked how one can argue for innateness without falling back on Platonic preexistence. Proposing a solution for Plato's geometry-solving slave boy in *Meno,* who had no learned knowledge of geometrical figures, Chomsky described an analogy for a Universal Grammar: "A modern variant would be that certain aspects of our knowledge and understanding are innate, part of our biological endowment, genetically determined, on a par with the elements of our common nature that cause us to grow arms and legs rather than wings."[17]

An archetypal answer would be the replicating transformations of the twisting helix, Goethe's continuous screw, whose innate form is the compositional model for polarized light, protein chains, and the DNA molecule, as discussed in Chapter 3. Although that form may be a universal generating model of subsequent transformations, how does one shift categories from a genetic model to a model of communicating messages in sentences? I think the key is to biologically model the propulsion of sounds and bodies by way of the form of carrier waves. So I will now begin to show that the moving proportion of a wave propels both living things and the moving forms of sentences. Here Chomsky provides the lead with a biological analogy. He quotes Konrad Lorenz in order to support the idea of innate organizations in the central nervous system (such as the perception of line, angle, and motion) that would produce "hereditary dispositions to think in certain forms":

> Adaptation of the a priori to the real world has no more originated from "experience" than adaptation of the fin of the fish to the properties of water: Just as the form of the fin is given a priori, prior to any given negotiation of the young fish with water, and just as it is this form that makes possible this negotiation, so it is also the case with our forms of perception and categories in their relationship to our negotiation with the real external world through experience. . . We believe, just as did Kant, that a "pure" science of innate forms of human thought, independent of all experience, is possible. (*LM,* 81)

It may be that these innate forms are innate because they are intimately related to such bodily extensions as fins, wings, hands, and fingers. Bodily articulations are a kind of syntactic articulation, and their formal motions, we shall see, constitute a syntax of moving forms that can begin to describe a poetics of things moving about in rhythmic space.

I want to suggest that the form of the fin, a product of bilateral symmetry, serves a stabilizing and propelling function that is analogous to a generator in a Universal Grammar. But because the phrase *Universal Grammar* is too universalizing for my purposes, I prefer to work with the idea of an anatomical syntax, which will be defined as a common form of articulating things in their bodily arrangements. The tongue, for instance, is like a fin, in that it helps shape and propel phonemes into shaped billows of air. I want to begin to describe how syntactical arrangements constitute an innate ability that, as Chomsky suggested about Universal Grammar, is on a par with bodily morphologies. Steven Pinker was thinking of Chomsky's biological argument for natural languages when he speculated, "The overall impression is that Universal Grammar is like an archetypal body plan found across vast numbers of animals in a phylum."[18] My effort here will be to begin to show how the symmetries of animal bodies constitute an archetype that helps us to understand how the syntactical shapes of spoken waves propel the grammars of various languages. Could language be a form of a phantom limb? Could languages be generated by syntactic forms that have been adapted from an innate code for propelling bodies through the *Urform* of an undulating wave? Is natural syntax a derivative of an animal's ability to perceive the angles and wavy motions of things?

Before dismissing this conjecture as far-fetched, consider the way in which a biologist discusses the evolutionary link between propelling waves and symmetrical limbs. One can understand the evolution of fins and limbs with respect to the propulsion of waves by following Carl Gans's studies of wave locomotion in snakes.[19] His diagrams of a sidewinder's lateral wave motion show how its bodily wave moves off angle from the direction of intent, just as a water wave moves an off angle from the wind's direction or an ocean breaker breaks its wave at an angle from the shore. Gans shows how snakes, as they evolved from water- to land-dwelling organisms, retained certain wavy motor patterns common to undulating fish. In addition, they retained a muscle control mechanism. Gans says that this governing pattern consisted of a program that "yields staggered waves of muscular contractions that move from head to tail. The site of maximum contraction of muscle fiber always corresponds to the site of maximum relaxation, or stretching, of the units of the opposite side" (*LWL*, 13). From head to tail the body muscles are alternately compressed and stretched along the body's length in a directed wave. Gans says of a fish's movement through the water, "The reaction to backward movement of the wave pushes the fish forward" (*LWL*, 13). To interpret that sentence one must perceive the backward wave of a fish's undulation as a "reaction" to the formal properties of water. The body wave creates a reciprocating water wave that alternately compresses and expands groups of water molecules. A fish form rides the water waves that its body wave creates. But notice a tacit idea that seems self-evident because it occurs everywhere: the carrier vehicle is the wave form that oscillates from two apparently discontinuous media: the wave form moves from the supple body of the fish and shapes the fluidity of the water. The wave form is an independent identity that moves the body, because both kinds of bodies, animal and water, are commonly shaped by shared symmetries of natural construction. The form of a wave seems to be a universal archetype for the shapeliness of things in dynamic media.

Here is Gans's synopsis of the evolution of fins (as you read, keep in mind Lorenz's a priori model):

> Quite early in their evolution fishes developed paired lateral fins, apparently for steering; fins that later served for propulsion. At some point these fins became modified into limbs, providing one of the adaptations that permitted the successful invasion of land—a fascinating story that we cannot go into here. It is interesting, however, that the basic movements of these limbs were controlled by a modification of the old sequence of alternating contraction waves—a sequence that stayed in the animal's nervous system. (*LWL*, 14)

Much later, among other species, paired fins evolved into different kinds of limbs for motion upon land. These limbs, Gans says, could lift the bodily trunk "during the entire locomotor sequence." In other words, the bilateral symmetry of paired limbs lifted the body against the downward thrust of gravity on land, whereas the formal sequence evolved to the natural advantage of "balance" for walking (*LWL*, 14). All this was governed by the "old sequence" in the nervous system of alternating contraction and expansion waves. Balanced limbs for walking were a landward distribution of the equilibrium governed by fins. My study is guided by the phrase "old sequence." A later chapter updates this idea of an "old sequence" with a new sequence called *central pattern generators,* a phrase many neurobiologists now use to describe the oscillating feedback of various kinds of bodily movements.

To restate Gans's description: the alternating wave format, retained in the nervous system, was a code that processed the flow of information and redistributed this alternating form of a wave to the symmetries of the limbs that propelled bodies in alternating gaits. It is an *innate* patterning code, whose essential neural patterning has stayed with animals through different stages of evolution. Lorenz's fins then would be interpreted as bilateral parts of a vehicle whose a priori coded structure produced a bodily motion by alternating contraction waves. Is this wave form not like Ammons's "deepest memory"? I think that Gans's "old sequence" is the same model of motion as Ammons's "old invisible presence." My point then is that the form of the alternating wave is preserved in innate codes within the nervous system and that this feedback format quite literally governs the flow of information. "Flow" is not a metaphor here; instead it is an archetype of in*form*ation that is sent rhythmically to the muscles and limbs.

When microorganisms and macrorganisms evolved from salt to fresh water, from water to land, from land to air, it seems they kept their innate informational systems more or less intact. They retained the old sequences of their instructional codes as invariants, even while they transacted within new and different environments. True, the systems of their interior worlds must have gradually evolved in the feedback with their new external worlds, but since their receptors still worked via wave information, the interior sequences and circulations must have continued to guide different kinds of limbs through new media. But what does this biological theorizing specifically have to do with language? If the code of the old sequence remains invariant while different species are guided by it through their wanderings, then all those different kinds of bodies may be seen as vehicles for carrying the code over long periods of time and over vastly different environs. From the code's point of view, bodies are just wandering vehicles for keeping its system dry. The fixity of the code's "memory" for perambulations would serve much later when the code of the old sequence was transferred to a new kind of information in human language.

When we propel ourselves or our thoughts through a medium, when we celebrate the free and voluntary action of walking on our own two feet, or when we think creatively for ourselves, are we not reenacting that old sequence of alternating waves? Is an alternating wave an elemental form of a motion? When information is distributed, it is arranged in sequential patterns of contrasting chunks, as in the composition of a tune. Could it be that natural syntax is a part of that old sequence, that syntax is a prototypical code that propels patterns of signs which follow the form of a rhythmic wave in distributing information?

Before broaching these questions, I want to think more precisely about the *Urform* of a wave that is propelling bodies. In what sense is a body plan an "archetype," as Pinker used the term? In what sense is a wave form an archetype, and how are waves and bodies related? In Chapter 3 I define archetypal shapes more exactly as body symmetries, specifically in the section on *hockers* from primary cultures. *Hocker* is an ugly word for an elegant symmetry. But for now, consider that Max Black defines an archetype as "an implicit or submerged model" that is not completely articulated as a finished scientific hypothesis of discovery.[20] He follows Stephen Pepper's use of "root metaphors" that stand for "world hypotheses," that is, commonsense analogies that gradually arise as large metaphysical systems.[21] The awareness of trope as archetype may range from inchoate figures to explicit models, such as a crane or an endless screw. I have in mind a wave archetype serving as a proto-type of physical form that may generate subsequent transformations in artistic and poetic media.

In this regard, consider Henry Moore's sculptural use of the term *archetype*: "Our attention is held by the contour of a particular hill, by the shape of a rock or a tree stump, or a pebble we pick up on the beach. These shapes appeal to us, not because of any superficial beauty, any sensuous texture or colour, but because they are archetypal. That is to say, they are the forms which matter assumes under the operation of physical laws."[22] This passage makes exactly the connection I seek between archetypal bodies as they are shaped by the physical pressure of close packing and random rubbings of sand or water or wind. From the sculptor's point of view, archetypes are those three-dimensional shapes that are formulated lawfully. There seems to be a necessary connection in Moore's mind between certain primal forms and the laws of physics. The work of sculpting seeks to reenact those shapes, as if the sculpted shapes were operating under the same strokes, gestures, and rubbings as those the sculptor claims to see in nature. In this sense, archetypal shapes compose the prototypical syntax of a natural language of interrelated compositions. I think that Herbert Read is exactly right to seize upon symmetry theory in order to explain Moore's necessary connection between archetypal shapes and physical transformations. To make his point, Read quotes D'Arcy Thompson: "In every symmetrical system every deformation that tends to destroy the symmetry is complemented by an equal and opposite deformation that tends to restore it. In each deformation positive and negative work is done."[23] I return to this principle again and again in this book. I study this compensatory principle, first articulated by Pierre Curie in Chapter 3. This concept, called *symmetry sharing,* is central to my argument for the physical syntax of language: syntactical units of language, considered as billows of sound or shapes of light, must be modulated by symmetrical deformations and compensatory reformations. A natural syntax must operate under these archetypal laws or rhythms.

As Frost said of the complementary forces working in the wave, "It is from that in nature we were from." But both Black and Pepper use the terms to suggest that the archetype not only precedes (as in "arche-") but that it is underneath, either "submerged" or at the "root." Where is the archetype? Is it underneath? Is it anterior? Is it outside? Beyond? In conceptual or hypothetical space and time? Where is it that we are from? Where are things flowing to and fro?

All Things Flow

As Alfred North Whitehead said in *Process and Reality,* flow is the first intuition of poetry:

> That "all things flow" is the first vague generalization which the unsystematized, barely analyzed, intuition of men has produced. It is the theme of some of the best Hebrew poetry in the Psalms; it appears as one of the first generalizations of Greek philosophy in the form of the saying of Heraclitus; amid the late barbarism of Anglo-Saxon thought it reappears in the story of the sparrow flitting through the banqueting hall of the Northumbrian king; and in all stages of civilization its recollection lends pathos to poetry.[24]

To understand the phrase "all things flow" is for Whitehead "one chief task of metaphysics" (*PR,* 41). Notice that the fleetingness of things does not necessarily mean that flow lacks form. Because I am trying to turn the argument from finny forms of transportation to the transportation of ideas by carrier waves, and thence to the transportation of ideas in sentences, and thence to the wisdom of poetry as Ammon's form motion, I want to recall that Heraclitus's famous flowing stream, into which one cannot step twice, is for some scholars an idea that is akin more to a form of equilibrium within process than to an unmitigated process. For instance, according to Kirk, Raven, and Schofield, Heraclitus's river is a form of *logos.* Like a "formula," they say, *logos* was a moving proportion that stabilized pure process into measure.[25] This ancient theory of the form of motion was a way of measuring the orderly flow of events in time, unrepeatable perhaps, but not without pattern. For Whitehead, too, philosophy and poetry seek a correspondent stasis that might slow process. Thus he quotes the old hymn about flow:

> Abide with me,
> Fast falls the eventide.

He says of these lines, "Here at length we find formulated the complete problem of metaphysics" (*PR,* 241). I consider that to be an amazing revelation on Whitehead's part. All of his philosophical works, all of his writings about logic and mathematics, all his panorama of life experience, can be "formulated" in a short sentence in an old song. Abiding and perishing seem to be his preoccupation in *Process and Reality,* his last great work. One sentence in a communal song asserts the formula of his metaphysical quest. But a rhymed formulation is what poems and songs do with their terse and formal wisdom. In the quoted lines, the patterned flow of the words reenacts the pattern of flow in the natural archetype of tides. Not just a "vague intuition" about the passing of all things, Whitehead's final philosophy of organic living is formulated in a song of tacitly moving proportions. Perhaps his original attraction was not just to the words on the page but also to the communal recollection of the hymn as it was sung.

Beneath the simplicity of the three-word refrain, the hymn's assertion is deep and terse. It depends upon Ammons's old invisible presence, perhaps the oldest pattern of alternating sequences on earth. The words *bide* and *tide* are made to rhyme as differences, because of the phonetic shift in syntax from *b* to *t.* The syntactic repetition forward of the phonemes is a physical repositioning that incites the recollection as a rhymed pair of words. Within songs and poems this syntactic reorientation of physical units is the compositional set that enacts the orderly measure within process. (In later chapters I shall discuss these kinds of syntactic transformations within poems.) In "Abide with Me" the rhymed consonants and words combine semantically into a reassuring moving proportion of recollection within process. Scarcely noticed, however, is Ammons's swell, the archetype of a wave form that carries the tide, and also time, in its alternating frequency, its old sequence. For *fall* is the action verb of nightfall as well as the falling tide. The unexpressed mover of the alternating archetype of the passing tide is the gravitational pull of the sun and the moon. Whitehead's metaphysics is deepened by the form of a propelling tidal flow, which, as in Emerson's image of wave motion, waits for no one.

As moving proportion, therefore, the *logos* as a formula for the flowing stream is not meant to be a mathematical or algebraic equation; instead it is the existential moving measure itself, the large-scale rhythm of things in their processional sequence. Is this archetype for moving proportion, this *logos,* not the form of the swell in Ammons's deepest memory? If the form of a motion curves or twists through the physical reality of space and time, then it may be a sufficient a priori format for generating components of natural languages.

Because the hypothesis about the flowing form of *logos,* as the *Word* or as language, is Heraclitean, Socrates decries the idea that letters, syllables, words, and statements imitate the form of streaming motion. In fact, several of Heraclitus's adages are preserved only because Plato quoted them in order to refute them. In *Cratylus,* for example, a very technical discussion of the origins and implications of language, Socrates questions the first principle that "all things are in motion and in process and flux, and that this idea is expressed by names."[26] Socrates provides counterexamples of Greek words that underlie the premise of processional motion (437b). For instance, the Greek word for "knowledge" indicates the stopping of motion. "Sure" indicates station and position, not motion. "Inquiry" means stopping of the stream! And "memory" indicates rest, not motion. Pure process may be undermined, but various words take attitudes toward motion.

Furthermore, in the *Sophist* the speaker says that "discourse is a weaving together of forms" (60a). He seems to mean only word forms, though Platonic ideal forms are not beyond the limits of allusion. In later chapters I want to study various tropes of weaving, because the physical action of twisting something is a first principle of symmetry theory, and the twisting of strands over and under is a first principle of topological knotting. In other words, lacing or braiding some strands over and under renders a helical screw in three space. Weaving and waving are moving proportions with similar compositional forms. But in the passage quoted, the actions seem to be just a trope. In *Cratylus,* however, the speaker discusses a most important thesis concerning the composition of language. The compounding together of noun and verb is the essence of discourse. After explaining that a name is the spoken sign of that which performs actions and that a verb is a word applied to action, the speaker says that their union is the basis for any assertion or statement. An example might be "Socrates was healthy." This is the origin for the crucial idea in logic that any real proposition demands a subject linked to a predicate, combined into a sentence, and hence the idea that the sentence itself is the basis for meaningful discourse.[27] Then Plato's speaker uses a trope, an archetype, of twisting or interlacing or weaving (it is variously translated), which is my point of departure for thinking of sentences as grammatical derivatives of more basic moving forms: "When one says, 'A man understands,' do you agree that this is a statement of the simplest and shortest possible kind? . . . Because now it gives information about facts in the present or past or future; it does not merely name something but gets you somewhere by weaving together verbs with names. Hence we say it 'states' something, and in fact it is this complex that we mean by 'statement'" (*Cratylus,* 262d).

An assertion must be couched as a sentence that "gets you somewhere." A sentence is couched as a figurative vehicle that gets you somewhere you want to go in the discourse. It is a kind of propelling vehicle, operating not by fins but by grammatical agents of action. Like Goethe's continuous helix, Ammons's wave, and Gans's "old sequence," this trope is a primary form of propulsion. Cratylus is a Heraclitean proponent of moving proportion. (As we shall see below, the historical Cratylus was a radical Heraclitean.) Plato's archetype of weaving "gets you somewhere" in the hypothetical space and time of the sentence's subject and verb. Most importantly, the trope of propulsion allows the most important idea about assertions: by way of the moving sentence, one moves from what one does not know to what one newly understands. This "statement" addresses one of the first issues in linguistic inquiry. By using the conventions of formal grammar, with conventional verbal meanings, that is, by appropriating something we already know, how do we generate something new? Any alphabet, any grammar, any code assumes a set of instructions for generating an indefinite number of transformations from a limited group of counters. It is the grammatical weaving together of noun and verb that figuratively gets one somewhere, perhaps into a hypothetical future of potential action.[28] This figurative domain of the sentence is the hypothetical world of space and time, in which forms of motion direct one to an inquiry about possible action. For instance, one might say, "Socrates was healthy up until the point when he was made to poison himself." Because of the magical figure of the motions of weaving, sentences now seem to propel us to a new somewhere, just as fins propel and stabilize the body. The magic trick is that the physical order of moving things about in real space and time, which I shall call *the order of natural syntax,* is appropriated as a figurative trope applied to the grammatical assemblages of subject and predicate, so that it seems as if the parts of speech are themselves doing the work of moving toward a significant inference. Grammatical dexterity in the composition of words can be mere nonsense, as in Chomsky's famous sentence, "Colorless green ideas sleep furiously." Nevertheless, somehow the moving form of a sentence does seem to get us somewhere. But what agency is doing the work of generating new patterns of ideas by way of the moving forms of sentences?

By introducing Heraclitus's stream, I want to suggest that universal wave frequency is the *Urform* for a number of archetypal variations: the stream of time, the stream of thought, or meanders of discourse. The curvilinear *Urform* of a propelling wave, in a direction, seems to be a viable model for a syntax of forms that gets us somewhere in our conceptions. For instance, in *The Grammar of Science* Pearson modifies Heraclitus's adage in the title of chapter 7: "'All things move'—but only in Conception." For Pearson, motion, like position and other strategic concepts of science, is not in itself a perception but a combination of the concept words *space* and *time* that makes motion a "*mode* of perception" (*GS*, 203). These terms compose the "grammar" of his title, and the title suggests his philosophy of science: "The whole object of physical science is the discovery of ideal elementary motions which will enable us to describe in the simplest possible language the widest range of phenomena; it lies in the symbolization of the physical universe by aid of geometrical motions of a group of geometrical forms" (204). I am after grouped forms of real motion that are not necessarily ideal but that are certainly read as hypothetical, in that they are revealed only in their symbolic transformations. There is no simpler form of motion in dynamic three space than the ideal form of a moving wave. For Pearson, although bodies may change, they do not move. Like the form of a motion on the surface of a pond, the wave seems to move, while the water itself stays in place. However, his grammar of the ideal forms of motion allows one to make it seem as if "matter" is in motion. If natural syntax is a set of physical transformations that orient shapes in space and time, for the purpose of making meaningful assertions about ourselves and nature, then Pearson's grammar of science and an *Urform* for syntax are fairly closely aligned. I describe these elementary forms of actual motion as those from symmetry theory: translation, rotation, and twist in Chapter 3.

But look more closely at the metaphorical use of "grammar" in Pearson's title, *The Grammar of Science*. Notice also the titles of these books and chapters in which the word *grammar* is used as a conspicuous metaphor: Owen Jones's *The Grammar of Ornament* (1868), Lancelot Hogben's "The Grammar of Size, Order, and Shape" in *Mathematics for the Million* (1937), Jeremy Campbell's *Grammatical Man* (1982). In these few examples, which range from ornamental art through science, through a mathematics of counting and measuring, information theory, and beyond, the use of grammar is recast in some broader sense of a limited array of ordering principles that one consciously and actually uses to generate the symbolic forms in the discipline, but none of the authors literally mean a *linguistic* grammar. Their consciously metaphorical use of the conceptual word *grammar* suggests that there are just a few limited rules of composition for generating an indefinite number of permutations from a few symbolic forms. That metaphor is well and good, but because I want to focus on the actual transposition of physical entities, I hope that you will bear with me as I begin to draw a distinction between symbolic grammars and physical syntax.

Off-Angled Joints and Limbs

If this *Urform* of wave motion is encoded in the old sequence, then the set of instructions must undulate a bilaterally symmetric body with four limbs in an alternating order. Thus, sensorimotor instructions and symmetrical forms are interdependent in the evolution of various forms of propulsion: swimming, walking, flying. But how does one get from the symmetries of animals' bodies to the geometrical forms that we can describe as compositional forms? I discuss this idea of an archetypal body symmetry in Chapter 3, in exploring the sun's physical effects upon animal bodies. For now, let me introduce one answer concerning jointed and off-angled limbs that I note throughout this study. As with the structure of other animal limbs, human limbs are jointed off angle from the torso that is to be propelled. Bent-back limbs get you somewhere; they propel bodies and ideas. The potential energy of bent-back limbs is also that of the anticipatory dynamics of a strung bow or lyre. Pluck either one, and it sends a significant message through the air waves. This off-angled correlation between body symmetries and jointed bent-back limbs serves a larger aesthetic purpose in succeeding chapters. For example, Rudolf Arnheim observed that among humans there is a simple but crucial dependency between symmetries of bodily shape and the creation of basic geometrical forms: "The lever construction of the human body favors curved motion. The arm pivots around the shoulder joint, and subtler rotation is provided by the elbow, the wrist, the fingers. Thus the first rotations indicate organization of motor behavior according to the principle of simplicity."[29]

The principle of simplicity is Arnheim's large thesis about a few forms of motion for art forms, and I shall follow a variant of it: that the employment of archetypes, as Henry Moore wrought them, is the composition of the simplest forms under the constraints of natural forces. For instance, Arnheim observes that when set in motion, jointed angles inscribe arcs, as an angled compass can draw a rotation. Angles, when set in physical motion, describe actual arcs of rotation. Although Goethe belittled the lever in his meditation on metaphysics, the off-angled levers or limbs that propel the trunk are the requisites for the act of symmetrical rotation. Further, if a series of rotations are strung out in a sequence, like the stretching coiled wire of a child's toy, one achieves an endless corkscrew. This stretched wire also shows how the simple transformations of translation along a line, rotation in a circle, and twisting in a helix are correlated symmetry operations. The lever, the wheel, and the corkscrew are similarly interdependent forms of propulsion. All are descriptions of acts. This nuts-and-bolts discussion about body joints may seem at once yeomanlike and irrelevant to the understanding of poetry and art, but in a moment I will apply these off-angled symmetries to the understanding of work by a master of symmetrical representation.

So the simplest lever of geometry is the two-legged compass, whose jointed angles can inscribe arcs. It is a tool that incorporates, in its pure geometrical form and intent, the prototype of bent-back limbs, which propel things toward motion. If one of the "feet" of the legs is fixed to a point while the other leg is rotated, an arc or a circle can be drawn. This basic tool of geometry is a prosthesis of formal motion that extends the interdependence of angles and arcs of rotation. Perhaps Plato's slave boy could solve geometrical problems because his nervous system was built upon the innate old sequence of formal motion that drove his symmetrically jointed arms, legs, wrists, and fingers in rotating and twisting motions.

This interdependence between protogeometries and body joints leads me to summarize a theory on the evolution of language as a form of gestural communication that developed with the making of sophisticated tools by the coordination of hand and eye.[30] According to one theory, language evolved at about the same time as humans learned the knack of making tools to make tools, or metatools. In this view, language, like a metatool, is a vehicle for planning ahead. Among metatools, there were stone tools that were used to make the delicate percussion flakings of spear points and scrapers. And there were burins, long, narrow-flaked chisels that could be used for carving and for scoring softer material such as bone and ivory and wood. Of this process J. Z. Young observed, "The achievement of a goal was sought by increasingly indirect methods" (*SM*, 508). The aim of learning itself is to "build in the brain a model that can be used to forecast the probable outcome of events (*SM*, 252). For Young, the skill in making these tools of indirect method was accompanied by the skill in communicating, or passing along, that ability, the knack of making metatools, to others in the community.

Language, like education at large, is supposed to make us ready, indirectly, for a future. In this way language became a metaskill of forecasting outcomes by way of increasingly indirect mediations. Later in his work, Young summarizes a biological theory of the emergence of language in this way: language "constituted in effect the invention of a wholly new biological phenomenon, the transmission of detailed forecasts by codes other than genetic" (*SM*, 497). In later chapters I plan to change this diction from that of forecasts to that of rhyming forward. I am suggesting that there is an a priori code, an old sequence, for formal motion in everybody's brain by which one learns how to get somewhere, even in a hypothetical future, for a future is always a fictive projection.

Bronowski's analogy is that language is a kind of instrument that can be turned upon itself to speak about language. Metalanguage, for him, is a kind of "machine tool":

> For deliberately to make a tool in order to make other tools (including fire) in the future implies a similar analytic process—an ability to break a plan into a sequence of steps, and to foresee the total effect as a coherent sum of its parts. The commentators on the Old Testament were so impressed by the mystery and the mastery of the machine tool that they made it a special creation: at dusk on the sixth, they said, God made the first pair of tongs so that man might be able to forge other tools (*SF,* 129).

In William Blake's magnificent rendering, sometimes called *Act of Creation,* the Creator uses His fingers as a set of dividers to separate the firmament from the moving waters (Figure 2).[31] This image also may be seen as Blake's version of the story of creation in the biblical book of Genesis. The King James Version uses the verb "divided" almost as much as "created." If tongs are the master tool, then dividers are their perfection. The primal set of dividers was probably a pair of fire tongs that a curious smith idly rotated in the ashes, only to find therein a perfect circle. But Blake, I believe, would have agreed with Michelangelo that one should not depend upon compasses in making art. As Vasari said, "Least of all will the eye approve the object when it is without rules or measure. Therefore the great Michelanagelo used to say that it is necessary to have the compasses— namely, sound judgment—in one's eyes and not in one's hands."[32] By means of the scale of bent fingers, Blake shows the origin of rigidly rational form in a geometric act.

Despite this digression about the archetypal relation between angles and arcs of rotation, I am still in pursuit of the moving form of the wave as the primal form. For waves are usually thought of as being wobbles or oscillations that result from uneven rotation. Blake renders the wave only in the chaos of clouds beginning to take shape and in the wind that is blowing the Creator's hair and beard in a horizontal meander. This minor symptom is Ammons's swell, an "old invisible presence." Commentators have noted that God is seen as "hovering" above the flowing waters in Genesis. Also commentators have said that "the Hebrew word for creation by God. . . has the basic meaning of divide or separate."[33] So if one studies Blake's geometry, and if one applies Arnheim's insight about angles and arcs, one can see how the dividers can be associated with the circle that surrounds the Ancient creator. The dividers are the Ancient's forked fingers. And the light rays that extend out in an angled direction are taking form in straight linear motion. Blake

Figure 2. William Blake, *Europe. A Prophecy,* pl. 1: Frontpiece: 1794 Relief etching and watercolor. 9 1/4 x 6 1/2 in. Yale Center for British Art, Paul Mellon Collection.

incised many of his works as engravings, so he would be a master of the rotator joints that Arnheim mentions—fingers, wrists, shoulders. All are featured in this picture. However, to achieve a rotation from the Ancient's dividers, to make an arc out of the angle, He must rotate His divider fingers with His wrist and elbow joints, out of the flatness of the two-dimensional rendition into a three space that is virtually outside the frame of this very flat picture. Blake liked to invert this picture into its mirror reversal so that arms and legs appear reversed. To reverse this image, Blake had to rotate the picture outside of its plane through the three space of his workshop.

Rotation is one of the basic operations of symmetry theory, and Blake was a master of a rotational "fearful symmetry." In the Conclusion I shall discuss that famous phrase from his poem "The Tyger." The poem is a progress piece that describes the process of constructing the tyger's body by actions similar to the geometries of this Creator. Here I am suggesting a Blakean archetype of formal creation: the prototypical forms of geometry separate out from the curvilinear wave of wind and water. To conceive this, imagine holding two gold wires with the fingers of one hand so that they are extended together in a straight line; then imagine rotating the wires with the fingers, wrist, and elbow of your other hand. The result will be a twisted sequence. If you turned the two ends together and joined them into a bracelet, you would have a primal geometrical ornament found in many cultures, often called a *torc* or *torque*. Translation, rotation, and twist (or torque) are all basic symmetry transformations—to be studied later on— that can be derived from the corkscrew form of a meandering wave.

Wave Form as Helix

Here let me briefly introduce the idea that the ideal form of a wave tends toward a three-dimensional helix, or Goethe's endless screw. First think of a real ocean wave breaking on a beach. Picturing its breaker pattern, one recalls that the crest does not break simultaneously all along the length of the beach. If the wave is breaking directly in front of the viewer, its crest is still rising and curling downwind. The break may seem to be curling from left to right along one's line of sight. If it is large enough, the wave may twist into a huge tube into which a surfer can ride. By the thrusting form of the wave, the surfer is propelled off angle to the crest along the direction of the twisting curl. The surfer is riding from my left to right along the direction of the twist. The surfboard is carried by the form of a wave just prior to that point when it crests and spills, for that is the point when it momentarily both completes and collapses its form as a corkscrew tube. The scud lines all along the curl are traces of the incipient corkscrew. In a section about the morphology of waves, René Thom illustrates a photo of this kind of breaker, complete with curved scud lines.[34] It is important to think about the shape of waves in three space because speech acts—as we shall see later—actually move as spherically shaped waves through the air.

As it happens, there is a freshwater worm, an oligochaete, which propels itself by contorting its body into a complete helix, making its body wave move from head to tail.[35] Charles Drewes and Charles Fourtner describe the "retrograde passage"(*HS*, 1) of the worm's helical wave that propels the body forward by means of a "helical vortex of water displacement" (*HS,* 8). This set of instructions in the nervous system of a simple worm directs the body to take on the undulant wave of a twisting three-dimensional helix. Locomotion seems to be based upon this a priori formal motion of a wave. The oligochaete is a living form of Goethe's endless corkscrew.

My elementary point is that this old sequence of helical propulsion follows the form of the wave. The simple symmetry of the oligochaete's linearity lets it contort that shape into a series of rotations that resolve into a helix along the line of the body. So even before studying the basic transformations of symmetry theory, one can see how their rudiments are required to understand the rhythmic pattern of formal motion: the patterns are *translated* along the line of the worm's body from head to tail; the alternate instructions to expand and contract the line of the body make the body *rotate* into a circular shape on a plane perpendicular to the body line; the continuity of the body requires that the translation plus the rotation results in a series of *twists* that propel the length of the body's helices through the three space of the watery medium. Drewes and Fourtner also show that the oligochaete does not just twist its body into a simple right-handed helix; if its helical motion were always just one way, it would veer around in circles. Its feedback code compensates with an alternate patterning. In order to propel in a fairly straight line, in order to translate its length forward, it twists one way through one sequence of body length, and then it alternates and yaws into a helix of the opposite side of its length.

It is natural that water shapes into the form of a wave because the worm-wave shapes it. But what is hiding under *natural*? Like an oligochaete, we too exist in a total environment of wave frequencies. Some know how to use those wave frequencies as metatools, which are employed as carrier waves for the projection of messages. But, for many of us, what is always before our eyes as total environment is usually beneath reflection. We sense and feel strokes, caresses, and brutal forces upon our bodies, but we question what those interventions might mean. So, in the next chapter, I will turn to self-reference and focus on our tacit fit into this total environment. Here I simply focus on the idea of a complementary fit. One kind of body wave can fit into the environment of another kind of wave because of the common shape of the wave form. For instance, the verbal compound "electromagnetic wave" means that the electric charge and the magnetic charge rotate and alternate off each other in a form of helical wave traveling in a direction. How simple it is to say in a sentence, such as this one, that nature tends to resolve itself into formal waves across the electromagnetic spectrum. But how much work, how much discovery and contention,

ranging from Michael Faraday to James Clerk Maxwell, went into that hypothesis about electromagnetic waves! Without trying to review that story,
I want only to introduce the more simple point that animals fit into the
forms of waves, as seen in their movements, because every interruption in
dynamic nature resolves itself into another set of archetypal waves.

As I quoted Thompson above, wherever symmetry is broken, a symmetrical form appears as the very form of the interruption somewhere else.
It seems natural to observe that one wave form, like an oligochaete's, necessarily creates another. But notice that two different media are intercon-
nected, the worm's body and the water's mass. *This is a crucial observation
that is usually beneath reflection: the form of a wave is a crossover format between two different media. Like the alternating charge of electromagnetism,
which may be its prototype, wave forms interrelate different media.*

Presumably all the great physical laws fit together by wave forms. How so?
Because the form of wave can roll over and through all things, as a wave form
moves over the surface of the water without much affecting the particular
contents of the individual components. Meditations about waves usually derive from the radical experience of watching waves radiate over a still body of
water. Throw a stone into a pond and watch the waves form and radiate. This
is commonplace, a universal form. Dent a ping pong ball, and it will shape
into the form of a circle. This phenomenon is called *symmetry sharing*, and I
shall discuss it at length in later chapters. Natural forms resolve themselves in
optimal geometries, depending upon the available energy. This principle of
simplicity is the thesis of Thompson's correlation between growth and form.
If an oligochaete evolved to fit its propulsion unit into the helical form of a
wave, its whole body is transmitting the old sequence in the form of a wave.

Since every break in a thermodynamic pattern resolves into the component forms of a wave, as in the protogeometries of Blake's *Act of Creation*,
across the created world, I can begin to show how the wave form can cross
over and resolve into waves of speech acts. *Speech* can be seen as a series of
interruptions that vary the pressure of air into periods of condensation and
rarefaction of air molecules. I like these peculiarly Newtonian words for the
construction of sound waves. The form of a wave, with its troughs and
peaks, push and pull, backward and forward, is the archetype for a moving
proportion of optimal form. For the poet Samuel Taylor Coleridge, furthermore, this old sequence of propulsion is the very form of reading—and
perhaps also something more, as seen in the following profound passage (I
quote it at length, so that one can see how Coleridge gets us to a new and
brilliant idea via the format of sentences):

The reader [of a poem] should be carried forward, not merely or chiefly by the mechanical impulse of curiosity, or by a restless desire to arrive at a final solution; but by the pleasurable activity of mind excited by the attractions of the journey itself. Like the motion of a serpent, which the Egyptians made the emblem of intellectual power, or like the path of sound through air; at every step he [the reader] pauses and half recedes, and from the retrogressive movement collects the force which carries him onward. Precipitandus est *liber* spiritus, says Petronius Arbiter most happily. The epithet, *liber,* balances the preceding verb; and it is not easy to conceive more meaning condensed into fewer words.[36]

In a sense the project of this book is an effort to understand the physical component of moving thought that Coleridge figuratively describes in this passage. This movement of the archetypal wise serpent is the propulsion of Gans's "old sequence," the propelling secret of Goethe's screw, the wave nature of Emerson's society, the weave of the Platonic sentence that moves us somewhere. In the next chapter I discuss this archetype of serpentine motion more thoroughly. But is this comparison of reading to journeying just an analogy, like Coleridge's simile, or is it a homology, as in the "the path of a sound through air?"

A reader and a speaker are similar to the pattern of the oligochaete's corkscrew—thinking, discovering, and propelling their purposes through three space. Let us retain too, for a later occasion, Coleridge's idea that it is the elastic back-and-forth quality between the body and the medium that allows for propulsion. Pluck a string on a guitar, and it reverberates back and forth; bend a bow string back, and the arrow leaps forward. As they are bent back and propelled forward, the messages elastically follow the form of a wave. But Coleridge wants more from this formal motion of reading backward and forward. Pausing amidst the necessary, hurrying, onward motion of the old serpent allows him to remember and to quote Petronius's sentence about free spirit. It is the reflexive pausing, and the grammatical positioning of the word *free,* that lets him position the existential concept within the *being-precipitated-forward* that is necessity. This fits my distinction between the propelling motion of syntax as a wave form as distinct from the stops and segments of a learned grammar: the formal motion of the precipitance forward, which I think is a function of natural syntax, allows for the grammatical positioning of the Latin construction that sets off freedom amid necessity. By means of this self-referential sentence, one learns a new definition of *freedom* as an abiding kind of reflexive pause and patterned break within the flow of being precipitated onward in time and tide. To bolster this concept, recall the idea of the flow of feedback between internal and external worlds described earlier. The natural philosophy of this environmental principle for the circulating flow of information between external and internal environs may be summarized in Claude Bernard's famous expression, quoted by J. Z. Young in his chapter "Homeostasis," about the self-maintenance of living systems through feedback with an environment: "La fixité du milieu intérieur c'est la condition de la vie libre" (*SM*, 93).

Order, Position, Shape

What if one were to think of the components of a formal wave as ways to begin to understand the physical elements of syntax in natural language? How can one say that the curvilinear form of a motion, a propelling wave that runs through all things, also formulates a syntax that gets us somewhere in speech acts? In what follows I introduce a simple table for defining the compositional elements of a wave form. I think of these elements as a syntax, but I shall reserve a thorough discussion until Chapter 4, "The Anatomy of Syntax." Let us assume that beneath every culturally specific grammar that composes the linear order of a prose sentence, there is the form of a spoken wave that is the moving proportion upon which are superimposed the stops and moves in a series of phonemes ordered by the grammar of one's language. In poetry, however, as we shall see, the syntax of shaped sounds first imposes its rhythmic momentum upon the grammatical units, and not the other way round, as is the case with a prose sentence. Here I want merely to introduce a table of trans-forms that set *position, order, shape,* and *direction* of motion as basic aspects of syntax. These are the commonplace components of any code in language, mathematics, geometry, or cryptography that locate the units in a discursive space. If the form of a wave runs through all natural languages as the carrying vehicle, then a common syntax can begin to be described by this elementary form. It will comprise the elements of composition for a common syntax of identifying the orders of things in natural languages. In this sense then, the hypothesis to be tested is that all natural languages have an *inherent* common syntax, but they have different *inherited* grammars.

Let us say first that position, order, and shape are sufficient deictic pointers for limiting the physical *location* of units in a code. I take these terms from Aristotle. In his forthright way, Aristotle noted that an atom was described by Leucippus and Democritus in terms of order, position, and shape. One recalls that Democritus, the materialist philosopher, was credited with being the first to think in terms of atomic knowledge. Aristotle had been summarizing the traditional sense of Being as the real unifying substance of things:

For they say [Leucippus and Democritus] the real is differentiated only by "rhythm" and "inter-contact" [touching] and "turning"; and of these rhythm is shape, inter-contact is order, and turning is position; for A differs from N in shape, AN from NA in order [arrangement], I from H in position. The question of movement—whence or how it is to belong to things—these thinkers, like the others, lazily neglected.[37]

Leucippus and Democritus are credited with being the earliest proponents of the atomic view of nature, the view that physical things are composed in patterned elementary units. (Against this atomic theory was posed the continuous wave theory of the Stoics, who are discussed in Chapter 3.) Despite Aristotle's disparaging comment about laziness in this passage, he here explains their rudimentary code for identifying characteristic ways of locating atoms in terms of orderly relationships with one another. If one of the major discoveries in the history of thought is that all of nature is coded, and if new sign systems, such as algebra, geometry, and calculus are developed over time to help discern the codes of nature, then Democritus and Leucippus took an early and important step in describing physical atoms in terms of the orderly changes of relationships in alphabetical letters.

Why did Aristotle feature movement? The aspect of moving form that he cared most about was the phenomenon of becoming, the potential in everybody's existential body that gets us somewhere. Put in terms of this Introduction, becoming needs the innate form of a wave to get us somewhere with feet or with other archetypal forms of propelling limbs. And the dynamic pattern of the distributions of atoms—in their characteristic rhythm, inter-contact, and turning—can also describe the locations on the form of a wave or the forms of a natural syntax. For instance, look more closely at the relationships between the letters in Aristotle's example. In order to interpret his meaning, one must act to apply a tacit framework of space and time. One must employ a few transformations of the letters themselves to distinguish between the kinds of relations arranged by the letters. One silently employs a transformation called *translation* to see that A and N have been moved forward along a line of sequence to achieve a new *orderly* group NA. I suppose Democritus was thinking of "inter-contact" as the contiguity of things in a series, like a domino sequence. One sees that I and H are the same letter, only one sees that each has been set in *rotation* on its own axis to achieve a new *position*. I suppose this rotation to be what Democritus and Leucippus meant by "turning." And to see *shape*, one has to perceive that A and N can be assembled out of the same three limbs, but at different angles. I shall use many examples to show how shape, helical

Diagram of Transformations

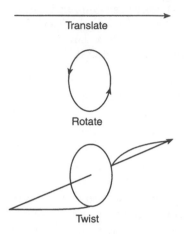

Figure 3A. Diagram of Transformations.
To achieve Order, translate a unit back or forth along a one-dimensional line.
To achieve Position, rotate a unit around on a two-dimensional plane surface.
To achieve (solid) Shape, translate and rotate and twist a unit through a three-dimensional space.
When a unit is Rotated on a plane, moving perpendicularly to a line of Translation, the resulting Twist into three space yields a Helix.

form, and "rhythm" relate in an aesthetic of curvilinear waves and weaves. But basically the point is that the two ancient philosophers are right: shape is never just static; a three-dimensional solid is paused in such a way that its moving rhythms are unapparent. Similarly, I think of symmetry as a kind of moving rhythm held briefly in stasis. In other words, the characters of *order, position,* and *shape* are atomic aspects that depend upon tacit formal acts of transformation which leave their primal characters invariant. The idea that groups of atoms or figures can be transformed while leaving their morphologies invariant is the essence of symmetry theory in physics. So *order, shape,* and *position* are spatial relations that locate things, whereas *translation* along a sequential line, *rotation* of oppositional pairs, and *twisting* of shape are instructions to act in measured rhythms. I diagram these categories in Figures 3a and 3b, and I will use them extensively in the next chapters as primal compositional elements that describe the locus of natural forms moving in patterns.

Aristotle	Location	Geometry	Dimension	Transformation	Democritus & Leucippus
NA AN	Order	Line	1D	Translate	Inter contact
I H	Position	Plane	2 D	Rotate	Turning
N A	**Shape**	**Solid**	**3 D**	**Twist**	**Rhythm**

Figure 3B. Diagram of Transformations: After Aristotle. Aristotle's transformations do not recognize an action through three space. But I have added that third dimension in order to meet the requirements of a physically rhythmic act, which requires motion in a real three space. Notice that to transform the order of NA to AN, one must also rotate the units. To transform the position from I to H, one must also translate the units. To transform the shape of N and A, one must translate the shapes into a new order; one must rotate the legs into their new positions, and one must twist the units into a flowing series to achieve a rhythm.

Because the topic of this Introduction is the form of a motion that may be seen in water waves, I want to recall that in his famous introduction to the *Metaphysics,* in which he summarized the philosophical theories that preceded his own, Aristotle began with Thales' first principle of water (983b). Aristotle noted that water was so old and so honorable that even the gods swore by its primacy. And before leaving Aristotle's *Metaphysics,* let me cite one other passage to help establish the lineage of these forms of motion that seem to me to have shaped the symmetries of off-angled limbs. In his summary Aristotle criticized those philosophers whose epistemology was based only upon sensation and upon "physical alteration." He named Democritus again, and he quoted Empedocles. Here is his quotation of Parmenides about the relation between physical alteration of limbs and the act of thinking:

> For as each time the much-bent limbs are composed,
> So is the mind of men; for in each and all men
> 'Tis one thing thinks——the substance of their limbs:
> For that of which there is more is thought (*M,* 1009b).

Here thought itself is theorized as a function of bent-back limbs. Although I have been arguing that the sensorimotor alteration of limbs evolved amphibiously from sea to land, I have also conjectured that it was the form of motion that shaped those symmetrical limbs. The passage may be interpreted as an early theory of bodily knowing. This elasticity is a kind of drawing back in order to leap forward. One word from morphology that defines this alternating movement is *palingenesis.* It is the elasticity, plasticity, and resistant *tension* of bent-back limbs that interests me in these ancient Greek theories of natural language. The figure of bent-back limbs would have been one of their main agents of propulsion. In Chapter 2, I study a few illustrations from primal art that feature bent-back limbs as designs of springing forward. Elasticity is considered one of the primary principles of physics:[38] as with bungee jumping, the more a thing is forced out of its resting place, the more it will return to it. *Attraction and repulsion, equal and opposite reaction,* are traditional phrases. Elasticity is here to be considered as a propelling concept that also involves a return to equilibrium and symmetrical forms.

Consider, finally, a few fragments of Heraclitus's famous adages about the elastic but unapparent unity that underlies the contradictory nature of the *logos:*

> An unapparent connexion is stronger than an apparent one.
>
> The real constitution is accustomed to hide itself.
>
> They do not apprehend how being at variance it agrees with itself [*lit.* how being brought apart it is brought together with itself]: there is a back-stretched connexion, as in the bow and the lyre (*PP,* 192).

The bow and the lyre are instruments composed of bent-back limbs. In *Philosophy of Symbolic Forms,* Cassirer showed how, among some primal cultures, language was seen to emerge from the formal structures of the natural world. Furthermore, he showed how among Greeks language and the world arose together as *logos* and *cosmos* (*PSF,* I, 117–19). Although language and world-being arose together, their relation was harmoniously polemical, counterstretched like the bow and the lyre. I surmise that both implements are bent back and plucked, one for purposes of war and the other for peace. But there is a deeper and more unapparent connection. The counterstretched elasticity of both tools lets sounds reverberate, one with the reverberating sounds of arrows in flight as they whir from bows and the other with peaceful music. Beneath both is the hidden connection, the reverberating archetypal form of a wave that hides itself. In these archetypal examples that I have been discussing so far, there is the assumption that language and world are mutually shaped by an underlying physical elasticity. The message of war or song was elastically propelled though air in the form of wave. Beneath both is the reverberating archetype of a wave form. Hence *logos,* for Heraclitus, seems to have been a formula of harmonically contrasted patterns whose character is the push-pull formality of flow. I am conjecturing that this back-stretched formula for *logos* is the transformer wave that can transfer over from one medium to another, or that can switch over from one kind of physical force to another, as electricity transforms into magnetism via an electric wire wrapped around a magnet. Just as rhythmic wave forms move from the plucked string of lyre to air through two different physical media, so the wave form moves in reverberating patterns from air to brain. Sender and receiver are unified by the *logos* of a reverberating wave. Further, I think that this archetypal transformer wave underlies the grammars of various languages and that its syntactic pattern of shaping sounds into morphemes is governed by the old sequence of propulsion in animal bodies. The form of a wave, as it extends along the electromagnetic spectrum and the sound spectrum, is thus sufficient topology for studying the composition of things.

Aristotle summarized his metaphysical forerunners by denigrating a Heraclitean sense of movement as all that can be known. He said that the most extreme of the Heracliteans was Cratylus, "who finally did not think it right to say anything but only moved his finger, and criticized Heraclitus for saying that it is impossible to step twice into the same river; for he thought one could not do it even once" (*M,* 1010a). In this kind of radical skepticism, one only goes through the motions, and with one of the smallest of bent-back limbs. Even though I suggested that the Ancient of Days rotated His angled fingers to describe a circle, I want to privilege the wave form of motion that enabled that bodily act. I have been suggesting with Kirk, Raven, and Schofield, and Whitehead, that *logos* is a formula for counterstretched stability and equilibrium within pure motion or process. I shall discuss the scientific and aesthetic formulation in later chapters.

But how is it that sentences take the form of a propelling wave? I cannot answer that question until I have reviewed some basic acts of symmetry theory in later chapters. If the form of wave action can be broken down into translation, rotation, and twist, then spoken sentences, with their puffs and billows of breaths, should have similar transformations that govern their eventual constructs. Translation is an action that is directed along a line. It is primarily one-dimensional. Rotation is an action in a circle. It is a two-dimensional operation of moving between opposite coordinates of up or down, or right or left. For Whitehead in his chapter "The Ideal Opposites," the concept words are permanence and flux, the issues of

"Abide with Me." Twist, as we shall see, is the rhythmic action that combines translation and rotation into a helical corkscrew.

To end this Introduction it is worth remembering that Aristotle's theory of language was primarily physical.[39] The form of the sentence followed the direction of physical becoming. If speech derives from morphologies of natural action, then a theory of language must also be a theory of indirect but natural action. As Frost wrote, "It is from that in nature we were from." For Bertrand Russell, too, language shares the physical forms of the spatiotemporal world:

> Our confidence in language is due to the fact that it . . . consists of events in the physical world, and, therefore, shares the structure of the physical world, and therefore can express that structure. But if there be a world which is not physical, or not in space-time, it may have a structure which we can never hope to express or to know. These considerations might lead us to the Kantian *a priori*, not as derived from the structure of the mind, but as derived from the structure of the physical world. Perhaps that is why we know so much physics and so little of anything else.[40]

The idea of structure to be tested here is that a spatiotemporal form of an undulating wave is a primary form of the physical world whose code is "in" the human mind. This wave, seen as an S-curved series, propels us in motion. Hidden, however, in every one of those abstract concepts of physics, such as "propulsion," or "ram ventilation," is a poetic or artistic trope. Embedded within those scientific abstractions, going beyond Russell's view, is word play and image play where new concepts and languages were born; especially within the grammar of subject and predicate there is plenty of creative play and twist. Our confidence in language also stems from its possibilities for creating hypothetical and fictional worlds that conflict ever so slightly with the structures of Russell's physical world.

I began by stressing the form of the impelling wave motion, and I quoted Emerson's assertion, "Society is a wave." A first reading of that assertion would suppose that Emerson meant the sentence as metaphor. But if it is archetype for a perduring model of the way the societal world moves, then it is worth considering that the form of the physical wave is the prime mover for the kinds of inquiry that propel society somewhere. It is worth conjecturing, I think, that the compositional code that has evolved over millions of years to regulate the ways in which animals maneuver their bodies through waves by means of alternating waves of their jointed and symmetrical limbs is also the prime mover for the ways that humans measure the world of their speech, as well as for other forms of physical communication. That is to say, it is the same body-based code of instructions, the old sequence, that can branch off into the syntax of language, of writing, of mathematics and geometry, and of American Sign Language. That, at any rate, is my effort here, to explore some of these compositions by means of the archetype of curvilinear form.

CHAPTER 1

Art
Frequency

Synopsis

The title of this chapter displays, in a fraction, a way of diagramming the relation between wave frequencies and artful communication. It displays a twofold distinction, and yet fusion, between an uplifted part that is artfully featured or sublimated, while the part underneath is tacitly suppressed. Also, it represents the underlying idea that the carrier part of a message is always a suppressed physical wave frequency. Furthermore, it exhibits the premise that the structure of communication is basically a compound, or conjunction, composed of a series of impulses sensibly received from the environment in waves, together with an intellectual component of the message that may be called an *idea* or an *image*. *Cognition* is a compound that brings together two dissimilar kinds of ingredients, a sensible component plus an intellectual component. If the gist of cognition is a dissimilar fusion of compound patterns, then *thinking* is essentially a metaphorical fusion of that which is above with that which is below the semiotic bar. This idea accords with a current hypothesis in cognitive science that thinking is a process of metaphor-making. If the brain transforms wave frequencies into images, then it is a pattern transformer. And metaphors themselves show signs of being transformational tropes.

Here then I study some tropes that enact this twofold character of thinking with signs, designs, and symbols. One who inhabits two different media is an amphibian. So here, too, I explore some art objects and tropes that self-consciously employ amphibious characters, even while they seem to be tacitly carried, below the bar, by the "old sequence" of a propelling wave frequency.

But first I describe some basic principles of wave frequencies and how they move in rhythmic proportions so as to carry messages by energy transformations. I describe three forms of wave propagation, and I suggest that these forms of wave movement line up adequately with the symmetry operations called *translate, rotate,* and *twist.* This alignment of wave propagation with symmetry transformations will be a basic part of my thesis: underlying the rules of linguistic grammar, such as the rules for nouns with verbs, is the constant pace and scale of a formal transmitting wave.

The Bow and Lyre, Back-Stretched

If you pluck stretched strings on a lyre, they will oscillate in frequencied sound waves. If you pluck a back-stretched bow, its string will reverberate back and forth. But how does this physical issue of bent-back limbs apply to literature? If this kind of oscillating frequency seems remote from poetics, let me recall a passage from one of literature's most famous scenes. When Odysseus finally returned to Ithaca in the guise of a poor wanderer, he contrived a contest among Penelope's suitors. Who could string his own old hunting bow? Penelope herself proclaims the challenge as a wager: whoever is strong enough to bend and string her husband's massive old bow will be her next husband. Its epithet is "back-strung" bow.[1] Then there begins one of the goriest action scenes in literature. Within Odysseus's great hall, now being transformed into the fateful precincts of Apollo the Archer, each suitor in turn tries his strength and humiliates himself. Then the disguised vagabond tries his hand:

> ... but now resourceful Odysseus,
> once he had taken up the bow and looked it over,
> as when a man, who well understands the lyre and singing,
> easily, holding it on either side, pulls the strongly twisted
> cord of sheep's gut, so as to slip it over a new peg,
> so, without any strain, Odysseus strung the great bow.
> Then plucking it in his right hand he tested the bowstring,
> and it gave back an excellent sound like the voice of a swallow.
> (*O*, XXI, 404–11)

At this moment of climax, Homer repeats the epithet of *polytropos* that he had applied to Odysseus in the very opening lines of the poem, which

Lattimore translates here as "resourceful," Fitzgerald as "skilled in many ways of contending," but Cook more literally rendered as "man of many turns." Here is Lattimore's translation of the first two lines of the epic:

> Tell me, Muse, of the man of many ways [*polytropon*], who was driven far journeys, after he had sacked Troy's sacred citadel.
> (*O*, I, 1–2)

In the homecoming scene, Homer explicitly brings together the bent-back limbs of bow and lyre in their back-stretched hidden unity. The twisted bowstring vibrated and sang, all in elastic preparation, for the fateful battle. At this moment of dread anticipation, Homer pauses to *reflect* on the physicality of the underlying unity that is going to send a terrible message, literally, by an elastic vibration. The arrows will whir as they are carried by the energy wave of elastic reverberation through the air, and the "twisted" bowstring will twang. In both examples of bow and lyre, the tense equilibrium of the stretched string is broken when plucked, and the air *reverberates*, "gives back" sound. Just as an arrow jumps from the bow, so the waves jump from the physical medium of the stretched gut to another medium, the air. Waves, we see, are the literal vehicles for carrying messages over from one medium to another. Waves are the common vehicles of transport from the physical world to the physiological body.

As Homer pauses to reflect about the impending final act, one remembers that the root of the word *reflection* is the physicality of bending back, either of bent-back limbs or deflected rays of sunlight. In the Introduction I quoted Parmenides' odd passage about thought as bent-back limbs. If one reads the passage hastily, the passage might seem like a simple formulation of behaviorism: thought as merely a knee-jerk reaction to other substances. But perhaps there is a more profound connection underneath the apparent idea. In what sense is thought itself physically bent back? Because the physical act of reflecting is to bend back and to spring forward, thought itself is a patterned combination of retrospection and anticipation. I shall be pursuing this combination throughout this study. Thought is, in some way still to be described, elastic. This is Homer's pausing attitude as he readies his audience for the action scene to follow. Consider that all figurative turns or tropes of *reflection* may be polytropical variants from sunlight. To reflect also means to return upon itself or to rebound. The *Oxford English Dictionary* cites two examples: "It is that violence, of which he is the author, reflected back upon himself" (1722 Wollaston, *Relig. Nat.*). And, "Each body will therefore be reflected with a velocity equal to that which it had before impact" (1799 J. Wood, *Princ. Mech.*). Even though *to reflect* has a Latin root, there nevertheless reverberates—beneath the varieties of languages, Greek, Latin, English—the physical wave that elastically sends the message. This book could be described as a way of regularizing into a syntax these polytropes, these many turns or transformations from one archetypal form, an elastic wave. Underlying this song of war, there is exposed here, if only briefly, the unity that makes Apollo, the sun god, both the archer and the musician. Homer's bow and lyre are implements dedicated to the power of the sun.

Still pausing upon Homer's connection between the bow and the lyre, I recall Simone Weil's meditation about the ancient Greeks' concept of the symmetrical reciprocity of force, which renders Wollaston's example timely and not anachronistic:

> This retribution, which has a geometrical rigor, which operates automatically to penalize the abuse of force, was the main subject of Greek thought. It is the soul of the epic. Under the name of Nemesis, it functions as the mainspring of Aeschylus's tragedies. To the Pythagoreans, to Socrates and Plato, it was the jumping-off point of speculation upon the nature of man and the universe. Wherever Hellenism has penetrated, we find the idea of it familiar. In Oriental countries which are steeped in Buddhism, it is perhaps this Greek idea which has lived on under the name of Kharma. The Occident, however, has lost it, and no longer even has a word to express it in any of its languages: conceptions of limit, measure, equilibrium, which ought to determine the conduct of life are, in the West, restricted to a servile function in the vocabulary of technics. We are only geometricians of matter; the Greeks were, first of all, geometricians in their apprenticeship to virtue.[2]

Simone Weil's monograph about the symmetrical reflection of force in *The Iliad*—"Ares is just, and kills those who kill"—is in my experience one of the most moving examples of literary criticism written about Western literature. Describing Homer's counterpoint of bitterness and poignancy for those who are turned to stone, to unfeelingness, by the grip of violent power, both as master and slave, conqueror and conquered, was her way of predicting in 1940 a baneful twist of the sword that would reflect back against the victorious Nazi invasion of France. One wants to believe in her conviction about the elastic reciprocity within violent action, that the indifference of those who grip force will be reflected back geometrically by the action of a physical counterforce. The Greek idea of social retribution is built upon the physical feeling of bent-back reflection. Force has an elastic or bent-back form.

Hers was a theory of literature and language and ethics that was based upon the physicality of reflecting force. It was not just a theory of literary language but instead a conviction about the hidden relations between literature and the form of an underlying physical action. My task is less arduous than hers, but it is related to her description of "geometrical rigor," and that is to suggest here that the syntax of language is unified by the underlying carrier of bow and lyre, and that a poetics of language can be served, neither by a geometry of matter, nor a geometry of technics, nor even by a geometry of virtue, but instead by an elastic geometry of form that describes the pattern of recurring rhythmic waves.

The timing of Homer's style of bent-back action in a plot is itself an awareness of the moving proportion between the destiny of the hidden plot and its eventuating narrative. Timing is an issue of sequential *order, position,* and *shape* of a figure in action, which are Democritus's categories for describing the formal proportions of atoms. This plastic sense of timing is so difficult to talk about that, to describe it, the best writers resort to those archetypes that recall the physicality of things being worked out in space and time. Here, for instance, is Robert Louis Stevenson:

> The right kind of thing should fall out in the right kind of place; the right kind of thing should follow; and not only the characters talk naturally and think aptly, but all the circumstances in a tale answer to one another like notes in music. The threads of a story come from time to time together and make a picture in the web; the characters fall from time to time into some attitude to each other or to nature, which stamps the story home like an illustration. Crusoe recoiling from the footprint, Achilles shouting over against the Trojans, Ulysses bending the great bow, Christian running with his fingers in his ears, these are each culminating moments in the legend and each has been printed on the mind's eye forever. . . . This, then, is the plastic part of literature: to embody character, thought, or emotion in some act or attitude that shall be remarkably striking to the mind's eye. This is the highest and hardest thing to do in words. . . . [3]

Whenever one wants to achieve this elastic anticipatory quality in a culminating act or attitude, leaning toward climax, one usually describes an image of embodiment that is full of recoiling countertension. Some physical derivative of a wave, such as twisted threads or such as music or a woven web, is liable to be employed to help describe this poised attitude. In Stevenson's examples the plastic part of the plastic arts are applied to literature.

Force Fields

In our era the most pervasive image of continuous recoil is of a tensely oscillating force field.[4] The natural trope most habitually used to represent the social uncertainties of the past century is a whirly continuum in space and time, seen as a series of S-curved waves oscillating and reflecting between two poles. According to Richard Bernstein, the metaphor of a force field pervades the "new constellation" of postmodernism.[5] Or as Gerald Bruns aptly paraphrases the mélange of critical controversy:

> Our world is a contest of narratives, a whirl of theories and practices. It is not one thing but an accumulation of forms of life both ancient and modern, near and remote, advanced and backward, alien and familiar, all arriving in California and spreading outward in bits and pieces in every direction. I am a modern, a creature of modernity, just insofar as I don't know where I belong in any of this; or rather, I float free among all the fragments of discourse and action that swirl though time and space like radio waves and electronic images. No one can pin me down.[6]

But what has been habitually bypassed in metaphorical discussions of these force fields and radio waves and electronic images? I think we misunderstand or suppress the actual physics that explains the force of the trope. Oscillation is not just a sign of free-floating whirls. Indeed, for Paul Davies in *Superforce*, "periodic motion, or oscillation, is perhaps the most widespread example of order in physics. Wavelike oscillations lie at the heart of all quantum motion; electromagnetic waves carry light and heat across the universe; planets, stars, and galaxies all involve objects moving on periodic orbits through space" (*SF*, 241). In other words, all the great laws of physics, even gravity as we shall see, manifest themselves in signs of periodic order that are experienced as the moving forms of waves. Waves are not symptoms of disorder, as the popular preconceptions would have it, but waves are omnipresent symptoms of certain dynamic laws of form that can be described by symmetries. Waves transmit semiotic symbols, of course, yet they always transport signs in rhythmic patterns of distribution. Even the flotsam and jetsam on a beach are randomly distributed by waves in patterns of likely shapes. Without understanding some of the physical forms of wave motion, one can get misled by the idea that these radio waves are mere metaphors instead of archetypes of periodic transmission. So I want to look more closely at the underlying shapes and rhythms of sound and radio waves. What is it, physically speaking, that makes the plucked bow sing?

Three Types of Wave Motion

In the Introduction I began to describe some of the formal character-istics of wave motion, such as the seeming paradox that it is the form itself that moves. In some early scientific accounts, the formal aspect of wave dy-namics had been slighting. For instance, Willard Bascom, perhaps the fore-most authority on the practical dynamics of surf and beaches, imagined the experience of the first person to think through a theory of waves.[7] Initially, the person idly watched a stick bobbing up and down, while the waves passed along underneath it: "But suddenly this man saw in his mind the fact that waves are only moving forms and that the water stays in the same place. The stick, and the water around it, moved in a slow circular os-cillation as each wave passed." Although Bascom claimed that this discov-ery of an oscillatory theory of waves was "roughly equivalent" to Newton's thought experiment about the gravitating apple, he still passed over the questions about the moving form. What is hiding under what Bascomb terms "only" forms? This is my pursuit, the primary moving forms of waves that are omnipresent in frequencies but that seem to have been dismissed into the subconscious of some sciences and humanities. But it is to this swell that the poet Ammons goes in his mind where all such waves "are."

These up and down motions, in which the stick just bobs up and down, are now called *transverse waves*.[8] Trefil and Hazen observe that if you "do the wave" in a football stadium, it seems as if a wave undulates around the field. But all the individuals are just bobbing up and down. "Longitudinal waves," however, are pressure waves that carry force. Trefil and Hazen use the domino effect for a visual example of particles that stay in place but that create an accelerating formal wave that moves along a horizontal axis by the toppling of each domino's neighbor, in what structuralists would call *a* metonymic series of contiguous neighbors. What they say about sound, as a longitudinal wave, will be instructive for later discussions about S-curved waves as the representations of carriers of messages:

> Sound is a form of wave that moves through the air. When you talk, for example, your vocal cords move air molecules back and forth. The vibra-tions of these air molecules set the adjacent molecules in motion, which sets the next set of molecules in motion and so forth. A circular wave moves out from your mouth, a wave that looks very much like ripples on a pond. The only difference is that in the air the wave crest that is moving out is not a raised portion of a water surface, but a denser region of air molecules (*TS*, 144).

Trefil and Hazen here provide a protogeometry of circular images and a jostling series that may be applied to Fodor's linguistic question about the motion of waves. The mouth cavity and the throat and the lips create these spherical waves that explode from a speaker toward a listener. More about this aural phenomenon will come in Chapter 4. It will become useful later to clear up one point now about the circular shapes of waves. If one thinks iconically about the ripples on a pond, matching up the written phrase "very much like ripples on a pond" with one's own visual trace, one 'sees' the ripples as being only slightly raised above the two-dimensional flat plane or pond surface. But the shapes of sound waves are roundly helicoid. As Ben Bova explains, it was Christiaan Huygens who first proposed that energy is propagated in *spherical* wave fronts: ". . . the waves of light coming from a candle are spherical: they expand outward spherically, like swelling balloons."[9] They are three-dimensional in the contours of their spherical shape, and they are one-dimensional in the elapsed time of their recurrence. The waves of a space-time model are four-dimensional and they "warp" in helicoid shapes.

In addition to transverse wave motion and longitudinal wave motion, there is also a third form of wave motion called *torsional* wave motion.[10] In order to review, I have redrawn Prout and Bienvenue's diagram as Figure 4. The main idea of "wave propagation" is that even though the particles of the medium are closely contiguous, it is energy that is being propagated, not matter (*AY*, 54). This is the conundrum of the Introduction: wave energy is being transported, not matter. Notice in the illustration that the particles are represented as if they move up and down, or back and forth, or around and about an axis of translation. Hidden in the idea of back and forth, up and down, and around and about, is the more basic principle that the particles return to equilibrium by means of some elastic quality of push and pull that is sometime identified as the Newtonian principle: for every action there is an equal and opposite reaction.

When the propagating energy is emitting in regular sequential intervals, it is called a *periodic motion* (*AY*, 15). When one transposes that periodic motion to a graph of regular troughs and crests, one calls the S-curved signs that delineate the curvilinear pattern *a simple harmonic motion* (*AY*, 19). So the diagram represents a series of oscillating periodic motions in which the particles are seen to be jostling off angle from the direction of the propagating wave energy. "Torsional" wave motion is one in which the particles "twist" around the directional axis of wave propagation. Sound waves move in this twisty way through solid matter, such as the bars of xylo-

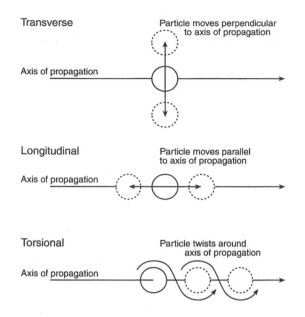

Figure 4. Types of Wave Motion. Redrawn from *Acoustics For You*, Figure 4.2, by James H. Prout and Gordon R. Benvenue, 1990. Permission Krieger Publishing Company, Malabar, Florida.

phones (*AY*, 57). And when the bars reverberate their tones, the wave forms themselves move out from the metal to the air in helical wave crests and troughs. It is important to understand, I think, that the moving wave forms are truly the transporting vehicles from one physical medium to another. The wave form transmits energy in patterned reverberations from metal to air to brain. There is no subject-object split in this theory of wave forms. If waves everywhere oscillate across from one physical medium to another, then they are sufficient transporters of rhythmic signals in all forms of communication.

These three forms of wave motion, represented in Figure 4, will serve as the basis for my eventual thesis about the physical transformations of syntax. I have chosen three transformations from symmetry theory to stand for certain transformations in the sounds of rhythmic speech. Translation along a line, rotation in a circle, and twist over and under, are the protogeometric commutations to be studied later. (See also Figure 3). One can see in Figure 4 how these symmetry transformations may be made to coincide with the three forms of wave motion. Although not a geometry of virtue, these geometric turns of energy are transformed into images by the brain. As such, these three kinds of turning waves comprise a group of *polytropes*. One can see now that these transformations match up with the actual movements of energy by wave propagation. I am proposing then that *these symmetry operations reenact the forms of a wave motion.* Because they occur everywhere that waves are formulated, they will appear in the sound waves of speech acts. They can be seen as a few basic units of commutation and composition that make up the syntax of a physical construction in space and time. That is, translate, rotate, and twist, are the few acts of transformation that minimally suffice to describe position, order, and shape within the momentum of harmonic waves. They reenact, but they do not imitate, the directions and momentum of wave motion. This distinction will become important much later when we decide that the speaking of a sentence has an actual momentum in direction, because of the recurrence of sounds that underlies the message. Just a few forms of physical composition suffice to be this rudimentary syntax.

Trefil and Hazen later say that an electromagnetic wave is an extension of the wave propagation model, but unlike the medium of water or air, this kind of wave "has no medium whatsoever, but simply keeps itself going through its own internal mechanisms" (*TS*, 151). What is hiding under "simply"? The not-so-simple truth is that nobody knows what domino prompted the energy of the first formal wave. Our scientific creation narrative is the evolution of the forces and the chemical elements from a Big Bang. According to that moment of creation, it would seem that all of the power frequencies in the universe have been expanding and flowing torrentially and inevitably since that instant, but nobody knows yet what was the occasion. Here is the issue of the moving form of a wave, baldly defined by Trefil and Hazen in their synoptic science text for college students: a wave is a "traveling disturbance that carries energy from one place to another without requiring matter to travel across the intervening distance" (*TS*, Glossary).

Three Modulating Waves in a Communication Signal

How could this formal motion have developed so that it could serve as a syntax underlying the languages of humans? Before suggesting one theory for such a development, let me sketch a model of the present state of communication. For it is easier to tell where something may have come from, once one has provisional agreement about where it is at present.

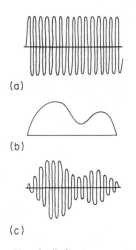

(a)

(b)

(c)

Figure 5. Three Forms of Wave Signals. "The propagation of wave signals: (a) a high-frequency carrier signal, electronic or optical; (b) a low-frequency analog signal, sound waves, for example, (c) the carrier signal is modulated by the amplitude of the analog signal. . . ." From Sobel, *Light.* With permission from The University of Chicago Press.

Common to the transmission of all messages over the electromagnetic spectrum is the form of the wave. Michael I. Sobel introduces the principle of transmission common to telephone, radio, and television:

> A wave "carries" information by the principle of *modulation.* For example, a radio wave with a frequency of 30 million hertz (cycles per second) can have superimposed on it a variation of its amplitude with a frequency of 440 hertz (from the sound of the note A in the middle of the piano keyboard); it then transmits the signal to a point where a receiver and speaker reconvert it to sound.[11]

The superimposition of the sound signal of the message upon the carrier signal results in the modulation of both waves into a third form of wave, with its own distinctive amplitude. (See Figure 5.) That is a commonplace of telecommunications, the present state of the art. This principle of modulation will be necessary for my thesis about the syntax of speaking language. Notice that the wave forms must have distinctive shapes in order to carry the message. Because the transmitting waves already have distinctive shapes in moving proportions through four dimensions, they are projecting their physical shapes, which will then be modulated as the essential syntax of the formulated message. These waves are not random slosh or noise. To the extent that the waves already have characteristic shape, positions, and orders of recurrence, they already have a physical syntax.

Furthermore, this modulating form of the wave explains how information can be transmitted by "wireless" waves. The message is being sent out from an antenna in all directions in three-dimensionally shaped wave fronts. Just as the light waves from a candle radiate outward from the light in three-dimensional forms, so, too, do radio transmissions swell outward in all directions from the source. It is the moving form of the wave in three situations that is modulating and propelling the signals. So now one understands how a sentence can be both a vehicle of phonetic transmission, transmitting by wave motion, as Fodor and Trefil noted, while underneath the grammars of languages there flows a "deep memory," an "old sequence," that is the *Urform* wave of propellancy. It is the ubiquity of waves, together with their direction of motion and their scales of moving proportion, that generate a syntax of universally commonplace forms. Notice that this physical syntax of wave forms implies a necessary movement of the waves *in a direction,* toward an unfluttered region into which the sound waves have not yet projected. The orderly compositions of wave energy always move in a direction, toward the unspoken. A spoken sentence always literally moves in a direction toward a presumed listener.

Let us retain Sobel's diagram as an initial way of distinguishing between an underlying curvilinear physical syntax and a superimposed grammar from one culture or another. The carrier wave (a) can also represent the physical form that transports the message (c) as a vehicle while a grammar (b) is the culturally specific set of instructions for composing sound patterns into their own learned arrangements. The carrier wave provides the physical form that enables us to think in terms of wave *vehicles* in the first place.

What is propelling these vehicles that allows them to steer in a certain direction? Consider, for example, that in *Philosophy of Symbolic Forms* Cassirer paraphrases Heraclitus when he says that *logos* is "helmsman of the cosmos" (*PSF,* I,119). This is one of the examples he used to show the simultaneous emergence of language and being. The Greek word for helmsman, "pilot", or *cybernos,* was also one of Socrates' most common metaphors for self-controlling guidance. Just as fins or legs or wings are jointed in different forms to propel through carrier waves, so syntax has developed different kinds of cybernetic tillers and rudders and steering oars to maneuver through the waves of the old curvilinear physical sequence. So I am suggesting that it is useful to distinguish between an underlying syntax of wave shapes that really carries and propels the message in a direction, whereas grammars can be seen as compositional rules that segment, stop, and divide the phonemic forms of waves into lexical arrangements. This distinction then allows the assertion that the generative shapes of natural languages have, so to speak, always been with us, as Chomsky and Gould have said, without evolving, whereas grammars of sophisticated languages did evolve and then did develop in their own histories. This distinction allows for an interactive compromise between those theorists like Dennett and Pinker who think that language evolves, and those like Chomsky and Gould who think that language follows physical "laws of form" that need not have evolved (*DDI*, 391–95).

Perhaps the most important feature of this kind of wave model for the propagation of physical energy in a direction is that it represents a moving pattern for the physical transfer of signs from the Past into the Impending. The underlying wave frequency that carries the message is the format for the transmission of energy. These waves are sending a physical patterning to the brain for its mapping of the world. So then, what is the form of motion? I think it is the ripple effect of a curvilinear space-time, for which we have not yet an adequate conceptual word or lexicon, and for which we invent metaphors, archetypes, poems, and pictures to make up for the lack of agreeable terms.

For there are many examples of art works and poems about wave theory that are based upon a tacit agreement that we shall agree to misunderstand the actual physics of the medium. In the poem, "The Railway Children," for instance, Seamus Heaney implies that, in knowing more than children about electrical transmission, we have also lost something of a miracle:

> When we climbed the slopes of the cutting
> We were eye-level with the white cups
> Of the telegraph poles and the sizzling wires.

> Like lovely freehand they curved for miles
> East and miles west beyond us, sagging
> Under their burden of swallows.

> We were small and thought we knew nothing
> Worth knowing. We thought words traveled the wires
> In the shiny pouches of raindrops,

Each one seeded full with the light
Of the sky, the gleam of the lines, and ourselves
So infinitesimally scaled

We could stream through the eye of a needle.[12]

The poem begins with a point of view open to children who can climb railway embankments and who can thereby get close enough to the lines to hear the sizzle of electronic transmissions. (The point of view is also like landscapes by Edward Hopper that portray the viewer's line of sight as if one were in a train upon an embankment, so that one sees only the top halves of row houses.) The telegraphic wires, like freehand lines, are curved where they apparently swag under the nil weight of swallows, and are presumably drawn up at the pole terminals, so that one reads and sees a series of concave arcs that sag down and draw up along the length of the span, as far as the eye can see. In their innocence and visionary ignorance, the children thought that words themselves traveled the telegraph wires. And the words were light seeded and gleamed with the light of the lines that carried them. That synaesthetic mix of seeing and hearing such words rendered the children so infinitesimally scaled (like swallow weight) that they embodied a small miracle of those poor in spirit who can enter the kingdom of heaven. Because the poem does reenact the miracle of that visionary seeing, its poetic lines, with their swagging confluence of "w" rhythms, merge with the telegraphic lines, and as in Wordsworth's poetry, we are reminded fleetingly of the vision that we may have forgotten.

Another example of a visionary misunderstanding of the physics of electronic transmission is remarkable because it was written in 1900, very early in the popular culture of electronic messages. When she was twenty years old, Zitkala-Sa wrote an autobiographical account for *The Atlantic Monthly* of her sudden transport as a child from a Sioux reservation in South Dakota to a Quaker school in Indiana. In the train, which she calls an "iron horse," she describes her tearful embarrassment at being rudely stared at by fair women and blue-eyed children:

I sat perfectly still, with my eyes downcast, daring only now and then to shoot long glances around me. Chancing to turn to the window at my side, I was quite breathless upon seeing one familiar object. It was the telegraph pole which strode by at short paces. Very near my mother's dwelling, along the edge of a road, thickly bordered by wild sunflowers, some poles had been planted by white men. Often, I had stopped, on my way down the road, to hold my ear against the pole, and hearing its low moaning, I used to wonder what the paleface had done to hurt it. Now I sat watching for each pole that glided by to the last one.[13]

Just as the relative motion of the telegraph poles is seen as "striding" so they are also seen as having been "planted," like sunflowers. They moan because the child believes their spirit, like hers, has been hurt by the transplanters. By means of this early use of relative motion and the projection of her own fears onto an inanimate thing, Zitkala-Sa describes the loss of her own animating spirit by means of a new tutelage. By turning her face away from the stares that fix her in the seat, she recollects her past and she anticipates the future by recurrence of poles until she reaches "the last one." It is a very early example of the moving picture of thought, one that works because she projects the illusion of motion on an orderly series of static objects. But because the child apparently does not understand the principle of relative motion, the adult writer is able to render the child's body as inanimate, while her thoughts race ahead. Part of the elegance of the passage is the restraint by which thoughts can be unexpressed verbally, at the time, but still can be carried by the archetype of the transporting vehicle.

Another "misreading" of the sound of the transmitting carrier is Henry Farny's painting "Song of the Talking Wire" (1904). Farny was a Cincinnati artist who, beginning in the early 1880s, nostalgically painted the traditional ways of Native Americans degenerating under the expansion westward of the "Fire Horse."[14] Farny wrote in a letter of 1884 that at an Indian agency in Montana he had seen a man named Long Day who would stand next to a telegraph pole and listen to its hum. Because Long Day wanted to become a medicine man, Farny wrote, he would listen at the pole and tell others that he "heard spirit voices over the wire" (33). Farny's written account lends itself to the possibility that the trainee might be misappropriating the white man's power, but the painting itself depicts nothing but the subject's resignation. At dusk, amidst a snowy landscape, a lone Indian, wrapped in a buffalo robe and backed by a luminous faint sunset, faces the viewer while leaning his body and head against a telegraph pole, poised eastward toward the danger of the message. Here the wires are rendered only faintly as disappearing lines, but poles are featured, retreating into the background. Here the title does the appropriating. It renders a typifying figure of speech into a translation of a new song of power. The White Man's "song" is of the Red Man's disempowerment. Depending on the listener, the song is of the future or of the past; but because the subject is featured in stark isolation, the effect is of hyperbolic pathos for those who mistranslate the mystery of modern telegraphic power.

For a later visual example, Charles Burchfield's "Telegraphic Music" (1949) enacts a visionary music with oscillating strokes of arcs that are the "lines" connecting a landscape of telegraph terminals.[15] (The word *arcus* is Greek for "bow," so the shape is all that remains of the elastic transmitter.) And the whole landscape pulsates in curvilinear waves: trees, clouds, birds, the ground, the poles—all are bird-winged arcs of vibrational shapes. In these examples a misreading of the nature of the transmitting signals allows one to focus on the moving forms themselves. All these works of art show minds trying to divine the nature of the transmitting lines. In focusing on that part of the message that is usually suppressed, they project onto the transporting medium their own visions.

Projective Arts

I began this discussion about force fields of energy by questioning the social application of wave theory to postmodernism. Let me state my chief presupposition about projecting the wave from the bow and the lyre onto social issues. Agreeing with many critics' misgivings about "Heidegger's silence," as Richard Bernstein called it, I still want to retrieve Heidegger's sense of "projection" from *physis* as a set of prejudgments about interpretation.[16] In *Myth of the State*, written toward the end of World War II, Cassirer exclaimed against the coinage of "magic words" in German philosophy that had contributed to the destining of that state.[17] Recall that all of the artful examples that I just discussed, in different ways, do enact or embody "magic words." Specifically, Cassirer paraphrased Heidegger's *Geworfenheit*, one's "being-thrown," in order to attack it:

> To be thrown into the stream of time is a fundamental and unalterable feature of our human situation. We cannot emerge from this stream and we cannot change its course. We have to accept the historical circumstances of our existence. We can try to understand and to interpret them; but we cannot change them" (*MS*, 369).

This kind of mythic fatalism, Cassirer said of Heidegger's destining, may have been a "pliable instrument in the hands of [Nazi] political leaders" (*MS*, 369).

How did Cassirer think of such projections of physis onto human life? He described a dialectic that was different from Heidegger's destining and Weil's reflective Nemesis. As I mentioned at the end of the Introduction, Cassirer quoted Heraclitus in *The Philosophy of Symbolic Forms* in order to discuss a "dialectic unity" in Greek thought between rationality (*logos*) on the one hand and a destining will to power on the other (*mythos*). This unity was an "attunement of opposite tensions . . . [back stretched] like the bow and the lyre."[18] He returned to this elastic image at the end of his life. Still seeking a trope for harmony, Cassirer also quoted the adage of bow and lyre as the last sentence of *An Essay on Man,* his summing up for Americans of the philosophy of symbolic forms. First and last, this dialectical unity between apparent opposites depends upon a feeling, a recurring kinesthesia of natural forces in counterpressure. Most arguments about the meaning of *logos*—as the Word, as a ratio for stability, as a synonym for *polemos* or conflict, or as a law of change and phase—depend upon the back-stretched feeling of countertense reverberation. But the physically periodic form of elastic wave energy is left tacit.

Forewarned, if not forearmed, about the myth of a necessitarian stream of time, I want to reconsider a kind of "projection" that is carried by the physical vehicle of wave frequency. For Heidegger, projection also meant to throw forward, out of the habitual forestructures of one's cognition, as an arrow is directed from a bow. When we think of the arrow of time, or the stream of time, one of several archetypes of streaming to be considered in this study, it is possible merely to see an arrow drawn on a sheet of paper to signify direction, as in Figure 1. But time's arrow is now seen as a function of a primal transmitter, the incalculable force of the Big Bang, which gave asymmetry and direction to the feeling of time passing. This model of the Big Bang is our archetype for all other physical models that employ waves radiating from a source in expanding shapes of four-dimensional waves. *Physis,* for the Heidegger of *An Introduction to Metaphysics,* did not mean mere physics or *natura*; rather, the Greek for him meant a fundamental "emergence" from concealed physical ground for the making of those significant Heideggerian terms of *techne* and *poiesis.* (Although the Greek *physis* apparently derives from the verb that signifies "to be born," the Latin *natura* is rooted in *natum,* whose meaning we retain in words such as *natal.*) Not wanting to be implicated in Heidegger's destining, I suggest that *physis* includes the underlying shapes of wave forms that *propel* or *transmit* or *project* a message, like the back-stretched projections from the bow and the lyre, into an emergent hypothetical future of meaning-to-say. In that regard, wave-curved patterns may be seen *retrospectively* as natal signs for a projection of meaning. In this sense, the physics of light waves, sound waves, gravitational thrust and counter-thrust, thermodynamics, such natural energies are transformed and socially constructed toward new ends by *techne, poiesis,* and *phronesis.* Just as the brain transforms wave shapes into images, so humans transform natural energy into various kinds of social power. But even the most subtle social power always depends upon the hiddenness of the natural carrier wave. Real light waves, for example, can be bent to new angles and physically projected as the illusions of Plato's cave.[19] When I project the idea of Plato's "table," when I read it or speak it or point to it or thump on it, I am not referring *to* an object, *to* an it. I am referring *with* radiant energy, riding different physical frequencies, at different times, upon which to project my meaning. In this sense, a symbolic projection is an imagined supposition about action, a hypothetical act, that nevertheless "rides" real force lines of S-curved waves as the transmitter of the intended meaning into the indefinite future of a proposition. If a future is a projective fiction, it is the very realm of the spatiotemporal assertions of sentences. There will be more about this in later chapters.

In order to point this issue of projection and transformation more directly in terms of my thesis about wave forms, recall that in the Introduction I began with the idea that Goethe was one of a number of natural philosphers who studied the ways in which one form of energy could be converted or transformed into another. Benjamin Franklin was the "modern Prometheus" who demonstrated that lightning could be converted within a jar as electricity. He demonstrated empirically that two very different forms of Promethean fire were aspects of one *Urform* of energy. When Michael Faraday, the sorcerer's apprentice of Humphry Davy, wrapped an electric wire around a magnet, he constructed an electromagnet that would later become the model for the dynamo, that transformer of a power unsurpassed until the advent of nuclear technology. In Heidegger's well-known essay "The Question Concerning Technology," he used the example of a hydroelectric plant on the Rhine River as a dangerous example of the "standing reserve" of technological power (*QT,* 4–6, 15–19). Written after World War II, this essay may have been a response to social critics like Cassirer who doubted his concept of "destining." That gigantic transformer serves as a model for him of the danger of a technological revealing of the energies of nature; it serves, too, as a way of discussing the transformations of *physis* into *techne,* even *poiesis.* Like the sorcerer's apprentice, he asks the question, "Who is master here?" Eventually this kind of study of different forms of energy allowed for the generalization in science that all forms of energy are interchangeable or convertible from one form to another. The point of my brief excursus is that the vehicle for all these forms of energy transformation is the periodic order of wave forms. When wave forms move from one form of energy to another, order is projected, transformed, expended, and stored in another locale. When air waves reverberate the eardrum, that form of resonant energy is converted in the brain into another form of energy, the electrochemical reaction of immense numbers of neurotransmitters.

Charles Olson featured the breath of the body as it projected the stuff of *poiesis*. For him, "projective verse" meant something more American than Heidegger's destining: "*(projectile   (percussive   (prospective*". [20] Olson asked, what is "the *kinetics* of the thing? A poem is energy transferred from where the poet got it (he will have some several causations), by way of the poem itself to, all the way over to, the reader. Okay. Then the poem itself must at all points, be a high-energy construct and, at all points, an energy-discharge" (*PV*, 19).

For Olson the new physics was not an extended metaphor. The energy or "field composition" (*PV*, 16–20) of the poem was not a metaphor. He meant that there is a real physical transmitter that projects through the poetic field. It is breath: "Because breath allows all the speech-force of language back in (speech is the "solid" of verse, is the secret of the poem's energy)" (*PV*, 19). This is exactly true. The breath of speech projects from one syllable to another all along the line. This is not a theory of artistic imitation or of production; it is a series of clapping plosives, felt by the ear. Olson says that this physical percussion breaks up conventional syntax as we know it in grammar. Because breath has solid shape, the physical syntax of poetry is a physical syntax of solidly breathed shapes all along the line. That at least is the thesis to be tested in later chapters.

Physis serves as the necessary vehicle for any ride upon frequency, which may be turned toward a conditional future. To what ends do I choose to appropriate natural force? How do I safeguard against the danger and violence, or the caring, that others might project upon me and I upon them? In this regard *techne*, *poiesis*, and *phronesis* are critical points of view on the projected possibilities of action, that is, the powers of fictions. In what follows I review the physical archetype of wavy oscillation from several projected points of view: as an Aristotelean metaphor of ratio and proportion, in Cartesian fractions of matter in motion, and in waves of aesthetic frequency.

Amphibious Metaphors in Ratio

Remember Odysseus' penance for having tricked and blinded Polyphemos and for having mocked his father Poseidon, the lord of the sea waves? In order to end his perpetual journeying, Odysseus must travel inland, carrying a well-shaped oar that acts "for ships as wings do," until he meets people who know nothing of the sea. In Hades, Teresias the seer instructs him:

> When, as you walk, some other wayfarer happens to meet you,
> and says you carry a winnow-fan on your bright shoulder
> then you must plant your well-shaped oar in the ground,
> and render ceremonious sacrifice to the lord Poseidon. . . .
> (*O*, XI, 127–30)

The same *shaped* thing is given different names by different kinds of workers. Although its shape remains physically unvarying, its designating word shifts with different trades in different locales. The *physis* of its Signifier, that which carries the message, remains the same physical shape, but the Signified changes socially with its lexical definition as it is embedded in the local grammar and diction. This transformation is another play on the idea of *polytropos*, man of many words. Note that the tool is in both trades a lever that deflects elements by means of energies of wind and wave. Without recalling this incident, Michel Serres cites an ancient Greek adage that divided the world according to the ways that farmers and sailors saw the same tool: one saw it as a shovel, another an oar.[21] But now both kinds of natural workers are dying out, replaced by automobilists. What the sailor and the farmer had in common was a tacit agreement naturally connecting them with their ground by a commonly shaped tool, but known technically by different names for different purposes. In this study, *physis* will be known tacitly by the three-dimensional shape of wave energy that carries the message and not by the dictions and grammars that characterize its eventual uses in languages.

Odysseus was helmsman, *cybernos*, of his ship. By means of the cybernetics of a steering thing, he rode the sea waves as automobilist. But he was forced to see that, despite his skill in contriving words and deeds, as polytropos, he did not shape his direction toward Ithaca entirely by his skill as navigator. One must pay compensation to the sea and the land. Plato's leading trope for wisdom was helmsman or pilot, and his image for a godless world was a helmless ship. "Ploughing the sea" was perhaps an old trope when Homer used it in *The Iliad* (Book III). This old turn, by which an oar and a winnow are seen as the same instrument for plowing furrows of significance in different frequencies, was used by Aristotle to discuss the ratios of metaphor. When one says, "The ship plows the waves," we learn that four things are related, not two. As Hugh Kenner interpreted the ratio, "The ship does to the waves what a plow does to the sea."[22] Donald Davie supple-

mented Kenner's total to six. In addition to the four, one must add the action of plowing and the action of sailing (*AE*, 41). Supporting Ernest
Fenollosa's manifesto that the line of language reenacts the direction of
natural force, Davie argued that the rhythmic acts of sailing and plowing in
nature are essential to the ratio. Following Aristotle, Fenollosa said that language duplicates the spatiotemporal power of nature by the directional
momentum of the sentence's subject, verb, and object. For him, too, the poetics of action is grounded in natural force. (I'll discuss Fenollosa's idea of
a sentence further in the next chapter.) But as we learned from the Greek
adage and from Odysseus' penance, there is a missing term that demands
the metaphoric exchange. There seems to be an unapparent ratio (*logos*) for
which there was no word in common by which to explain Odysseus' steering or guiding device, but for which there was a material shape and which
prompted the need for metaphor. But an English word roots the commonality: the "tiller" of the soil is preserved as a synonym for a farmer, and the
"tiller" of the sea is preserved as another name for the stick or oar that connects to the rudder. Perhaps most important, because unnoticed, is that
both leave unintentional wakes as frequencied traces of their winnowing.

Implicit in this count of ratios is the act of imagining in the first place.
The ship plows the waves. To imagine with metaphor is not just to see the
other fellow's point of view, but to find common ground by projecting
one's self into the other's work shoes and tools, whether oar or winnowing
fan. As old as critical theory, the imagination was, for Hannah Arendt,
Kant's aesthetic sense of social judgment seeking consensus:

> that the capacity to judge is a specifically political ability in exactly the sense
> denoted by Kant, namely, the ability to see things not only from one's own
> point of view but in the perspective of all those who happen to be present;
> even that judgment may be one of the fundamental abilities of man as a po-
> litical being insofar as it enables him to orient himself in the public realm, in
> the common world—these are insights that are virtually as old as articulated
> human experience.[23]

But by gathering up words into linguistic ratios, by reifying the physical force of ground and sea into linguistic ratios between words, one sublates a missing trace that really unifies the action. The ship plows the waves.
The two kinds of natural workers transformed the latent energies of soil
and sea to their own uses with a common tool expressed in different terms.
We usually forget that each of us is *polytropos*, a transformer of winged
words. Odysseus was forced to learn the hard-and-long-way-round that his

wily ride on the sea was at the expense of the sea and the wind frequencies. So his compensation, his sacrifice, is that he must learn that he cannot remain a rationalist, a tricky economist, one who thinks only in the calculating ratios and exchanges of linguistic deals.[24]

Serres laments that we have all forgotten that agreement with nature: all of us, not just language theorists and busy industrialists. We have forgotten that all language, all sign systems, are carried by the suppressed vehicles of natural frequency. In a well-known essay about metaphor, "White Mythology," Jacques Derrida doubles back on the trace of that conviction, the Aristotelean paradox of metaphor's missing term.[25] His brilliant preoccupation, his dazzling pivot, has been "the concept that erases the trace," the concept that sublates a curved furrow or wake. In this essay one finds Derrida's most extensive study of *Aufhebung*. For example, he discusses the missing term of a poetic metaphor for the sun that, to me, mediates two frequencies: "sowing a god-created flame." About this nameless act of sowing and kindling, he says: "Let us leave open the question of this energy carrying absence, this mysterious break, that is, this gap which creates stories and scenes" (*WM,* 40). By his paradox, the "absence" in language (for which the metaphor was imagined) is made equivalent to the "energy carrying" modifier, so that the heuristic word *gap* in one category of linguistic experience can substitute for the gaping transmission of physical energy in the category of experience that prompted the need for metaphor in the first place. And yet the model of a wave-carrying force field implies that there is no real Cartesian split, because there is no occult action at a distance for which a paradox is required to leap across a logical or linguistic gap or an empty space in physics. Waves transmit from air to metal and from lyres to brains; waves travel from the sun to membranes that we feel as light and heat waves. The premise of a patterned field of force, defined as an extensive manifold of shaped waves in expanding series of shapes, means that there is no gap. The Cartesian split can be left to the history of ideas. Later Derrida adds:

> Metaphor here consists in a substitution of proper names having a fixed sense and reference, especially in the case of the sun. This referent is the origin, the unique, the irreplaceable (so at least do we represent it to ourselves). There is only one sun in this system. The proper name is in this case the first mover of metaphor, itself non-metaphorical, the father of all figures of speech. Everything turns on it, everything turns to it (*WM,* 44).

If there is a difference worth preserving between metaphor and archetype, where the former fuses two distinct categories of thought into a new illogical but illumining figure, and the latter refers to enduring tropes like the rotation of the seasons, then the sun is archetype. In the quotation, the main source of the earth's physical energy moves as the unique source of all figurative substitution. If we believe that the sun's energy is the origin and referent of all linguistic displacement, then we have a claim why humanists may defer the danger of *physis*. We are relegated to symbolic heliotropes. We imitate the hermeneutic circle around the sun. In this polytropical system, everything repeats itself in endless turns around the name, ever deferring by circling its source. We fear the Word (*logos*) as we fear physics. How can we protect ourselves against a violent force that may be wreaked upon us, if we close off the study of physical relations?

Here is another rationalization of the sun's energy. Heraclitus said, "All things are an equal exchange for fire and fire for all things, as goods are for gold and gold for goods" (*PP*, 198). For Greeks, the main heavenly fire was the sun. The adage has been misinterpreted by Kostas Axelos as being the progenitor of Marx's universal ratio of money: "Heraclitus had already grasped in all the splendor of its universality, the dialectical process of the conversion of all things into money and of money into all things."[26] This interpretation inverts Marx's use of the adage. It reifies the concept by subverting the simile. The sun is the converter of all things into life forms. Its wave energy is the origin of exchange and transformation. For Marx, the sun's rays are real vehicles for the transportation of light from objects into images. For instance, in the famous chapter on commodities he explains their fetishism as a sublimated system of social relations between productions, not people:

> This is the reason why the products of labour become commodities, social things whose qualities are at the same time perceptible and imperceptible by the senses. In the same way the light from an object is perceived by us not as the subjective excitation of our optic nerve, but as the objective form of something outside the eye itself. But in the act of seeing, there is, at all events, an actual passage of light from one thing to another, from an external object to the eye. There is a physical relation between physical things (*C*, 83).

Does anyone doubt that the light rays that reflect into the back wall of one's camera are real? A photographic camera has neither grammatical code nor syntactical code inbuilt. Like the eye, its arc lens reflects and bends the inherent shapes of characterizing light waves that have glanced off the object in patterns. The light waves pass through the medium of air into the dissimilar medium of glass, but the waves retain their shapes. In passing,

Marx exposes the dilemma of seeing natural wave frequencies. As transmitters, waves are always imperceptible, because one does not see the impulses that excite the optic nerve; one sees them visually sublimated, as an
object like a table outside the eye. "There is a physical relation between
physical things." This is not naive realism; his is subtle realism about *physis*.
Marx's comparison of optics and social relations needs the physical relation
between sunlight and our eyes in order to highlight his moral economy
about the supersession of the material base by a universalizing superstructure of capital. But is Marx's shift in point of view, introduced by the phrase
"in the same way," really the same way? In the next section, I take up the
fractions of base and superstructure as denominators and numerators.

The sun's fire converts most things on earth. As the propelling force of me-
tabolism, the sun's recurrent radiating waves comprise the "universal" energy
that we convert into alimentation, technologies, poetries, and harmonic acts—
techne, phronesis, poiesis. It is the origin of exchange, to Heraclitus. Marx
wouldn't have made the mistake of confusing a simile for the actual reality of
sunlight. Money doesn't really convert one thing into another, though we may
habitually believe it can, like the magical stone in the story "Stone Soup." When
he tried to eat gold, Midas learned that exchange and transformation are not
the same process.[27] Although gold may reverse all principles, and language can
reverse course and conceptually erase its trace, one need not fall into a universalizing linguistic ratio that substitutes money for physical energy.

Marx's sense of simultaneous perceptibility/imperceptibility is a shrewd
example of a physical base superseding Hegel's *Aufhebung* of transcendental
consciousness.[28] Wilden reminds us that the term means in German both to
suppress and to conserve, with the base meaning of "lift up." He also calls it
an *amphibious* word. If all words are *amphibious*, in the sense that they combine a materially transmitted signifier that is supplanted by a superstructural signified, then *Aufhebung* is one of those words that calls attention to
its pre-structural *physis*. It is also a good term for the description of the
brain's basic act of transforming a physical carrier wave into an image. It is
characteristic of cognition, for as we convert a carrier signal into an image,
we habitually suppress the vehicle in favor of the tenor.

Habitual Recurrence and Prejudice

The back-stretched bow and lyre are haptic analogies that revert to the
suppressed base of natural recurrences in wave forms. Without such re-
minders we habitually sublimate these frequent occurrences of natural

forces. In our era some have been schooled to repeat that all natural recurrence is merely an economy of symbolic thought. Recurrence seems something that only happens in our minds with frequency. Here, however, is Whitehead's simple definition of scale or "measure" by way of physical signs that recur in nature:

> The general recurrences of things are very obvious in our ordinary experience. Days recur, lunar phases recur, the seasons of the year recur, rotating bodies recur to their old positions, beats of the heart recur, breathing recurs. On every side we are met by recurrence. Apart from recurrence, knowledge would be impossible; for nothing could be referred to our past experience. Also, apart from some regularity of recurrence, measurement would be impossible. In our experience, as we gain the idea of exactness, recurrence is fundamental.[29]

A century of symbolic indeterminacy and positivism have prejudiced many from knowing these periodic recurrences of natural frequency. But recurrence will be important later in Chapter 4, as we think through some issues of rhyming recurrence. Perhaps I should say "habituated," but *prejudice* is a hermeneutic term that I will also feature in this chapter.[30] Hermeneutics is seen by Jurgen Habermas as a kind of interpretation that "has abandoned an ontologically grounded natural law" as a basis for practical philosophy.[31] In agreement with some others, I want to help redeem natural measures in an aesthetic judgment about waves, without reifying natural "laws," as fields for being and doing.[32] But because of hermeneutic prejudices we find it unnatural to found our principles upon recurrent wave frequencies. How so?

We ordinarily deny natural wave frequency because we cope with regular events by unconsciously sublating them through language or other automatic skills. Habit is the mind's accommodation to frequent recurrence of events. Habit is the converter for the economy of thought in language. I generalize *tiller* as a concept only by way of repeated instances of use, but I cannot be reminded of each instance of use in my experience, else I would be swamped by all those particular memories. Frequency, my awareness of periodic recurrences of like events in nature or elsewhere, is the occasion for habit. Habit requires frequency. Under this kind of recurrence, frequency becomes only the displaced vehicle that makes the habit. Yet the awareness of periodic frequency is suppressed in a general concept, the word *tiller,* with the trace ignored. Our most basic habit is that we suppress the natural wave frequency that sustains the skill.

How can this be? The exchange with displacement of *Aufhebung*, the habits of *seeing-as* and *hearing-as*, only seem to require the suppression of natural frequency. As William James said, "The moment one tries to define what habit is, one is led to the fundamental properties of matter. The laws of Nature are nothing but the immutable habits which the different elementary sorts of matter follow in their actions and reactions upon each other."[33] Nowadays physicists would speak of probabilities instead of habits for particles of matter, but the measure of predictable *periodicity* is the same. John Dewey rephrased it: whenever we speak of "natural laws," he thought, we might just as well think of "natural rhythms" (*AE*, 149).

Seeing-as and hearing-as, and writing and speaking, are carried by the natural frequencies of light waves and sound waves, but mere repetition of a meaning makes it seem as if habit is only the *routine* recurrence of symbolic thought. Consciousness of something always requires a focus from the tacit awareness of frequencies of light and sound to the projected meaning. In this use, the signifier is first of all a natural frequency, converted as a vehicle to transmit a message. As soon as I see-as or hear-as, I see it as a sign of something else. E. H. Gombrich, the art historian who taught the ways that habitual norm shapes optic form, said that we don't ordinarily focus on recurrent order; rather we attend to a "break spotter," a sign of something amiss in the regular pattern.[34] If inquiry is predicated upon a turn from what we already know to what we don't know, then the shock of the new experience is the physical converter from Saul to Paul. But surprise changes habits into new ones. What an odd surmise presents itself: to learn something, a skill perhaps, is to close down other alternative choices that were open hitherto.

The transmission of natural wave frequency always was what it was in its own natural periodicity, prior to its transformation by the brain into a sign of something else. Some have taught that alienation of the sign—its conversion and displacement (from natural periodicity), is the radical of universal exchange, of social contract, of property.[35] This kind of habitual alienation therefore bespeaks the social dilemma, a nostalgia for the natural world so close and yet so far; and, as Freud said, a duty to maintain the threatened breakage of civilization, to suppress our natural aggressions by the ego's discontented exchange for security.[36]

The large-scale shift from theories of the State based on a natural law of an "order of things" to the folk history of the State as the arbiter of law, roughly from Enlightenment to Romanticism, from Kant to Hegel, is part of a hotly contested history of social presuppositions and prejudgments that is not my task. Bernstein, for instance, discusses some of the issues in chapters on Michel Foucault and Derrida (*NC,* 142–98). Cassirer also tells some of that story in *The Myth of the State.* But from Hume to Peirce and from Heidegger to Gadamer—to choose a few philosophers—pragmatics and hermeneutics have claimed that habit, custom, prejudice, and belief guide human action.[37] Although I am featuring a pragmatic turn to practical action at this point, I am not celebrating it. Instead I am seeking the natural wave rhythms beneath the practice, rhythms that nevertheless generate the patterns of the practice. If habit is a recurring pattern of thought, then, along the way, we must look closer at this idea that thought itself is a recurring pattern. If such habits are grounded in natural frequencies that have been suppressed and converted into other vehicles, then it helps to try to turn the focus upon these automatic vehicles. A difficult task this, for habitual *unreflection* must be the economy of the rhythmic unconscious. Everything about natural syntax rides on this sublation. *Automation, ritualization, reification, industrialization, prejudice,* are all terms that describe a scooping out of particular contents in favor of routine formats. This unreflection amounts to an "apathetic fallacy," the inverse of the pathetic fallacy, "treating living creatures as if they were inanimate."[38]

A Review of Fractional Ratios

In order to highlight some approaches toward the transformations of wave frequencies into habitual images and word-images, I want to review a few fractional ratios that have diagrammed the symbolic thinking of the century. All of them address the question: How are physical events related to mental events? Here for instance is a version of Saussure's diagram of the sign:

$$S = \frac{Sd}{Sr}$$

I will be clearest if I can maintain a rigorous non-slippage that the Signifier, beneath the bar, is, for my uses, that part of the message that physically but tacitly transmits or transports by a natural wave frequency. Also, for my purposes, the fractional display can be said to represent a communication model for the transmission and reception of shapely waves via skewed semantics. The fraction necessarily represents an off-angled point of view upon the several parts. This kind of off-angled perspective lets one see two other points of view from the aspect of a third. It is a semiotic version of the Golden Rule, in which the doing is seen dialectically, now from the point of view of the Sd, now and again from the point of view of the Sr. But both cannot be seen simultaneously. Most importantly, the diagram clarifies any prejudgments for interpreting the syntax of the transmitting wave frequency.

Consider Whitehead's recurrent heartbeats, and our own. Locke said of heartbeats and circulation that it is not in our power to stop them. To arrest the recurrent frequency of their transmission is not in our power, if we wish to live. The term *heartbeat* is a habitual economy of thought that suppresses the numbers of beats that measure the extent of our lives. Like the visible convulsions of Saint Vitus's dance, Locke said, we are under the invisible necessity of this kind of recurrent motion.[39]

Yet through habitual repetition we remain unconscious of these kinds of automatically rhythmic frequencies. We live, we enjoy, we die through habitual variations in pressured natural frequency. With respect to certain kinds of frequent recurrence I am not a free agent. For instance, I cannot choose to avoid the recurrent frequencies of light and sound in motion as I write and speak, or read and listen. I can choose various codes: Morse, German grammar, semiotics, algebra, geometry, logic, trades jargon. But I cannot sidestep the necessity of these moving proportions projected from *physis*. And yet we habitually forget these projectors. We are so habituated to see-as and to hear-as that one discipline can speak as if its tool were a plow, another an oar, another a stylus.

In the use of a sign as word, the physical world is not primally a referent; it is known only tacitly by transmission and transport.[40] The word *sun* is referred to among its group of associated words in a local lexicon and grammar, but the sun is experienced only tacitly by means of its imperceptible carrier waves. One usually turns from the suppressed periodic order of things—its physical order of syntax—to the order of grammar and semantics. So when I refer in English to the Signifier, of course I am already in the hermeneutic circle of the transcendental Sun. Although the articulation of the issue arose in the aftermath of German Romanticism, the crux began with Descartes, who lamented the unknowability of the object by the subject. This review aims toward a different point of view about the scandal: we are suffused in an environment of natural wave frequencies all along the electromagnetic spectrum, which we habitually deny and simultaneously appropriate, as those who continue to live under a high power line, or above a radioactive waste dump. This is another point of view within Bernstein's "new constellation" of forces, his metaphor that would replace Cartesianism. Totally in it, we habitually act as if we have a point of view apart from it. We all uplift ourselves for an automatic ride upon the technological vehicles of naturalistic wave frequencies. This recurrent wave frequency is the underlying transporter of habitual recurrence in symbolic thought.

To extend Saussure's diagram further, let it also represent a ratio between part and whole, the vexed ratio that began Schliermacher's discussions of the hermeneutic circle (*HT*, 12–13). Consider that the figure and ground of Gestalt psychology can be redrawn, again with the connotation that underneath the bar is the suppressed signifier, the ground of natural frequency:

$$\frac{\text{Figure}}{\text{Ground}}$$

Other familiar variations of the era:

$$\frac{\text{Text}}{\text{Context}} \qquad \frac{\text{Tenor}}{\text{Vehicle}} \qquad \frac{\text{Superstructure}}{\text{Infrastructure, or Material Base}}$$

Although I do not know whether Chomsky may have ever drawn his terms *deep structure* and *surface structure* as fractional diagrams, his idea was still that a sentence has an underlying pattern that is essentially different from the superficial or surface grammatical pattern. More pointedly, the fractional ratio of Marxism reflects the ideological crux within the structure of any sign. For instance, V. N. Volosinov located the class struggle as a dialectic within the sign between the material base and a supra-class.[41] For him, every sign is a conditioned ideology, and signs become habitually worn-out concepts when they are removed from the polemics of class. In their discussions of material base and superstructure, Marx's followers habitually presuppose Marx's primal inversion, his standing on its head, of Hegleian *Aufhebung*. Perhaps not *polytropos*, Marx, according to his followers, in a single turn exposed Hegel's transcendental spirit as an illusion. Marxists say that they think with their feet on a material ground, not with minds but with heads. Whatever one may think of Marx's political beliefs, his one inversion was an important turn in an exposure of real physical relations that underlie social relations and constructions: "There is a physical relation between physical things."

Instead of featuring a stuffy materialist base of things however, I am substituting a physical wave form of natural frequency as the repressed signifier, beneath the bar. Here for example is a less familiar diagram, a physical formula for waves in frequency. Frequency is measured as the number of waves that pass a point in a given time. The frequency (F) of an object in harmonic motion equals the inverse of its time period (T): $F = 1/T$. Or:

$$\text{Time} = \frac{\text{One}}{\text{Frequency}}$$

This fractional formula of moving proportion gives me the title *Art/Frequency.* All of these fractions are implicit ratios, but this formula for physical frequency, with its inclusion of time, more accurately represents a ratio of moving proportions through space-time. The necessary visual image for this formula and for all spatiotemporal warps is a harmonically drawn curve.

The horizontal bar of Saussure's diagram, which separates the Signifier from the Signified, has also come to represent a Lacanian barrier of psychoanalytic indirection, the "father of all figures of speech," to quote Derrida's discussion of sun out of context.[42] According to Henry Sussman, much of this indirection derives from the Kantian bar, where Knowledge = Transcendental/Empirical.[43] Crossing the bar can also represent a logical discontinuity by which any sign mediates by conflation. This discontinuity has been given different names in different disciplines: for instance, category mistake or logical typing.[44] In most of these assumptions, the social relations between parts and wholes being illogical, the sign's compound structure must be an arbitrary relation. But if the brain is more truly a pattern transformer from wave impulses to images, then the compounding of the sign is not completely an arbitrary social construction.

All of these diagrams represent fractional ratios between a numerator and a denominator. That is, the figure stands to the ground, not in radical discontinuity but in a part/whole relationship of reciprocal frequencies. A fraction stands for an inversely reciprocal ratio, diagrammed as levels. My aim thus far has been to show how habit and recurrence are correlated, under an economy of symbolic thought, but that natural wave frequency is habitually barred and harnessed at the same time as the rhythmic vehicle for transporting our ideas and desires. But what if we considered that there is not a radical discontinuity between categories (drawn here as fractions), but rather that there is a recurrent continuity between them? Figure 6 is

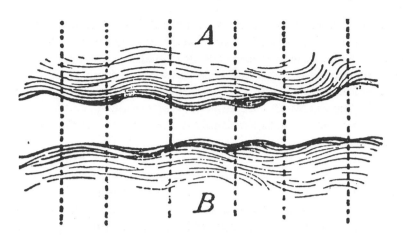

Figure 6. Diagram of Speaking and Thinking. Source: *Course in General Linguistics* by Ferdinand de Saussure, p. 112.

one of Saussure's. It represents "language as a series of contiguous subdivisions" that demark two different kinds of waves, where A represents "jumbled ideas" and B represents similarly incoherent sounds.[45] Then he says, "Visualize the air in contact with a sheet of water; if the atmospheric pressure changes, the surface of the water will be broken up into a series of waves; the waves resemble the union or coupling of thought with phonic substance." Notice that the energy waves of idea and sound, the intellectual and the sensuous component, are in themselves imperceptible. One must substitute the archetype of water waves to imagine any other kind of wave on the electromagnetic spectrum. Despite Saussure's brilliance as a teacher of linguistics and semiotics, his illustration is misleading, however, because of his prejudice about the supposedly chaotic slosh of water waves. His hypothesis is clear: without language, ideas are jumbled; and without ideas, sounds are incoherent. But I would like to hold out the alternative possibility that physical waves are not mere slosh, but that they have characteristic patterns; and that ideas themselves are not arbitrarily patterned only by the patterns of grammars of a language, but that thought is always only patterned thought, and that ideas are shaped by the characteristic patterns of their projecting wave forms.

For Saussure, linguistics works on the "borderland" of thought-sound. If you perform a figure-ground reversal upon Figure 6, you can see the white space between the two waves as the borderland or the bar of the fractional ratio that I am pursuing. In that view, a white wave shaped by the upper and lower wavy lines, there is not radical discontinuity between upper and lower levels, but rather an amphibian reciprocity between two kinds of frequent recurrence. These fractional linguistic ratios are amphibious depictions; they ride between different levels of thought. As Stephen Land said of Locke's account of words: "The amphibious nature of words, their necessary mediation between the mental and the physical, is at once the essence of their communicative virtue and the source of a problem."[46]

The use of the word *amphibian* to describe the nature of language reveals an unapparent relation between steering oar, winnowing fan, and language. All are amphibious things that deflect wind and waves. This archetypal deflecting tool, deeper still, reminds briefly of the old sequence of alternating motion with fins and feet, as amphibious animals paddled through sea and earth.

Amphibians are creatures that live between two boundaries, usually sea and land, or air and water, like a duck, a duck-billed platypus, or even humans who use physical transmissions to signify ideas. So the *amphibious* quality of words is an archetype derived from creatures whose bodies glide between frequencies with their own moving proportions. I have been making the observation that waves must be the primal amphibians, because they can jump from water to air, or from metal to air, or from electricity to magnetism, or from air to eardrum. For an example of an amphibian who lives in and rides above wave frequencies, see the pattern of recurrences in Figure 7, a photo of a Ming Dynasty silk painting that has been called "Lohan Riding on Lotus Leaf." In the next chapter I shall take up the Chinese Tao in relation to water. But in this illustration of the lohan, is there a harmonic measure or ratio of moving proportions that is based on waves? From the previous section, carry over the diagram of figure and ground as a leveled fraction, as well as the amphibious analogy of language mediating between two frequencies. Notice that the boat serves as the mediating bar, Saussure's borderland, between two kinds of waves: water and wind. The hat, the cloak, and the prow billow in synchrony with the waves of wind and water, making palpable the reason why waves are almost always archetypes for rhythmic recurrence. Could this Chinese picture depict a world in which the *Urform* of curvilinear waves must shape everything in it?

$$\frac{\text{Lohan}}{\text{Lotus}}$$

Figure 7. "Lohan on Lotus Leaf," or "Traveling on a Raft." China, Ming Dynasty, ca. 15th century. Painting on silk. With permission of The Walters Art Gallery, Baltimore.

A lohan is one who has deferred immortality for a time in order to help others understand everyday ways of living. Lotus is the cosmic flower of Buddhism.[47] For instance, there exist many pictures and statues of Buddha standing on a round pedestal representing a lotus. Sometimes, the Buddha's lower garment is leaf-like in its folds, suggesting that the whole torso is the emergent bud. So figure and ground are continuous. Continuously the lotus buds into a flame. With its roots underwater in the mud, it floats upon a pond, aspiring amphibiously. This image lets one visualize the earlier archetype from the Greek tradition, mentioned by Derrida, "sowing a god-created flame." The source of metabolic exchange, fire, is here seen as flame that aspires to the origin. In the original Indian concept of Buddhism, according to Heinrich

Zimmer, the lotus "was the exclusive sign and 'vehicle' of the goddess Padma—mother, or yoni, of the universe." Sometimes in the tradition, either a monk or a Buddha stands on a lotus but also holds a lotus blossom as the world in his hand. In some statues he holds a thunderbolt, representing the shock of cosmic energy. For Zimmer, the design of the lotus also combines male and female triangular designs, which together represent reproductive energy through the four directions of the cosmos, as lily pads cover a pond.

Notice the twisty triangles in the silk painting (Figure 7). The lotus-boat in the silk painting has been abstracted so as to make the vehicle insufficiently recognizable either as a boat or a lotus leaf. It doesn't look much like a lotus or a boat or a raft. In what ways can it be imagined to be a lotus design, which might represent the "vehicle" that transports all things on earth? The painter has made the conveyance look like a blank sheet of paper, amphibiously curved by wind and water into the twisted triangular shape of a curvilinear boat, riding between curvilinear waves, with front and back almost indistinguishable.

What can this design of *twisted* paper mean? Throughout this study I shall be calling attention to the deforming morphologies of a *twist*. Twist is always a sign of an apparent move from two dimensions to the third dimension of shape. A real sign from the environment must always be emitted from some kind of three-dimensional shape "in the round," so an environmental sign is a curvilinear twisting wave that apparently carries or transports as vehicle the patterns of waves over into the patterns of human thought, like the reverberating sound of a gong.

Other kinds of animals also receive these physical signs from the environment, and they transform them presumably into images of their own signification, but humans further transform physical signs from the environment into symbolic lexicons and grammars. By studying this image, thought itself may be seen as a new twist in the patterns of nature. Perhaps thought exists only in the hypothetical dimension of symbolic discourse. If thought is in part prompted by a moving proportion of natural vehicles of twisting waves, then a new thought may be a new pattern of moving inferences. Perhaps Saussure's diagram, which unifies wind waves and water waves, ideas and phonemes, could be modified a bit, so that "jumbled ideas" might be replaced with patterned ideas. By thinking of waves as being something more than the slosh of water, I can suppose that thought exists not as some arbitrary and independent entity within a symbolic envelope, such as the conceptual word *tiller,* but as deriving from a constant pattern, such as the cosmic lotus, twisted though it may be.

When the boat or raft is seen as paper, then there is a categorical switch that allows us to ask: Is there a rhyme or consonance between the paper of the boat and the paper of the book in the lohan's lap? [48] And the paper that will enframe it all? Think of the construction as if it is formalized by the semiotic bars drawn earlier: Boat is to Sea Waves as Book is to Lohan:

$$\frac{\text{Boat}}{\text{Sea}} :: \frac{\text{Book}}{\text{Lohan}}$$

The lohan reads the sea (as the ship plows the waves). I calculate the linguistic ratios only to diagram the four-part harmony of a visual metaphor, which is more accurately a visual archetype. These fractional proportions prompt the guess that both paper boat and paper book are likely means of conveyance to different realms. My visual diagram, with its double colons, is a static proportion, but the sentence, "The lohan reads the sea," with its subject and verb, points out that metaphor is a moving proportion in apparent action. That is, for Ricoeur and others, metaphor itself is usually a sentence that asserts an odd kind of "category mistake," some kind of predication that asserts an untruth on the face of it but that allows the discovery of an unapparent unity connecting the subject and predicate into an idea, for which there is no conventional word, concept or lexicography.[49] The ship plows the waves. Books then, and meditation about them, help allow the lohan to transport concepts of Buddhism. This then is not a helmless vehicle, Plato's fear. As Toynbee observed: "The disorderly motion of a rudderless ship, which stands in Plato's eyes for a Universe abandoned by God [*Statesman,* 272d], can be recognized by a mind equipped with the necessary knowledge of dynamics and physics for 'programming' a computer, as a perfect illustration of the orderly behaviour of waves and currents in the media of wind and water."[50] Notice that here waves have an orderly composition and are not just chaotic. But where is the hidden tiller? Where or what is the steering oar, the winnowing fan, the deflecting rationale, the cybernetics? The (blank) book is a sign of the missing term, that is, the mental tiller that steers the boat in the right direction with wind and water. The actions of reading, sailing, meditating, and acting are rhymes of a sort. But they are

not automatic, not habitual associations, for they must be studied to un-
fold their moving proportions in time.

For instance, the poet Denise Levertov writes:

> As you read, a white bear leisurely
> pees, dyeing the snow
> saffron,
>
> and as you read, many gods
> lie among the lianas: eyes of obsidian
> are watching the generations of leaves,
>
> and as you read,
> the sea is turning its dark pages,
> turning
> its dark pages.

Quoting just the last stanza, Rudolf Arnheim comments that the common relation in the metaphor must be the perception of rhythmic turning.
He notices further that the rhythm transforms both page and sea, for the
common element of turning conveys "a sense of elementary nature to the
pages of the book and of readability to the waves of the ocean."[51]

By means of the formal device of repeating nature's frequent curves,
the Chinese artist lets us see the blank book as a recurrence in kind of the
blank-paper boat. By way of the book does one read the order in the sea?
The twisting operation at each end of every plane suggests the ambivalent
figure-ground relationship of turning waves; in this case, since all the
things wave, they may be all of one kin.[52] The turning back upon their origins of these S-curved forms create a useful figure-ground ambiguity for
any animating artist. As Gombrich says of waves at large, "What is part of
the 'figure' in one reading is part of the 'ground' in the next, but there is no
real break or border which would help us to settle any interpretation" (*SO*,
137). The barred difference between figure and ground is lost as the semiotic "border" twists back and forth in indifferent waves. One sees how figure becomes fractionated as ground to the next level (sea, boat, lohan,
book), but with no radical discontinuity, no permanent semiotic bar or
borderline.

The rhythmic pulsations work in opposition to the equilibrium, the
balance, of the hovering lohan. By way of Buddhism, perhaps, he has
crafted self-command, and he has cultivated an imagination that projects
himself dangerously into the direction of another's plight. No automobilist, this helmsman tills the waves of natural frequencies. This lohan does

not arch forward anticipating truth. He reposes in everyday frequencies. Presumably any place he might seek to rest will be just like this, a locale to frequent, with its own local recurrences that interfere with the overall rhythms of things. But what else is natural scale but the interactions between local growth and universal frequencies? Balanced on the crest of the present, the lohan moves into a more violent Impending, where the waves are steeper. Amidst a whence-whither continuity, where waves of the past, depicted on the left, seem smoother than those on the right, the lohan steers and mediates between this locale and another in slouched skill, that is, in danger always of foundering on the habitual *pharmakon* that one has found the right way.

Simplicity reproduces in indefinite waves. Both prow and stern are twisted or torqued into apparent triangles by forces of wind and water. So it is not singly by way of a book that one finds one's way, but the book exponentially sums up or economizes in itself the all-encompassing reciprocities between natural forces, seen as waves. "Pen and sword in accord," says the Zen adage. Pen, till, plow—all sow in recurrent waves. All are comprised of similar strokes of physical gestures or deflections. They suggest a common syntax of wave forms.

Point of View of Push-Pull

S-curved waves seem to demand shifts in points of view, if only because one may be now on the crest and then in the trough of a frequency. S-shaped curves seem to be natural shifters. The form of a wave amphibiously carries energy from one physical medium to another, for waves can shift between different physical media. Consider that image as if it were shaped by the three groups of waves in Sobel's diagram (Figure 5). As I look at the image, I know that I am imperceptibly receiving actual light waves from the image, which are carrying the message. As Marx indicated, even though I infer that light waves are real, I cannot perceive them as such. But the straight lines and arcs that were formed by the strokes of the artist's brush are sending their shapes as impulses to my retina, which I transform into the represented image of figure and waves. This image of an unruffled lohan, with the book in his lap in his boat on the waves, requires me to see it as an illusion: everything else in the painting is twisting in apparent motion. Because the ends of the triangles rhyme in recurrent forms, they allow the conjecture that the waves in the painting are also the media for the twisting strokes of the artist's brush, which we might not ordinarily attend to. I can either see the waves in the picture as the illusion of waves in nature, or I can see the waves instead as strokes of the brush which lend themselves to the illusion that everything is billowing and not still. In other words, the image can be read as a metapicture, which draws attention to the physical waves that run through all things but that are ordinarily imperceptible as such.

The paper-lotus, in precarious balance, drew attention to the physical structure of the boat. Although the boat seems ready to swamp, it is levitating like a magic carpet. It is sublimated, uplifted. I thought I knew how a boat floated until I saw the picture. It shocked my prejudice. Working as a break spotter, it turned me to an odd interpretation of its physical world. The lohan nonetheless floats with that calmness that always characterizes Buddhist immortals in Chinese and Japanese and Indian painting, who may be resting ambiguously on rocks or clouds, on dragons or vacuities of (silk) space. So while there seems to be a discontinuity between the physical nature of the fragile vehicle and the lohan's compassionate expression, there is no discontinuity between such fractions in everyday use.

Those of us whose crafts are reading and writing may tend to forget how much of physical balance is not metaphor in the other arts, but habitual feeling of muscular skill, from which all of any method turns—the gesture of a paintbrush, the curve of the dancing hand. In that context then I notice that there is no calligraphy on the pages of the lohan's book. Why blank? Perhaps because all of the message is inscribed in the supple brush lines of the painting, which occasionally we admire as nature's wild waves, while occasionally we might speak of the artist's craft in handling calligraphic paint stokes with grace. But we do not see both simultaneously, not material brush stroke and its illusion. This phenomenon of perception Hans Hofman called a "push-and-pull" effect, which also describes the complementarity that one cannot focus on both brush stroke and intended representation of depth simultaneously. C. H. Waddington noticed that this push and pull creates the essence of painting, the illusion of moving pulsation:

> Every patch or stroke of pigment, by its colour, texture, or shape, at once pushes itself into the background, and at the next glance, pulls itself into the front. The surface of a painting, therefore—regarded purely abstractly, without any reference to anything it may 'represent'—pulsates in space, sometimes the reds striking out towards you, sometimes the blues. It is a transposition, into colourful painterly terms, of the spatial ambiguity which was exploited, in black and white geometrical language, by late constructivists like Albers. This ambivalence of near and far, this 'movement of the surface,' has remained one of the characteristic features of the new painting.[53]

This complementarity of focus between Signifier and Signified, between front and back as depth levels, is the radical of reading recurrent images of figure and ground as apparent oscillation, as motion. But it is also the scandal of the twentieth century's physical indeterminacy. It prompts the overused metaphor of oscillation between two poles. In fact, Jerome Bruner reports that Neils Bohr once was reminded of this kind of visual indeterminacy when he was puzzling about quanta of action.[54] Bohr's son had stolen a pipe. As the father was meditating about the discontinuity between Love and Justice, that he couldn't make them meld together simultaneously, he remembered the figure/ground reversals of Gestalt pictures, when first you could see the vase and then the face, but not both images at once. "And then the impossibility of thinking simultaneously about the position and the velocity of a particle occurred to him." Reading that passage startled me. The radicals of ethics, cognition, and physics combine in Bohr's narrative of the sequence. In all three cases, the discontinuity depends upon switching point of view to another fractionated level of scrutiny. The whole phenomenon cannot be seen simultaneously, but each must be projected in space and time, first from one vantage and then from another. When one switches points of view to solve for different ends, then the optic relations of perceptibility and imperceptibility, mentioned by Marx, are simultaneously switched too. In sum, the twentieth century's metalanguage about objects is that the sign of an event must be seen as having two inseparable aspects, like the front and back of a sheet of paper, as Saussure put it. And each aspect must be seen antecedently or consequently, that is, narratively in experience. All three codes depend for their interpretation upon the tacit awareness of a transmitting wave's substructural counterrhythms, which force us to see un-simultaneously.

Perhaps the most significant point about the push-and-pull effect may be that it records an essential pattern of thought. Thought itself is patterned. Thought cannot exist independently from the patterns of nature that the brain uses to transform impulses from waves into images. Although I may bracket an apparently independent concept like the word *tiller,* the word is still and always tacitly connected to the transmitting waves that carry the message of the letters *t, i, l, e, r.* And though the word has no necessary connection with its associated object, its shape is still relatively constant. And its function as deflector of waves is also constant. For instance, the tongue is a deflecting blade that shapes airwaves into speech acts. So I can assert metaphorically, "The tongue tills the airwaves." Instead of being enveloped in symbolic indeterminacy therefore, the Gestalt switch in point of view—the transformation from a sensuous to an intellectual part of the sign, or vice versa—allows the notion that thoughts themselves are composed of patterns that ontologically connect by fitting into the patterns of the world's wave spectrum.

Although set up simultaneously, any opposing pair in an oscillation resonates successively. To interpret a set of opposites in space, one must use time. On the other hand, to solve for opposites that are composed as if they were before and after, as if antecedent and consequent, one must use spatial brackets. One cannot set space against time as if they were symmetrical opposites, like left hand and right hand. (The amphibious word *space-time* misleads if one does not recall that the complex delineates a strange curvilinear twist or fold.) For example, if one wants to use the force field as a metaphor for the dialectic of temporal history, one projects from the point of view of spatial antitheses. As Walter Benjamin wrote, "Every historical state of affairs presented dialectically, polarizes and becomes a force field in which the conflict between fore- and after-history plays itself out. It becomes that field as it is penetrated by actuality."[55] Or if one wants to privilege a simultaneous conflict set in a space, one necessarily projects from a tacitly temporal point of view. For instance, here is Freud's explanation of the neuroses engendered by the counterrhythms of the life instinct versus the death instinct: "There is as it were an oscillating rhythm in the life of organisms: the one group of instincts presses forward to reach the final goal of life as quickly as possible, the other flies back at a certain point on the way only to traverse the same stretch once more from a given spot and thus prolong the journey."[56] In all of these examples, there is some natural force field, not yet understood, which backstretches and turns us and projects us elastically.

Some of the trouble of thinking about oscillating complementarity arises from setting up an equation between two opposites that are only apparently symmetrical. Here I have used instead a set of reciprocating fractions of figure and ground. Anthony Wilden in *Man and Woman, War and Peace* warns against double binds that send choices oscillating indecisively between two apparent opposites. Wilden's opening chapter is called "Oscillation, Opposition, and Illusion East and West: the Ideology of Error." His leading example is the yin-yang symbol of ancient China. He recalls that some understand this symbol to be a three-way relationship, not just a dyadic balance between traditional opposites of female-male, dark-light. That which shapes the distinctive oppositions is a third thing separating and relating yin and yang: the curvilinear boundary between. What makes for illusory oscillation is the curvilinear S that in Gestalt parlance pivots figure and ground into rotating inversions of one another. Wilden argues that this is an illusory oscillation between equal opposites. For in his interpretation of Chinese society, the yin-yang does not really symbolize "the complementary interpenetration of opposites," because it sublimates the ideological domination of male over female. The visual symbol of equality binds the minds and choices of women into a paradoxical submission within an invisible hierarchy. For Wilden, many of the so-called "symmetries" between traditional opposites mask dependent hierarchies of inverse restraint: nature vs. society, man vs. woman, capital vs. labor (*MW*, 22–23). The habitual inversion comes about because of the prejudice that men outrank women, culture outranks nature, capital outranks labor. His main point is then that the use of a root metaphor of binary opposition at one level is usually a strategy designed to project an ideological hierarchy of control upon one of the members of the pair. So oscillation is for him the sign of a double bind. The projector of the illusion is signaled by the intermediate boundary. Wilden sets up a "Three-Way Rule" that should be studied: "The minimum number of connections required to establish a relation is three: system, environment, and the boundary mediating between them" (*SC*, 7). In short, for Wilden, the closed circle of yin-yang can be seen as a vicious circle where self-referential dependency is habitually projected and inverted as ascendancy over an other referent. I do not wish to quarrel with his strategy. In Figure 1, however, I projected an alternating group of splayed half circles, seen as S curves, apparently moving not in a vicious circle of eternal return but as a continuous series in direction. The half circles can be seen as correspondences in space, or as sequences in asymmetrical time; or together they can be seen as an oscillating force field.

Perhaps then I can see the illusion of a paper boat, not because the illustration sets up opposites, like light versus dark, but because it conflates oppositions as *levels* of meaning, in figure-ground nests, twisted into before and after. The four-part harmony of the lohan's visual fraction plays amid the complementarity of different spatiotemporal figures and grounds. The S-curved waves "pulsate" as bimodal lines that call attention to the reciprocity of separation and relatedness, especially in terms of yin and yang, but always as a Gestalt possibility of balancing and switching from figure to ground. In order to see-as frequently, you switch habitually from pigment to meaning, from stroke to the illusion of motion. But it is the push-pull of natural frequency that forces Marx's perceptible imperceptibility of an optical twist.

A boat sails amphibiously between wind and water. A poem waves between light and sound. We act counterrhythmically between chance and necessity. The spectrum seems to be dangerously bimodal, whose sign of interference is an oscillating wave of furious formalism. This metaphor of push-pull draws attention to what we habitually erase everyday. Our era's custom is that we simultaneously appropriate, while we deny, the natural wave frequencies that allow us to switch to meaning. In cognition, in physics, in ethics, in aesthetics, we switch points of view. *Polytropos*, I turn from apathy to sympathy, from third-person objectivity and distance to first-person empathy (*I-Thou*), from justice to love, from the Chorus to Oedipus and back again in the reversal of roles that is peripety. At one time we see people as objects, and at another we see and judge others as we would be judged. This duality turns us from atomism and objectivity to mutual parts of a sympathetic process in motion, a field theory.

If habitual unreflection is the economy of the recurrent unconscious, working through us while we continue unaware, with cant its worst language use, then poetry and art can turn up and trace, to-and-fro, the signs and tools of frequency, tiller and plow, bow and lyre. Art and poetry turn us about on our untested premises to the possibilities of a surprising promise. Their counterrhythms seem to be the form of the natural continuum, like the double helix of DNA, which inverts opposites in a succession that splits and doubles chance and necessity in a new projection for life. Artistic recurrence enacts experiences that cannot be drawn up under one concept that erases its trace.[57] Art reenacts nature's alternate recurrences, by kindred foldings and unfoldings of meanings that cannot be suppressed by one concept or a rational equation. So I would not assert a ratio about the lohan on lotus that the book is the way. Art pushes you in and it pulls you out, first beach surge and then undertow, now simultaneously in space, then narratively in time. An ensemble of rhythmic wave frequencies, the range of related meanings suggested by a work of art unfolds slowly like a lotus.

Natural Forms of Carrier Waves in Traditional Cultures

Synopsis

In this chapter I review some archaic stories and images whose subject is the original creation of being and language. According to some traditional accounts, when beings came into being, so did languages. Some of these creation stories say explicitly that language came into being as natural forms of waves. In these primal accounts it is apparent that language was made up of a rude but adequate natural syntax of protogeometric gestures, strokes, and plosives of speech that were to be used for making and carrying formal distinctions around and about the natural world. Here are gathered some traditional examples of natural language, in which *language* means a limited set of physical forms from nature that transmit or carry the idea of natural force or energy—as the animating principles of life and language.

The artists, storytellers, and poets who used this early natural language tried to duplicate in their own physical gestures, strokes, and sounds this rudimentary syntax of a few naturally curving forms. Just as nature seems to be composed of only a few rudimentary periodic forms, so, too, does the syntax of drawing and singing the arts of nature make do with only a few kinds of formal strokes. So this chapter about the emergence of language and beings, in some traditional cultures, allows the suggestion that theirs was a poetics of *methexis,* of reenactment of natural forms, more than *mimesis,* or imitation of nature. This distinction illustrates how the physical world comes into play in signification through the enacting of a carrier wave, not in reference to nature. The poet-artist's syntactic strokes that carry the message would rhyme with, as they reenact, the patterned forces of nature. *Consonance* here describes this kind of rhyme or match between a syntax of a few physical forms and representations of meaning.

Because it is easier to *see* a syntax of forms by way of visual thinking than it is to *hear* the syntax of speech, I feature visual illustrations in this chapter, saving an extended discussion of poetry till later. But some poems do show up here as examples. To help unify this task of making the case for carrier waves in primal art, I study the superheated image of the Plumed Serpent, an *amphibious* beast who combines a bird's air waves and a snake's ground waves into one huge caracole. So this section amplifies the question about the amphibious character of language, studied in the previous chapter. The trope of the bird/serpent was widely shared in East Asian, Mexican, and Native American creation stories. Like fusion music or fusion food, the image of the bird/serpent, re-described here, brings together Eastern and Western images in some old and in new ways, though perhaps not to delight all tastes.

By studying several different kinds of primal art forms, I shall explain how it is that within these creation stories, and their plastic embodiments or images, there seem to be just a few basic strokes of composition across the arts. When theorists speak of the syntax of music, are they speaking in mere metaphor, or is there an element of true patterned kinship in the idea of syntax? If there are basic physical strokes of composition that derive from the forms of the carrier wave, then it may be that the physical units of syntax in a sentence, those parts that carry the message in rhythmic waves, are of a kind with the other basic units of artistic composition. A discussion of poetic syntax, however, will be taken up in later chapters.

Wind and Water Waves

In the beginning, Eurynome, The Goddess of All Things, rose naked from Chaos, but found nothing substantial for her feet to rest upon, and therefore divided the sea from the sky, dancing lonely upon its waves. She danced toward the south, and the wind set in motion behind her seemed something new and apart with which to begin a work of creation. Wheeling about, she caught hold of this north wind, rubbed it between her hands, and behold! the great serpent Ophion. Eurynome danced to warm herself, wildly and more wildly, until Ophion, grown lustful, coiled about those divine limbs and was moved to couple with her.[1]

These are the opening sentences of the poet Robert Graves's *The Greek Myths,* and they represent his poetic reconstruction of the fragments of a remote Pelasgian creation story. Graves tells how Eurynome transformed herself into a bird and then brooded the Universal Egg and bade Ophion to coil about the Egg seven times. From it were hatched all the things in creation. But Ophion later claimed himself to be the author of the universe: "Forthwith she bruised his head with her heel, kicked out his teeth, and banished him to the dark caves below the earth." This account is but one version of the great battle between the bird and the snake for dominion of the media. Graves's opening story reorients the mythology of male primogeniture in favor of the dancing goddess. It imagines Eurynome dancing with the coiling traces of her own relative wind as she sped south. One sees also how the four directions are beginning to emerge. Her curvilinear turning or wheeling back-and-about, while dancing forward, will be my prime trope in this chapter, as ophic or dragonic S-curved waves are construed as the twisty traces of creation. Especially in their male format, such wakes are not to be considered as the causes of creation, which would be a misformulation. Even physicists do not yet know what caused the Big Bang, our creation myth nowadays. All they can do is trace the structures and frequencies of the forces that ripple from it. For we all live within these ripples of wave frequencies as our total environment.

In Graves's poetic reconstruction one notes what may be seen as a Derridean compound of emergence and suppression: the sign, the deity, and the fictive trace of a transcendental signified, all emerge together in creation myths.[2] One learns from Graves's reconstructed myth that Eurynome's inversion is the primary trope or tropical turn and also that the natural trace is the sign and not the cause of her primal act of creation. The trace did not turn Eurynome. Grown cold, she danced to warm herself, and she created a thermodynamic current, an S-curved wave, in her wake.

One now knows that all such thermodynamic motions, and most curvilinear frequencies on earth, whatever their stripe, occur in the energy-change from hot to cold.[3] Air waves, dust devils, water spouts, tornadoes, striations of molten runnels cooled in rock, all are rippling wakes traced back to thermodynamic transformations: simply the ripple movement from hot to cold. All these physical whirlings will be seen later as designs of the Plumed Serpent in American lore. For humans, even cryptohuman forces like lonely Eurynome, the feeling of energy loss—cold, hunger, thirst—is a consciousness that new heat must be metabolized to create kin, and kinlike or kindly patterns. Cold, lonely, with longing ever unsatisfied, she was the first romantic. If stories are invented, in part, to court this longing in ourselves, then perhaps Eurynome reflects the reader's desire.

Eurynome's aesthetic act of dancing created a wake through the sea and the air. Not a "Big Bang" (which now seems a term characteristic of the Cold War fears of the 1950s), her act of creating the Universe was the physical enactment of dance. The Goddess of All Things turned the trace into Ophion, who is now known as the North Wind, so when he tried to claim authority for the creation, he tried out the first ruse, the first fraud. It is out of crafty inadequacy and the desire for primacy that the tricky snake would say that the trace was the creation. He coiled into the primal antithetical turn. Creation stories, like most stories, usually follow a narrative and visual pattern of back and forth, over and under, a weave of space and time. For the story must turn on a first principle, a bootstrap principle that is necessarily both transformer and transformed, seen from wheeling points of view and phase.

Because Ophion lied about her, she apotropaically kicked him into the ground, and there he became the first of the autochthonous gods. In much mythic story, some to be discussed here, the serpent serves as an agent of the substructural ground. He serves as the sign of the repressed vehicle. In much myth and rite the underground serpent is usually set against an uplifting bird, whose feather serves in many rites, myths and fairy tales, as totemic agent of the transcendental signified, the flight of meaning, the message from the gods of the air. So the feathered serpent, to be studied below is an archetypal Sign/Design, incorporating the opposite functions of substructural ground and transcendent air. "Be ye therefore wise as serpents and harmless as doves" (Matt., 3:16). The feathered serpent antithetically balances and interweaves these opposing forces. In its repression and transcendence, the tropical *Aufhebung* discussed in the last chapter, the feathered serpent denies while uplifting. The bird is uplifted and the serpent is suppressed beneath the semiotic bar. This chapter then sets forth some of those opposing rhythms of nature as they are symbolized by archaic curvilinear waves. Here, too, I begin to explore the signs and symptoms of an archaic syntax of helicoid forms of creation. For first acts of creation, like Eurynome's, must use a minimal set of natural forms with which to pattern the rest of the kindred shapes and orders and positions of creatures. Lonely, she danced upon the waves.

"Revelation of the Helicoid Sign": The First Word of the Dogon

The creation story that for me best features S-curved natural rhythms is told by Marcel Griaule of the Dogon religion.[4] There is not space here to sum it up, but it will suffice to introduce their idea of the first language as being one of thermodynamic shapes in nature. For the Dogon, the First Word of creation is a Water language of helicoid forms. S-curved undulations are the first forms or Words of Nature. Their syntax of helicoid signs will make my case about their prescience in using these forms to describe their world.

Here is part of the story. The God of creation, Amma, created the stars, the sun, and the earth from lumps of clay. Amma desired to have intercourse with earth, but did it violently, with the result that the Thos Aureaus was born, the Jackal who symbolized the blunder of Amma (*CO*, 17). Afterward the twins Nummo were born. The essence of God, they are the Water that is the energetic "life force" of the earth. The Water twins saw that mother earth was naked, so they made two strands of fibers that covered earth front and back:

> But the purpose of this garment was not merely modesty. It manifested on earth the first act in the ordering of the universe and the revelation of the helicoid sign in the form of an undulating broken line.
>
> For the fibers fell in coils, symbols of tornadoes, of the windings of torrents, of eddies and whirlwinds, of the undulating movements of reptiles. They recall also the eight-fold spirals of the sun, which sucks up moisture. They were themselves a channel of moisture, impregnated as they were with the freshness of the celestial plants. They were full of the essence of Nummo: they *were* Nummo in motion, as shown in the undulating line, which can be prolonged to infinity.
>
> When Nummo speaks, what comes from his mouth is a warm vapour which conveys, and in itself constitutes speech. This vapour, like all water, has sound, dies away in a helicoid line. The coiled fringes of the skirt were therefore the chosen vehicle for the words which the Spirit desired to reveal to the earth. He endued his hands with magic power by raising them to his lips while he plaited the skirt, so that the moisture of his words was imparted to the damp plaits, and the spiritual revelation was embodied in the technical instruction (*CO*, 20).

Here one confronts the primary premise of natural language thinking: there is a necessary relation between the structures and functions of the human body, the functions of the earth, and breathed speech. The carrier vehicle for understanding the mythopoeic triad is the curvilinear line that prolongs the common properties of Water as a first principle and which extends to infinity. From this passage I take a title for this section: "a helicoid line." The properties of water are the visual shapes of undulant motion and sound. Cyclones, whirlwinds, reptiles, twining plants that form the fibers, the "plaits" by which humans reenact them—these are the undulant lines of earth's reproductive power. In archaic art coiled unduloids are always organic sign/designs of the grand continuum of nature. This wave form is the thermodynamic syntax of nature, and it is a rhyming language of similar shapes, a first principle of form. Of primal syntactic importance is the idea that the plaiting of strands over and under one another is the analog of helicoid lines in nature. As Griaule summarizes the syntax of this first language, "Its syntax was elementary, its verbs few, and its vocabulary without elegance. The words were breathed sounds scarcely differentiated from one another, but nevertheless vehicles" (*CO*, 20). Here the syntactic order is rude but nonetheless universal. Speech is carried by vehicular breaths of plaited spiral shapes, and it dies away in a helicoid line. I shall have much more to say about the topology of twists and plaits and braids as the essence of syntax in later chapters.

What kind of twisty "vehicle" is this? What is meant here by "vehicle"? I am after a pattern of wavy-twisty syntax in speech, spoken out loud, that propels and carries as vehicle the different grammars of many languages. In order to compare this archaic story about breath with a very recent poem, consider the archaic syntactic patterns in Robert Creeley's "Breath":

> *Breath* as a braid, a tugging
> squared circle, "steam, vapour—
> an odorous exhalation,"
> breaks the heart when it
> stops. It is the living, the
> moment, sound's curious
> complement to *breadth,*
> *brethren,* "akin to BREED . . ."
> And what see, feel, know as
> "the air inhaled and exhaled
> in respiration," in substantial
> particulars—as a horse?
>
> Not language paints,
> pants, patient, a pattern.
> A horse (here *horses*) is
> seen. Archaic in fact,
> the word alone
> presumes a world,
> comes willy-nilly thus back
> to where it had all begun.
> These horses *are,* they reflect
> on us, their seeming ease
> a gift to all that lives,
> and looks and breathes. [5]

This is not the place to try to think through this poem. Here I simply notice the idea of kinship among etymological definitions of sound-alike words, and the poet's braiding of *breath* with other consonantal *b*-word-sounds that are akin to it in life. It is archaic in fact, this breathed word that presumes and summons a living world. Note too, in passing for the nonce, that the grammar is torturously wrenched to get at the breath of the word and the world: "Not language paints." This poem reenacts the syntax of a more elemental "pattern."

The great theme of nature myths, where animals are created first, like archaic horses, is that all living things are organically kin. They are related in a common syntax of living and breathing properties that lets them understand each other's functions by their helicoid rhymes. Later I shall call this rhyming and patterning set of gestures by the term *consonance*. Natural language has a helicoid syntax. This Dogon code about nature lets curvilinear forms render speech acts. The trailing helicoid fibers mean that all things are born out of and traced with the same kind of natural language.

In the Dogon creation story, as related by Marcel Griaule, plaits are the common elements of the next two stages of creating the Word. (Creeley's word is *braid*.) The second Word is learning the helicoid skill of basket weaving. Then a woven basket later serves as the model for the building of a clay granary, the Third Word, which in turn serves as the Axis Mundi for the eight sacred seeds that sustain the culture. I return to the trope of weaving in the concluding chapter. One notices that the Word, *logos,* here and throughout, is not to be construed as the letter or the law, but is more like a harmonic principle that weaves the scale and rhythmic unity of various forms.

These helicoid woven forms are clearly the first syntactical units of natural composition; their protogeometry shapes the variety of thermodynamic occurrences in nature. It is the visual shape of rhythmic frequency. This undulant form is the first Word of nature; it gives shapes to events just as plosions of vapor from the mouth convey and constitute speech. Hard to envision perhaps, but on a frosty morning watch your billows of breath. Then plosives can be seen, as hot breath takes shape in cold air. While it is impossible to see the in-taken breaths, they are just as important as the explosives, because they invisibly mark the patterned intervals of space and time that demark the sounds as rhythmic series. For the Dogon, the visual representation of "inner speech" is a curvilinear snake, drawn only in the innermost sanctuaries.[6] It is also consonantly linked with good rain, with drums, horns, and especially flutes. Sounds jump from one medium to another by waves. So all helicoid occurrences, and the instruments that reenact them musically, are sacred manifestations of the first act of organic creation.

In the sacred context of this helicoid primal thinking, one better understands the lowered voice of Ogotemelli, the Dogon storyteller, as he whispered of the violent inception of jackal: "It was no longer a question of women's ears listening to what he was saying; other non-material ear-drums might vibrate to this important discourse" (*CO*, 18). Speech is breathed by these awe-full helicoid forms. So Griaule interprets Ogotemelli's whispers about those things that are better meditated through inner speech than spoken aloud. He seeks to avoid the primal sound frequency that transmits the thought through air to ear drums. For people who live closer to nature than we do, speech is always vibrational in natural forms. They do not forget that any message is always transmitted by physical means, like bird calls and drum beats, that the signifier is always carried by rhythmic physical waves. It is not a good idea to decry, as did the colonialists, the "bush telegraph." Sometimes better left unsaid, it is banal to forget these matters. Turning back to examine first principles is a fearfully sacred task because the creation is counterfraught with turbulent action.

In this creation myth, according to Ogotemelli, language is neither good nor bad:

> Its function was organization, and therefore it was good; nevertheless from the start it let loose disorder.
>
>   This was because the jackal, the deluded and deceitful son of god, desired to possess speech, and laid hands on the fibres in which language was embodied, that is to say, on his mother's skirt. His mother, the earth, resisted this incestuous action. (*CO*, 21).

But the jackal triumphed, and his incestuous act "endowed the jackal with the gift of speech so that ever afterwards he was able to reveal to diviners the designs of God" (*CO*, 21). Jackal is therefore the first back-and-forth mediator, whose theft of speech both conceals and reveals to diviners the designs of God, which are themselves turbulent helicoid codes. The first language is a natural language of curvilinear forms that were woven thenceforth into crafts. To divine the future, the designs of God, one follows the shapes of the helicoid signs in nature from the present to the implied repetition of waves in metonymic series.

I consider this Dogon example to be diagnostic evidence that S-curved wave forms were seen as a helicoid syntax. Language and being are unified by this primal pattern of nature. In the course of this study I offer many circumstantial accounts, both aural and visual, of S-curved waves as essential forming ideas of organic nature. But this is the most explicit statement that I know, from a native speaker, in which the helicoid sign is embodied as the first Word of creation, a word with its own intrinsic curvilinear design. The shape of the helicoid word is a syntax sufficient to weave all that is to follow. It is a common form of natural syntax.

There are many accounts of the thermodynamic shapes of things in narratives of creation. For instance, in several North American stories, Buzzard or Hawk creates hills and valleys out of the primeval mud by flapping wings and drying out the earth in updrafts and downdrafts. Similarly, in a Jicarilla Apache myth, Black Hactin is an animating energy that first created living things. First he created animals and birds and plants, but they wanted another kind of companion, so Black Hactin created humans: "Black Hactin sent Wind into the body of man to render him animate. The whorls at the ends of the fingers indicate the path at the time of the creation of man" (*PM*, 267). Rhyming with Nummo's speech, this fingerprint, our ultimate signature in criminal sign systems up until the advent of DNA, "dies away in a helicoid line," to quote Griaule. Like Eurynome's trace and like the Dogon helicoid forms, the animating Wind whorls through humans, manifesting its curvilinear design even at the extremities. This curvilinear or zigzag "line of life" is depicted on many artifacts of early American cultures. If a first principle of knowing is pattern recognition, then certainly a primal recognition is that several kinds of vibratory waves may be seen as primal helicoid forms. One sees that oscillation and pulsation are the signs of life in art and nature. To animate an art form, so that it seems to speak and move, one crafts it as if it were pulsating rhythmically.

The Helicoid Line as Recirculating Form of Life

Because my pursuit here is a biologically based syntax of just a few forms of transformation, I begin with a visual figure whose form is odd but symmetrical. Figure 8 is a photograph of a Stone-Age Sicilian tomb door of about two feet by three feet. When I first saw the stone door, propped against a wall in the national museum at Siracusa, I assumed that its strong bilateral symmetry made it look more like a face mask than anything else. What does the outline represent? The strangeness of the image is that, on first sight, one does not recognize any one being in the creation. There seems to be no such beast for which the image stands as imitation

Figure 8. Tomb Door from Casteluccio, Sicily (3000–2500 B.C.E.) Stone, approximately 2 ft x 3 ft. Permission Archaeological Museum, Siracusa.

or mimesis. I became aware gradually that the image was a conceptual diagram, and not just the imitation of a face, only when I understood that the image incorporates female and male generative forms. And the forms are both external and internal.

Its severe stylization, moving toward the extreme of ritualization, its disembodiment, allows the speculation that this is a symbolic diagram of the life force and its circulatory rhythms. Because I want to consider the flow and circulation and transformations of fluid wave forms as a biological basis for rhythmic syntax, I note that fluids here are imagined to carry the generative or reproductive principle. Note the deeply incised groove for the creative force, a duct up from the phallic geyser and through the channel, which is all the torso is, and spiraling into the breasts, where it emerges as milk. Bearing in mind what was said about Ophion as trace, use your finger to *trace* the lines in one direction, with your finger as a moving point. The spirals may be followed in the opposite direction as well, with the fluid from breasts moving down to create the fluid of testes, giving birth to the round disks from the spirals. Depending upon the point of view from which one starts the trace, whether from up-to-down or down-up, the life force flows from disks to spirals, or vice versa, either as sources or as receptacles, senders or receivers. This image probably represents a recirculating model of the forms of life. Although the symmetrical image is static, it represents some kind of model of moving form, of circulation, and of rhythmic proportion. Here one notes in passing that the physicality of a *trace* resides in the actual reenactment of the aftermath of an original motion. In the reenactment the trace is not just a word. My own first tracing of the outlines of the image exists as a quick ball-point pen sketch that I drew on the back of a hotel bill.

One of the reasons why myth and rite are so formulaic, so translatable across so many different cultures, is the traceable preoccupation with just a few physical forms of the beginnings and endings of things. So the animating curvilinear line of the Dogon extends to infinity, and the animating life energy of the Black Hactin whirls to digital extremities. This tomb slab about life moves toward the formulaic, for it is an abstract symmetry of formal relations about conjugal force. Its schematic outline of conjugation moves it almost entirely out of the realm of realistic portrayal and into a symbolically conceptual space by means of a linearity that depicts *fittingness*. It fits complementary male and female forms into a third proportion, a new form. To define is a simultaneous act with the twofold effect, that of separation and relation. To "delineate" is the visual version of to "define." Thus the more a drawing moves toward the starkly linear, away from the illusion of realism, the more conceptual it will become, as in Art Deco poster designs. So in this linear diagram, death is set off by life not by absolute difference but only by implicit phases of change. Here death and birth, end and origin, are conjoined by the semblance of circulatory movement. Two complementarities create a third thing, a conjugation, a creation. Many ancients thought that the life force and the death force in nature, those forces of construction and of destruction, are the same force—only seen from variant points of view and phase.

Perhaps I seem to be far away from a syntax of language, but look closer at the incised lines that stand for the illusions that make the image. (Although the incisions look like drawn lines in the photograph, that is a distortion of the transfer from a three-dimensional shape to a two-dimensional plane.) To conceive the patterns as the circuitry of reproductive life is to infer that the grooved lines etched into the stone must stand for grooved lines of force. To make the interpretation, one's eyes must trace the lines as if the grooved lines were tubes for moving fluids. This implies also a rhyme between stroke and image. Then, too, I have translated the visual image into a verbal concept by saying the image stands for reproduction. More truly, I should say *it stands for* a certain kind of recirculation. That means I have accepted the straight lines, the curved lines, and the spirals as force lines that represent certain generative functions of bodily channels. By seeing them, and by tracing them as grooves, I have transformed the static lines into an image for the movement of fluid forces. Three kinds of related linear strokes suffice to depict sexual regeneration of bodies. A few forms of protogeometrical lines constitute the syntax of the drawing, as they stand in the place of "lines" of life. Put the other way round, the body contours, both inner and outer, fit into larger patterns of a zoological life force, whose natural language is a few forms of lines. Later in the next chapter I shall quote Erasmus Darwin and Charles Darwin, who say essentially the same thing: life begins with just a few protogeometrical forms—points and lines and arcs and tubes.

The radical oppositions of male and female, down and up, are here resolved into a masklike form of bilateral symmetry about life and death. Two complementary forms create a third thing. Their differences require conjecture because at first glance the completed thing looks so complete unto itself that it does not seem to need interpretation, only description. Even though complete, it is not classically beautiful. Symmetrical and grotesque, its very oddity seems to ask one to figure what made so strange a thing. But there is no superb grotesque shape without some regularity in its proportion. Some may decipher the phallic symbol first, others the vulvar torso; most interpretations resolve upon conjugation of the two complements.

Nevertheless, the conjugal third thing still looks like an archaic ritual mask. Note the language of formal joining in this passage: "Throughout the anthropological literature, masks appear in conjunction with categorical

change. They occur in connection with rites of passage and cultural ceremonies such as exorcisms. They are, as well, frequently associated with funerary rites and death."[7] These are the important opening lines of A. David
Napier's *Masks: Transformation and Paradox.* A mask, then, is usually an
implicit sign of transformation, which in transgressing boundaries, creates
a new entity out of the two boundaries, one that metaphorically conceals
and another that reveals. If the brain is in part a transformer from wave
input to image, then a mask draws attention to the phase of metaphorical
transition. It incorporates in itself the ambiguity of contraries, since it
crosses from one state in transition to another. Victor Turner has called this
transitory state in rites of passage and in narrative a state of *liminality.*[8] And
the way that "categorical change" is associated with the definition of Form
will be central to my discussion of harmonious forms.

The interpretation of many of these art forms begins with an awareness
of some strangeness in their proportion, some dissymmetry, which has been
called a *break-spotting* capacity in perception.[9] Instead of looking for symmetry, Ernst Gombrich asserts, one looks for dissymmetry: "Thus while the
Gestalt approach fastens on our perception of order I would draw our attention to the reverse, our response to disorder." The break-spotter is designed
"to search the environment for discontinuities of stimulation." Usually the
break-spotting is forced upon one, either by nature or by art, when one is
forced to see or hear an interference, or "noise" in the parlance of information theory, not simply as noise but as a significant interruption, a possible
sign, that requires conjecture. An interruption for me may be merely noise,
but for you it may be a significant break-spotter. The archetypal interruption
is a lightning bolt, itself the heat trace of a static discharge, seen as a sign of
something else, especially when accompanied by the frequencied roll of
thunder. In poetry the analog is metaphor, which is a category mistake that
combines two separate categories of thought into a conjugal third thing, a
sport and the amphibious possibility of a new sort. Thought for poets, artists,
and some cognitive scientists, occurs at a metaphorical junction. I shall say
much more about junctions as the chapter goes along. In both the visual
form and in the literary form, one looks for the invisible relationship or code,
that transforms the combined categories into a third meaning. By way of the
Sicilian example I have tried to isolate a first principle in the forms of conjecturing about poetic language and being: conjecture with a few archetypal
forms has a family likeness with conjunction and conjugation.[10] The opposition between two opposing categories creates the symptom, but in this example there is no distinction between sign and design. More explicitly, the

relative fitness of complementary figures creates a third thing: in this case, the formal design of a curvilinear life force.

A drawing of only the upper part of this tomb door is reproduced in Marija Gimbutas' remarkable survey *The Language of the Goddess*.[11] The image has been drawn as a door in its original place, near Casteluccio, situated at the entrance of a rock tomb. Because the image, as it is drawn, is partially blocked by rocks in front of the door, she saw only the "eyes" and the upper part of the torso. She saw the image as a face. Brea reproduces a photograph of the whole image (his Plate 33). But for Gimbutas the spiral is part of a widespread visual language that would have interpreted it as the protective eye of the goddess. What she calls *language* in her title is a pictorial "script" for understanding the energy of the goddess (xv). Her script is what I would call a *syntax* of a few forms of drawing that suffice to render a protean life force. And her panoptic thesis is that this visual language of goddess worship was the first religion of culture. Perhaps her chapter most appropriate for this section is called "Opposed Spirals, Whirls, Comb, Brush, and Animal Whirls." How to resolve the variant interpretations, in which one construction thinks "eye" and another thinks "breast," I admit not to know. An expert interpretation would depend on a knowledge of the ancient Sicilian culture that contextualized the making of the image—the rites and stories of burial that narrated the act in their own regional grammars. For example, Brea places the linear scrolls of this slab door, and similar kinds of images on pots and jewelry, along a widespread Bronze Age trade route whose navigators knew their way from the eastern Mediterranean all the way up to the tin mines of Cornwall (96–97). These formulaic icons accompanied the earliest carrier vehicles of trade routes of exchange and circulation.

In the next chapter I shall discuss how a way of representing body parts in many primal cultures, as a kind of ball-and-socket joint, allows for the modular use of the same abstract circles to stand for breast, eyes, haunches, and joints. So perhaps an interpretation of the image as a face mask is not at all inconsistent with my hypothesis about a limited syntax of forms of transformation. Perhaps the image is after all a visual pun, for which the abstract face is the symmetrical exponent of the symmetrical two bodies that generated it. Then the tomb door itself can be seen as the head and the protective eyes of the goddess. Using the figurative trope of substituting the container for the thing contained (such as "the White House denied"), one supposes that the cavity of the tomb rhymes with the body of the goddess. If there are rhymes among the symmetrical face and the body, among the face and the body of the tomb, and the features of the earth at large, then this syntax of recirculating lines allows the interpretation that the stone image is a monument to the opening and closing rhymes of life and death. Though made of stone, it reveals elastic life. This carved slab seems to be a zoological admonition about how death is set off by life, not by absolute difference but only by recurrent phases of change in the apertures and portals of the earth's permeable body. So here, too, death and birth, end and origin, are conjoined by the semblance of flowing lines, seen as fluid helices, as in the example of Emerson's wave quoted in the Introduction, which neglects individuals while featuring the form of the governing body.

In closing this section about the Sicilian stone door, one may conjecture about its age and about the early date for thinking about abstract lines as abstract concepts. The ability to think in the abstract is evidently very old. I mean the holding or fixing of an abstract concept in visual design. Brea places the image in the Early Bronze Age, just after the transition from a copper culture. Gimbutas dates the carving at about 3000–2500 B.C. If it is indeed that early, it is the oldest "text" that I know of *addressing* the principle of generation of life through transmission of sexual fluidity between both sexes. At least two millennia older than Hippocrates, it represents not just a picture of body organs in copulation, but also an abstract principle of the transport and exchange of a curvilinear life force, because it interrelates ducts for milk and semen.

It is a remarkable feat to draw the slab door as if it were not just a face but also as a conceptual diagram. It is not the picture of the face of any *one* object or thing. It embodies nobody. The representation of a moving principle of sexual circulation is a sophisticated example of thinking with abstract incised lines. It means that language, at the very least picture language, is no longer in the barter stage of symbolic exchange and transformation. This image traces a going-over-into-opposites by curvilinear circulation. To meditate on this stone door as an artifact in a history of symbolic transmission by codes is to decide that an abstract language has indeed come of age, and most early, too. The image doesn't point to Mama and Papa. Although it points, as does all language as *deixis* (discussed in the next chapters), it refers to Nobody—nobody in particular. By refusing to refer to an object in real space and time, it is perforce removed to a symbolic space and time, to a proposition space or a hypothetical space, or a fictive space, or even a transcendental space—to give it anachronistic labels. By absenting realistic bodies, it begins to present an abstract concept. Nobody knows the Bronze Age word in Sicily for generation, or even if there was one. It may have been the name for the local goddess of all things. But the image does not refer to her so much as it *reenacts,* at a symbolic remove of physical tracing, the circulating life force. It is a diagram of the carrier waves that carry the principle of generating life, which does not refer to an *it.* Its physicality lies all in the transmission of carrier principles. The image is deeply carved, and to be interpreted, its etching must be traced. One follows the lines of physical signifiers of artful strokes, but one is thinking of the very abstract concept. So the physical act of tracing the gauged lines, as if they were ducts, reenacts the circulation and recirculation, perhaps as if it were the dialectical or reciprocal Way of the goddess: perishing and becoming; life passing over into death; and inanimate matter becoming animating fluids that presumably may have been the goddess's ancient principle of the earth's rejuvenation through children. In fact, wherever in primal art one finds some enactment of rhythmic periodicity through living form, one usually finds, as Gimbutas shows, that the message is the replenishment and renourishing of species.

This oddly symmetrical image uses abstract lines and disks and spirals to achieve its end of recirculation. The lines mean that one should reenact, not just imitate. Trace in a straight line, trace an arched curve, and then spiral around. In symmetry theory, to be discussed in the next chapter, these kinds of transformation are injunctions: translate, rotate, and twist. The artist's few lines are as abstract in their syntactic use as the message is an abstract concept. To be completely anachronistic, the lines of composition are like the Constructivist works of Barbara Hepworth, Naum Gabo, or Louise Nevelson. But this Sicilian work is not purely abstract art, like Gabo's see-through wire-line sculptures. The most telling point for my hypothesis about a few basic physical forms of syntactic transformation is that the image is a prototype, an archetype, for a sort of conceptual thinking that will come only very much later in history of symbolic codes. It is a prototype for drawing a circulation diagram. It would be anachronistic to say that it is like a wiring diagram for, say, electronic circuitry. But because it uses just abstract lines to represent recirculating force, it serves as the earliest model that I know of for the symmetrical outlining and inlining of a feedback system that sends and receives moving signals. Because the artist of the stone door and the designers of wiring diagrams use the same few linear units of composition, they can represent the same very abstract concepts.

If this hypothesis about syntactical similitude among a few basic strokes of composition is beginning to seem probable, then I have begun to make my point. But do we agree that this conceptual diagram is a visual model that does not need verbal language for its conceptualization? Although I use language to explain my conjecture and have learned much from the discussions of Brea and Gimbutas, I am still supposing that the artist who carved the lines so deeply used a sufficient visual syntax for articulating anatomical parts and functions. Inviting the viewer to trace the lines in order to visualize the concept of circulation, the artist depends upon a physical syntax of what we know about positions, locations, and shapes of bodies—in other words, an anatomical syntax of bodily order. If one does not need verbal language for the visual concept of life's recirculation, then I have begun to tease out a distinction between a physical syntax common to all grammars of various languages. In one sense, then, this early use of an anatomical syntax is also a "generative grammar." Broadly speaking, any grammar is a code that lets one reproduce, from a few alphabetical elements, an indefinite number of variations. So the phrase *generative grammar* is slightly redundant.

One more illustration (Figure 9) will sustain the conjecture that the curvilinear forms of the life force and the death force pivot back and forth as homeostatic points of view upon the same cycle of forceful energy.[12] This Scythian buckle depicts naturalistic forequarters of a catlike predator and a horse-like victim, but the hindquarters of each beast must end in stylized S-curved waves. S curves, or their topological variant, spirals, constitute an elementary syntax in many archaic cultures for depicting joints, as one can see from the shoulder joint of the cat. They serve as stylized "reversing gears" for back-and-forth reticulated motion. They are bent-back sign/designs of transformation, like the bow and the lyre. This pivoting semblance in S curves is perhaps their most important possibility, for to move back and forth elastically from antecedent to consequent, from retrospection to prediction, or vice versa, is the essence of sign/design-making and interpretation.

What is the element of patterning that allows the inference that these separate events are part of a group? These formal designs are signs of propulsion; they serve to remind how all animals get somewhere in space and time. In much archaic art of course these ball-and-socket joints are conventional designs. The patterns are just habits and customs. But in this example the S curves that attach tiger and horse to ground may not be only ornamental but also they may be conceptual. There are myriad forms of Aphrodites, mermaids, horses, bulls, dragons from the sea, some of which I shall discuss later,

Figure 9. Scythian Buckle, *Fantastic Tiger Tearing at a Horse,* gold. Courtesy of The State Hermitage Museum, St. Petersburg.

that are partially naturalistic and partially conceptual in the sense that, centaurlike, they amphibiously signify life's chthonic origin in the rhythmic forces of nature by ending in S-curved waves. But in the buckle the unnaturally reversing S-curves allow the inference that the horse will live again and the predator will die in its turn. The pounce of the predator, its *élan vital,* is correspondent in concept as it is complementary in form to the death throes of the horse. The animating energy of the line of life circulates both ways in curvilinear extensions.

According to Cook and others, the spiral in nature is a "growth under resistance" to some counterforce.[13] It is therefore a model for an elastic propellant in nature. According to Cook also, the femur in the ball-and-socket joint, the weight-bearing bone of many mammals, is spiral shaped at its knob (*CL,* 222–23). Hunting cultures would have noticed this form as they severed the joints. Hence the ball-and-socket joints, drawn at the haunches of many animals by archaic artisans, would have been representations of leaping and pouncing power. This power explains Ogotemelli's comment,

"The joints are the most important parts of a man" (*CO*, 51). I take up joints and junctures as physical parts of a formal syntax in a later section.

This rendering of metabolic opposites—the prolific versus the devouring—is not isolated. There exist many other archaic metalworks of the liminal moment of the pounce of life and death. All seem to represent these opposites as the ratio and rationale of being and perishing. In Greek myth, after all, Harmonia is the daughter of Aphrodite and Ares, Love and War, creation and destruction. A poetic example of the pounce is described by Homer, the poet who gloried in symbolic metalwork. Testing his wife on his homecoming, Odysseus, in the disguise of a wandering beggar, recounts how he had met her husband over twenty years ago. Circumspect Penelope in turn asks him to describe her husband. He responds:

> Great Odysseus was wearing a woolen mantle of purple,
> with two folds, but the pin to it was golden and fashioned
> with double sheaths, and the front part of it was artfully
> done: a hound held in his forepaws a dappled fawn, and
> strangled it and the fawn struggled with his feet as he tried to
> escape him. (Book XIX, ll.)

One might reflect that Penelope—who herself had given him those clothes and had pinned that pin upon him— knew that Odysseus' name meant "the bearer of pain" because he had been scarred on the thigh by a boar's tusk in his youth. But she means that reciprocating sense of his significant name as a carrier of terrible force, both the agent and the patient of suffering. Poetry, more than any other art form, seems to feature these elemental feelings—pain, suffering, perishing, loss, and the irresolution of how to cope with them. And we know to fear Odysseus, for he is about to unleash a terrible revenge upon the suitors. He bends back the massive bow of suffering, to which Homer's lyre is the accompaniment.

Few would deny that the principles of archaic crafting implicit in the gold buckle and in the funerary slab are grounded in physical forces of nature. Put differently, the curvilinear tropes seem to invite the conjecture that as sign/designs of reciprocating power they invoke the implicit forms of artful crafting. An S-curved design is a sign of the reciprocal harmony between the forms of the craft and the forces of nature. Herbert Read once described aesthetics in a way that helps me understand the rhythmic forms of wave patterns as underlying syntactic structure. He asserted that for him

the aesthetic experience was not a superficial phenomenon, an expression of sur-
plus energy, a secondary feeling of any kind, but rather something related to the
very structure of the universe. The more we analyze a work of art, whether it be
architecture, painting, poetry or music, the more evident it becomes that it has
an underlying structure; and when reduced to abstract terms, the laws of such
structure are the same whatever the kind of art—so that terms such as "rhythm,"
"balance" and "proportion" can be used interchangeably in all the arts.[14]

Read wrote before the term *law* came under scrutiny for its juridical
connotations. For instance, he mistranslates Lao Tzu anachronistically
about law: "To live in harmony with natural law—that should be our one
sufficient aim." I would want to substitute for Read's term *law* the generative phrase *common syntax*. Then we would be speaking of the same few
rhythmic principles that shape the structures of the earth and the technical
compositional elements of a craft.

For instance, in *Primitive Art* Franz Boas cautioned against the then-
prevalent theory of will-to-form or will-to-power arguments—that early
artists succumbed to seizure from an Other in creating their forms. He
adopted a pragmatic position that technical expedience more often than
not determined the designs of crafted things. For example, he said that the
reason why spirals are omnipresent in primitive utensils is that in both basket weaving and pottery one started with a long single strip of material and
began the basket or pot at the bottom by coiling the line about itself in a
resulting concentric spiral that gradually builds up from the bottom to en-
close a space within.[15] Nevertheless it is difficult not to imagine the potter
or weaver imagining the beginning and ending of his or her composition
as rhyming with the prototypical creation, that is, in a radical trope or turn.

If the techniques of crafting things rhyme with the larger rhythm of
earthly creation—as in Eurynome's dance of creation, the Dogon weave,
the Sicilian carved "face" mask, the Scythian buckle—then all these creatures that call attention to the craft of their structures as first principles of
creation are amphibious creatures.

Bird and Serpent

In her book *Inanna*, a remarkable translation and analysis of the
Sumerian forerunner of Ishtar the fertility goddess, Elizabeth Williams-
Forte speculates that snake-weaves derive from the tangled pattern observed when snakes are mating.[16] She illustrates a cylinder seal from about
2600–2500 B.C., which represents a snake weaving its body over and under

into a lattice. The single snake has a bird perched on its head to complete the creation myth. Snake and bird, according to Williams-Forte, may also represent, respectively, their homes in the roots and leaves of the tree of life. And a drawn lattice is the Sumerian pictograph for plant (*I*, 144). I consider this kind of union between snake and bird in studying plumed serpents below. But note that, for now, the snake here apparently represents the subterranean aspects of creation, while the bird represents the aerial aspiration of the union. Together they signify a generative cosmology of antithetical design. Do they deny while uplifting? Do they together sublimate a life force that weaves together two frequencies, earth and air?

John Ruskin observed that in Greek myth these two orders of animals are always associated with Athena. In his chapter "The Bird and the Serpent" Ruskin claims that birds are the very voice and spirit of the air. These qualities are "woven by Athena herself into films and threads of plume; with wave on wave of following and fading along breast and throat, and opened wings, infinite as the dividing of the foam and the sifting of the sea-sand."[17] Notice the *consonance* of wave, wing, weave—all introduced by a "W," a sign with a precise symmetry, whose shape repeats the up-down and back-forth trace of waves themselves. It is an ancient visual sign/design of moving junction.

The "living hieroglyph" of the serpent fascinated Ruskin without repelling him:

> There are myriads lower than [the serpent], and more loathsome in the scale of being; the links between dead matter and animation drift everywhere unseen. But it is the strength of the base element that is so dreadful in the serpent; it is the very omnipotence of the earth. That rivulet of smooth river—how does it flow, think you? It literally rows on the earth, with every scale for an oar; it bites the dust with the ridges of its body. Watch it when it moves slowly:—A wave, but without wind! a current, but with no fall! all the body moving at the same instant, yet some of it to one side, some to another, or some forward, and the rest of the coil backward; but all with the same calm will and equal way—no contraction, no extension; one soundless, causeless march of sequent rings, and spectral procession of spotted dust, with dissolution in its fangs, dislocation in its coils. Startle it;—the winding stream will become a twisting arrow; the wave of poisoned life will lash through the grass like a cast lance. It is a divine hieroglyph of the power of the earth—of the entire earthly nature. As the bird is the clothed power of the spirit of the air, so this is the clothed power of the dust; as the bird is the symbol of life, so this is the grasp and sting of death (*QA*, 68).

Ruskin was perhaps the last English naturalist for whom the whole earth was sacred and every image a parable, including that of the snake. In another essay entirely devoted to ophidians, called "Living Waves," Ruskin described their undulant motion more exactly.[18] He was proud that he had been the first naturalist to describe a snake's actual pattern of propelling itself. He used the same idea of alternating contraction and expansion of muscles to propel the body, as does Gans, whom I discussed in the Introduction. But being Ruskin, always whimsical, he derived his model of propulsion from contemporary models of glacier flow. A matter of debate at the time, the movement of glaciers was not self-evident. Ruskin favored the idea that glaciers creep by a plastic quality of expansion and compression of ice molecules.

Having heard T. H. Huxley, in a lecture on evolution, say metaphorically that a snake is a lizard with its feet dropped off, Ruskin began his lecture by drolly suggesting that perhaps a snake could better be imagined as a twisty honeysuckle with a head grown on. Because I am in pursuit of the idea that a few syntactic forms from nature suffice to compose archetypally the rest of articulating bodies, let me recall an identical kind of antithetical dialog in the composition of body parts, but this time in haiku. First, Kikaku:

> Darting dragon fly . . .
> Pull off its shiny
> Wings and look . . .
> Bright red pepper pod.

To which Basho replied:

> Bright red pepper-pod . . .
> It needs but shiny
> Wings and look . . .
> Darting dragon-fly!

I guess that Basho used the identical Japanese words and syntax to parallel but to invert the body articulation from fission to fusion.[19] One can see in the English translation that Basho has inverted the order of Kikaku's composition, just as Ruskin reversed Huxley's evolutionary order of things. Reversing the syntactic order of the words rhymes with the syntactic articulation of transforming body parts. More about the interrelations of body syntax and grammatical syntax comes in succeeding chapters. But now I think that we can agree that, if one premises a primal syntax of a few primal strokes, then there can be seen a necessary relationship between the emergence of language and being. To repeat Ruskin, "the links between animation and dead matter drift everywhere unseen."

Within this section of "Birds and Serpents," it is enough to notice the juxtaposition and occasional fusion of bird and snake as counteractive vehicles or living waves that carry the mythic messages of life and death and superlife. In this compound the snake is usually associated with the suppressed and chthonic signifier, Ruskin's "base element," while the bird uplifts the ethereal signified. Although animated nature may run to extremes, in archaic art it occasionally meanders through apparent opposites seeking equilibrium.

But since an S-curved wave is a propelling vehicle that can jump from one medium to another, the search for absolute meaning of an S as form will become eventually lost in one of the two main turns that the form sustains, one of which is a dizzying spiral downward to terror, as in Alfred Hitchcock's *Vertigo*. The archetype of the spiral surely involves a vertiginous loss of consciousness, even toward death.[20] In a late work Ruskin brilliantly illustrated that the decorative motif of the Greek Fret was characteristic of the labyrinth.[21] The crafting of the labyrinth by Daedalus, the Minotaur lurking in the bottom at the center, the love of Ariadne for Theseus, the retrospective thread that allows them to escape, and the story and the labyrinthine form seem formulaically fused in a curvilinear fret. Plot and narrative, design and eventuation, are reciprocally interwoven.

Yet Ruskin ably shows how the difference between the form of a thing and its eventual meanings is "adventitious." To use Susanne Langer's way of describing a "genetic fallacy," it is

> the error of confusing the origin of a thing with its import, of tracing the thing to its most primitive form and then calling it "merely" this archaic phenomenon. In a philosophy of symbolism this mistake is particularly fatal, since all elementary symbolic forms have their origin in something else than symbolic interest. Significance is always an adventitious value.[22]

Her wise admonition might well serve as epigraph for the chapter; signification is what gets tacked onto the primal carrier patterns. I shall argue in a later chapter that grammar is also almost always retrofitted back upon a primal syntax. I shall have more to say about the semiotics of whence and whither, for in the construction and use of any sign there is a habitual suppression of the physical wave form and a shift to adventitious uses and meanings. No instrument is bound by its original inventor's intention to a single use, not even a law of force, even though rules be set down for its eventual use. So an S-curved wave form, as syntactic carrier, will have no necessarily inherent meaning, but its shapely pattern may sustain a family of rhymed meanings in primal art that can be associated with its eventual uses.

Ruskin's own examples graphically show how the form of an ornament can become of symbolic import, and more to the point, how the form of a meandering line is an inexhaustible design for symbolizing:

> Of course frets and running lines were used in ornamentation when there were no labyrinths—probably long before labyrinths. A symbol is scarcely ever invented just when it is needed. Some already recognized and accepted form or thing becomes symbolic at a particular time. Horses had tails, and the moon quarters, long before there were Turks; but the horse-tail and crescent are not less definitely symbolic to the Ottoman. So, the early forms of ornament are nearly alike, among all nations of any capacity for design: they put meaning into them afterwards, if they ever come themselves to have any meaning. Vibrate but the point of a tool against an unbaked vase, as it revolves, set on the wheel,—you have a wavy or zigzag line. The vase revolves once; the ends of the wavy line do not exactly tally when they meet; you get over the blunder by turning one into a head, the other into a tail,—and have a symbol of eternity—if, first, which is wholly needful, you have an idea of eternity!
>
> Again, the free sweep of a pen at the finish of a large letter has a tendency to throw itself into a spiral. There is no particular intelligence or spiritual emotion, in the production of this line. A worm draws it with his coil, a fern with its bud, and a periwinkle with his shell. Yet, completed in the Ionic capital, and arrested in the bending point of the acanthus leaf, it has become the primal element of beautiful architecture and ornament in all the ages; and is eloquent with endless symbolism, representing the power of the winds and waves in Athenian work, and of the old serpent . . . (*FC*, 404–05).

Notice that Ruskin's finishing stroke of a sweeping pen, like the Dogon plosives of speech and like the whorl at the end of one's fingertips, notice that all die away in a helicoid line. Even at the exhausted end of things, the detritus falls away into an insignificant curlicue. The Be-All and End-All of energy dies away not in the power of waves but a trickle. Even the tired flick is a physical flicker. Without meaning, adventitious, unintentional perhaps, the form lingers, like the exhausted line of beauty in late works of Art Nouveau. All of Ruskin's techniques, beginning with the direction to vibrate a stylus, necessarily harness physical forces from nature. Notice the transformations therein. If you rotate an object like a vase, and you set a stylus or pen to draw a straight line up and down, perpendicular to the circle of rotation, then the resulting trace yields a series of meandering curves. The line of straight-line translation, when combined with the line of circular rotation, yields as trace a linear series of waves. Those three strokes are the primary syntactical parts of a wave. (Recall from "Art/Frequency" the three motions of molecules in the motions of waves—horizontal, longitudinal, and torsional; in the next chapter, I return to those allied terms of transformation from symmetry theory: translate, rotate, and twist.)

For Ruskin, vibration and oscillation can be merely automatic, yet they can also become "the primal element" in symbolic art. For instance, he says that *labyrinth* means "rope-walk" or "coil-of-the-rope-walk" (*FC*, 408). That is to say, following Boas, the clue to its structure is the mechanical coil of the rope, turning upon its origin outward in a spiral. The key is the turn back upon itself that involves expansion outward and forward as it coils back. The explanation is tropical, the turn back and forth upon means and ends in a compound of space and time. From the labyrinth onward, all symbols of S curves ride on the wavy counterturn backward that also expands forward. What forces are being personified when Eurynome dances and separates the sea waves from the sky waves? Most important, what forces are being represented when she turns back over her shoulder and twists her own trace into a creation? How do you represent a universal beginning as a twist? Solve that twist and you have the beginning of a three-dimensional rhythmic shape, a reenacted curvilinear trace of the formal creation of all things.

Amphibious Dragon Rhythms

The undulant form of a dragon, half visible, half invisible, and doubling back upon itself, has personified the beginning of creation in several ancient narratives. Other than recalling Eurynome's trope, I shall not stress the history of that image in creation stories; instead I shall describe its wavelike nature as an elastic element for personifying rhythmic power in creation. In most creation myths, nothing becomes something by a curvilinear separation of light and dark, water and sky, chaos and earth. Nothing becomes something by halving itself into opposition, a fundamental paradox. In several ancient cultures dragons are represented as having features and components that personify curvilinear rhythms of natural power. It follows that those who try to interpret dragons follow a curvilinear trace of hiddenness and exposure. A dragon both conceals and reveals, like a mask, metaphor, or sign/design. Their interpretation is intertwined with curvilinearity.

Perhaps the earliest drawing of a dragon, a footed snake with a humanoid head, apparently dated from the third millennium B.C.E. It is a simple brush drawing on a jar, which traveled to this country with the breakthrough "Chinese Exhibition" of the early 1970s. Here is its description in the catalog: "This appears to be the earliest representation of the dragon of Chinese myth. In later times it is described as amphibious and benevolent, but sight of the whole animal being dangerous to man, it is represented partly hidden in cloud or water."[23] Here the term *amphibious* literally means inhabiting both water and air. Any wave form that jumps from one physicial medium to another is by definition amphibious, such as a light wave that moves from air to water, which makes a stick in the water seem bent. But it can also mean metaphorically its artistic representation, as being represented as both half visible and half invisible. In this sense its composition accurately reenacts the twofold nature of the sign, with a visible superstructure and an invisible base, discussed in the Chapter "Art/Frequency." Any composition whatsoever is a compound, constructed of the invisible but material form of a carrier wave, plus the adventitious illusion of its meaning.

If one could rotate the Chinese jar, as Ruskin did his hypothetical vase, one could see the other parts of the dragon. But in understanding dragons and the codes of all curvilinear form, the key is the turn. Dragons should never be represented so that they can be seen fully all at once. A partial image signifies their amphibious nature, as halving may represent doubling. In this case a physical turn of the jar may underscore a figurative trope. As George Rowley has shown, most Chinese representational art, whether a "dragon scroll," that one unrolls from right to left, a hanging scroll, or a pot, invites one to follow a path or gradual sequence of visualized interpretation.[24] The sequence of interpretation, in other words, follows a spatiotemporal pattern in the art work of rhythmic intervals that invite attention in their appropriate place and phase. So even in a static dragon painting there is an implied temporal way of reading it rhythmically in a narrative sequence. For instance, in the scroll that Rowley has photographed in four sections, one follows a gradually unrolled sequence of islets, hummocks, rocks, and trees amidst a lake landscape (*PCP*, Figure 15). Although Rowley does not mention it, to be in the place of the observer enjoying this passing illusion of seeing different islets appear before one's eyes, is to understand that the observer can only be in a boat that is imaginatively carrying one's body about the lake. A point of view is tacitly implied, but its moving platform is an invisible but physical vehicle that is carrying the viewer through the medium. This attitude of the observer being an implicit part of the landscape is here intimated by the slight movement, drift even, of the propelling vehicle, so that the gradual motion of unrolling the scroll is displaced by the empathy of drifting.

While it is commonplace for Westerners to think of the rhythm of a few prescribed natural elements such as a stream, Rowley wants to explain the Chinese understanding of the rhythm of a rock or the rhythm of bamboo or the rhythm of birds in flight, indeed the characterizing rhythms of all things in nature:

> Furthermore, the qualities of plastic organism were avoided in their favorite theme of the dragon, who seemed to epitomize their delight in abstract and fluid rhythm. Indeed, this characteristic rhythm might be called the mother of all Chinese rhythms, a kind of yin-yang dualism of reverse curves, concave and convex, stretched to attain the utmost in speed. It is the rhythm behind all the freest rhythms in nature, rhythms of flame, of swirling water, and of flying clouds. It is the rhythm of growth, each branch being made of responding curves, and it is the rhythm of erosion, present in most mountain wrinkles. One might ask if it is not also the rhythm of the muscles moving around the bones of the human body. That is true, but the test of rhythm is quality, not general shape; the curves of the body imply guided plasticity and power, while the curves of nature suggest the flux of natural forces (*PCP,* 42).

Yet if the strokes of the brush are applied by the supple joints of fingers, wrist, elbow, and even shoulder motions, then the rhythms of the body's structures may rhyme with the different rhythms of natural landscape.

Consonance and Geomancy

Dragonlike undulating intervals are for Rowley a basic principle of Chinese design, one that is fundamentally involved with the Taoist concept of rhythmic relationships, of allowing yin and yang, originally shade and light, then female and male, and most abstractly positive and negative, to play out their convexities and concavities in time. Perhaps I should say *convolved* rather than *involved* because his preceding passage, plus the following, set forth a unity in much Chinese art that Rowley called *consonance* or a *rhythmic likeness*:

> To realize such a dynamic conception of unity the Chinese resorted to many devices of which the two most effective were the synthesis of opposites (yin-yang) and the likeness of parts which may best be described by the term consonance.

The relationship of yin and yang was the most basic single principle in Chinese design. [If one visualizes the tao chi tu, or yin-yang emblem, one can see a single S boundary delineating the black and white tear shapes.] Since k'ai-ho was simply a mighty yin-yang, it was inevitable that this dualism should have been applied to everything from the association of thematic material down to qualities of brush stroke. ["On the basis of the eternal flux of nature Tsung-ch'ien, an eighth-century writer, described creation as k'ai-ho, an 'open-join' or 'chaos-union' process" (*PCP,* 48).] This is coherence through the interdependence of opposites. Although this principle played a role in Chinese painting corresponding to the Greek principle of unity and variety, the two principles differed fundamentally. Unity is order derived from the chaotic variety of nature, whereas yin and yang are opposing forces that need one another for completeness. According to the Chinese, the substantiation of such coherence was to be found in the dualism of forces throughout the universe, whose interaction is the source of life—heaven and earth, male and female, birth and death. By analogy, every possible dualism was supposed to express this cosmic dualism (*PCP,* 50).

In the first few illustrations of the chapter, I have already described a reciprocal dualism of curvilinear life force, and I shall follow an apparent dialectic of opposing forces throughout this study. But Rowley demonstrates that dragon intervals signify a dynamic dual code that subsumes male-female, life-death, prey-predator into one natural symbol of oppositional interdependency in space and time. Nature is not chaotic in Chinese thought. Its amphibious code of composition is manifest in dragon rhythm. Rowley also reminds us as Westerners not to think of absolutely rigid oppositions. Instead, for early Chinese artists and scholars, all is mutual interdependence of different particulars, where in the natural flux of varietal things, one mode will eventually go over into its opposite. So for other art historians of Chinese painting it is useful to think of structural patterns of thought underlying these images as a kind of "complementary bipolarity" that summon natural phenomena and their inverse image: emptiness-fullness, stillness-movement, softness-hardness.[25]

The Mustard Seed Manual says that dragon rhythms are images that depict the designs of weather: "Masterpieces of dragons were direct representations of the power of Heaven arousing life with rolls of thunder and flashes of lightning through thick clouds, dispensing its power by wind and nourishing rain. In landscape painting, the dragon itself is invisible but the weather suggests its presence."[26] Visible signs of weather mean the code of an invisible dragon force. But sometimes the dragon is rendered half visible and sometimes the weather waves are rendered half visible. For

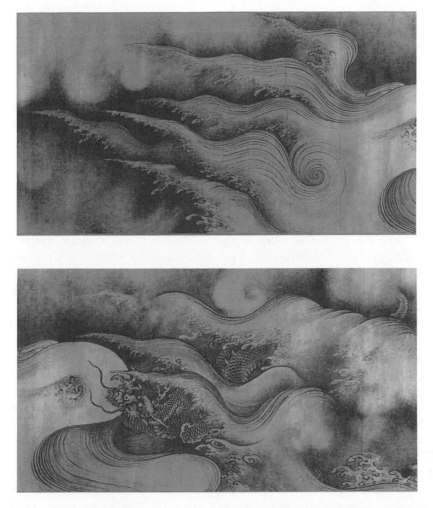

Figure 10. Chen Rong, *Nine Dragons* (section), Southern Song dynasty, China, first half 13th century. Permission Museum of Fine Arts, Boston, Francis Gardner Curtis Fund.

instance, in Figure 10, a section of Chen Rong's *Nine Dragons,* the dragon emerges from the wave half visible, half invisible, in such a way as to allow the inference that the dragon is here meant to personify the abstract forces that are the rhythms of water waves. In that sense this rendition is very much like the combination of natural and of stylized force in the Scythian belt buckle of Figure 9.

Kenneth Clark compared the lines of force in Chen Rong's dragon scroll to the apocalyptic turbulence of Leonardo da Vinci's water drawings. For Clark, Leonardo's visions of a deluge of wind and water

> are the most personal in the whole range of his work. They express, with a freedom that is almost disturbing, his passion for twisting movement, and for sequences of form fuller and more complex than anything in European art. They are so far from the classical tradition that our first term of comparison might be one of the great Chinese paintings of cloud and storm, for example, the Dragon Scroll in the Boston Museum. Yet, as with his landscape, closer study shows that Leonardo's scientific attitude has given his drawings a character fundamentally different from Chinese painting. By profound research into the movement of water he has learnt to give his lines of force a logical as well as expressive significance.[27]

I do not think that Chinese dragon rhythm can be thought unscientific, especially considering the anachronistic way in which Clark imposes the phrase "lines of force," popularized by the Victorian Michael Faraday, upon Leonardo's method. When in his manuscripts Leonardo repeatedly drew wave forms in grasses, in braided strands of hair for Leda, in elegant flower arrangements, as well as in deluges of rain and flood, he probably was seeking in the helix or vortex the generating archetypes of forms of construction as well as of destruction in nature.

Dragon rhythm, then, is a metaphorical expression for the "consonance" or resonance in a design, the rhyme between the kinds of strokes an artist uses, his laws (*li*) of organization, and the representation of the landscape, such as the upthrust of a certain rock formation. An extension of this metaphor was "lung-mo" or "dragon veins." Rowley cites two observations about this widely used principle of interconnection. *Lung-mo* could imply rhythmic wrinkles or indentations, as in "Wang Meng who used dragon veins (*lung-mo*) abundantly, making them like winding snakes, or in Wu Chen who painted them like straight lines." Or "Make the dragon veins slanting or straight, complete or in fragments, hidden or visible, broken or continuous, but all bristling with life, then you will make a real picture" (*PCP*, 68).

In Chinese painting, according to Wen C. Fong, one is always supposed to notice the calligraphic brush stroke: "Known as the 'trace' of the brush and ink, the subject of a calligraphic work is the brush as an extension of the calligrapher's body" (*BR*, 5). Unlike the Greek tradition of mimesis, or "imitation of nature," Fong says, the Chinese painter did not seek to achieve the illusion of representing nature by "concealing the pictorial medium."

Instead, the physical movement of the painter's body incorporates the strokes of nature (*BR,* 4–5). This idea of stroke is, again, one of *methexis,* not *mimesis.* In *The Mustard Seed Manual,* furthermore, the stroke is illustrated by a simile: "Each brush stroke should move and turn, with abrupt stops (*tun*), sinuous as a dragon" (*MSM,* II, 130). When a student learns to paint calligraphy, one is taught these various strokes as a kind of syntax.

If these techniques are intended to match physical stroke with rhythms of the earth, then dragon rhythms imply that rhyming patterns are the ways to achieve animation in a picture, but also they suggest that all the earth is animated. For example, Lin Yutang, who helped interpret Chinese ways to Americans in the 1930s, stressed that the dragon is not merely antediluvian or mythic: "To the Chinese, the mountains and rivers are alive, and in many of the winding ridges of mountains we see the dragon's back, and where the mountains gradually descend and merge into the plain or the sea, we see the dragon's tail. That is Chinese pantheism, the origin of geomancy."[28]

In this primal kind of hermeneutics, to divine is to interpret all natural forms as being alive; so to see the world as the body of the amphibious dragon is to retrofit its body parts back upon the structures of the earth. But the point of divination, of course, is to be able to tell which way things are going. And if you can trace the rhythmic lines of the world's body—the lay of the land, the weather, the bends of a river—you have some leads. This half visible, half invisible, troping is not quite the same as tracing the conceptual diagram upon the tomb door in Sicily, where a death mask represents a recirculating life force. Here the dragon reenacts the animating "turn" itself, rather than being a conceptual diagram of the turns. To find a way of seeing ordinarily inanimate things as being subterranean but living parts of the great dragon requires building a principle of artistic composition back into the body of the earth, and thereby reanimating rivers and mountains by seeing their declensions in space as elements of the dragon's code. For example, here is a translation of the calligraphic commentary for an illustration called "Method of painting an overhanging rock that hides part of the waterfall":

> Mo-chieh (Wang Wei) said: "When one is painting a waterfall, it should be so painted that there are interruptions but no breaks." In this matter of "interuptions but no breaks," the brush stops but the spirit (*ch'i*) continues; the appearance of the flow of water has a break but the idea (*li*) of it is uninterrupted. It is like the divine dragon, whose body is partly hidden among the clouds but whose head and tail are naturally connected (*MSM,* II, 205).

For ancient Chinese, water rhythm was the most common archetype for interpreting the Tao or Way. The term *ch'i,* or spirit, Rowley says, was often used to describe the Way. But certainly no word can describe it. For instance, Lin Yutang, further extends the word to other compounds of spirit or force: "*Yuanch'i* means 'vital force' in the universe and in an individual, and one should do well to nourish or cherish it. A literary or artistic masterpiece is supposed to have stolen the secrets of nature, thus leaking out the *yuanch'i,* and it is a thing not to be attempted too often." [29] One takes note that it is often better to let the connections of natural force remain tacit, only partly visible underneath the semiotic bar. Since the spirit or force of Tao energizes all things, it is effaced by the faces and shapes of the myriads of created things. Yet by way of self-effacement and self-government and least resistance, Lao Tzu seems to say, one may catch a half-glimpse into the government of the Way. In *Tao Te Ching* the most common trope for describing the non-contention of the Way is the form in which streams and rivers flow downward to larger rivers and thence to the sea.[30] By seeking the drain of least resistance, the Way is submissive yet rhythmically overwhelming, as yin may go over into yang. There is an unsettling rhyme here between natural "law" and political "law"—between a natural flow and a political force of quiet resistance in a long revolution. But the point is, of course, that by following the least resistance of the Way one can interpret or divine which way things go in the world. The Way doesn't tell the future, but if one follows the strokes of nature, one can get some leads. Here is one watery example of many: "Highest good is like water. Because water excels in benefiting the myriad creatures without contending with them and settles where none would like to be, it comes close to the way" (*TTC,* 64). Here is another poetic adage that also depends upon gravitational flow for its interpretation: "The Way is to the world as the River and the Sea are to rivulets and streams" (*TTC,* 91). The syntax is of proportion and ratio, but as we saw in the last chapter, the sense is of moving proportion, which here gathers in volume and immensity from little runnels, where minnows live, to river deltas, where civilizations flourish: "A large state is the lower reaches of a river—The place where all the streams of the world unite"(*TTC,* 122). Notice that the archetype depends upon capillaries and streams and deltas, but the conceptual geomancy is the interpretation of branching diagrams or Y-shapes as models of physical interrelations. The way is figured as a meandering archetype that flows necessarily down, and political geomancy is a divination by way of dynamic flow. In the next chapters I shall show how gravitation, that which makes things flow in a direction, is inherently linked to symmetry theory. Wisdom literature is prescient because it anticipates the way that things will lead.

The Way is usually described in terms of subtle hydraulic resistances. "Turning back is how the way moves; weakness is the means the Way employs" (*TTC*, 101). Another instance, a fable of Chuang-tzu, tells of an old man who fell into a cataract but emerged unscathed downstream. Did he have a way?

> No, . . . I have no way. There was my original condition to begin with; then habit growing into nature; and lastly acquiescence in destiny. Plunging in with the whirl, I come out with the swirl. I accommodate myself to the water, not the water to me. And so I am able to deal with it after this fashion. . . . I was born upon dry land . . . and accommodated myself to dry land. That was my original condition. Growing up on the water, I accommodated myself to the water. That was what I meant by nature. And doing as I did without being conscious of my effort so to do, that was what I meant by destiny.[31]

Accommodating himself to the helicoid whirls was another way of saying that the Chinese learned to fit within a concept of rhythmically alternating waves of nature epitomized by yin and yang, flux and reflux. Intervals of accommodation or rhythmic fittingness, as we shall see throughout, is the leading possibility of the trope of an S-curved wave, which allows the possibility of fittingness with its opposite but only in fluid succession. Joseph Needham has admitted that so dominant was the concept of wave motion in Chinese thought that, as opposed to atomistic thinking, it inhibited the advancement of scientific knowledge in China.[32] In the same passage he notes, too, the "dialectical quality" of the thought that everything carries the seeds of its own disintegration (*SCC*, 6).

Perhaps the essential teaching of Needham's chapter is that for the Chinese the universe was conceived as a sine-curved continuum with waves—standing waves, acoustic ripples, dietetics, hygiene, the sun and the moon—all participating in the rhythmic unities of natural forces in their places and phases. Needham points out that among Westerners only the Stoics adumbrated a wave theory similar to the Chinese (*SCC*, 11–12). Given the ubiquity of waves in the Chinese design of nature, it is possible that every act of human creation may evoke the latent rhyming of the half-invisible dragon. The design itself may be seen as an amphibious code for those rhythmic forces where all things seem consonant with one another. In this sense of consonance, then, the relationship between the making of a thing and its meaning as the pattern of a living dragon is not entirely adventitious as it would be in much Western art. For the syntactic forms of the strokes rhyme and turn back upon one another as the re-patterning of the ridges of mountains and the windings of rivers. The myriads of creatures, from small black and white fish chasing each other in rivulets to emperors and empresses in their deltas, all turn and twist in their natural channels of wave frequency. In European epistemology there is the futility of the hermeneutic circle, but in much wisdom literature and traditional art there was and is a patterned and elastic twist back and a propulsion forward, which is not so much a model of the Way but a bodily gesture or moving posture.

T'ao t'ieh

Dragon rhythms were not mere images for Chinese artists; they were seen as primal forms for the wavelike convergences of all things toward a primary direction. Perhaps this is another semblance of the twisting interdependence of opposing cosmic dualities, which is Rowley's large theme. That is certainly one way of describing the *t'ao t'ieh,* the dragon mask of ancient Chinese design (Figure 11). William Watson has shown that wonderfully grotesque combinations of this most basic Chinese demon mask become more astonishing when one realizes that the monster can be seen not only as one full-face being but also as two monsters nose-to-nose in profile. But he says that both aspects cannot be seen simultaneously.[33] This then is another example of the Gestalt principle of push and pull. Although Watson does not call this monster a dragon but rather a prey-predator duality of perhaps tiger and ox (*SAC,* 27) (and therefore consonant with the stylized lion and horse of Figure 9), it is often combined with dragons, serpents, cicadas, birds, and other stylized animals in the most ancient decorative motifs. When combined with a fret motif of squared spirals (*lei-wen*) or thunder calligraphy, one recognizes that the motif begins to rhyme with other examples of transforming S-curved waves.

For Gombrich, the spectacle of two profiles confronting one another in *t'ao t'ieh* may be part of an archaic maker's grotesque response to the psychological anxiety about depicting dragons in the first place. His section "A Great Dragon Force" is devoted to the notion of "protective animation," as, for example, the archaic desire to ward off evil by painting eyes on boats.[34] He mentions that, in much folklore, demons and witches hate to be confused; hence the tangle of knots, mazes, and riddles—for instance, from Maori to Scandinavian art—are protections against monsters as they themselves pause and turn back to decipher the design.

The makers of dragons make them seem enchanted by the codes of their own construction, as they twist the dragons so the dragons themselves

Figure 11. Diagram of T'ao T'ieh. By William Willetts, *Foundations of Chinese Art.* Permission Thames & Hudson Ltd. Imagine this symmetrical creature as being two profiles split longitudinally along its dorsal side with the two halves rotated forward from the tail so as to meet the plane of the page, while being hinged at the snout.

can turn back upon themselves and decipher their own presuppositional codes. How does one unravel the presupposed principle of order in the apparent tangle of the decorative field? Sometimes, for instance, the dragon is represented as amphisbaenic, with a head at either end. Surely part of the anxious pleasure of making and deciphering demon masks is involved with the incipient threat that one is confronting the half-exposure of certain underlying principles of natural order. Johan Huizinga says that much of the sacred play or ritual of archaic making is intended to recreate some portion of the "cosmic order" in which play is in a sacred space set aside for a competition of dual forces that helps induce cosmic events.[35] He claims that all such ritual making is *methetic* rather than *mimetic;* it helps the larger action to happen rather than imitates it (*HL,* 15). The sounds and strokes embodied in the Signifier, the carrier of the message, are reenacting the primal strokes and gestures of the natural world, and they are designed to evoke a responsive chord in nature.

For Boas and Levi-Strauss and Gombrich, among others, one of the most fascinating riddles about t'ao t'ieh is that they occur profusely in archaic cultures of the Far East as well as the Americas. Whether by diffusion theories of migrating symbols or by means of the psychology of split representation in humans, how does one account, in different cultures, for

this conceptual symmetry of two opposing parts uniting to form a terti-
ary full face?[36] This theory of migration, and the questions of body symmetry seen as anatomical syntax, are discussed more fully in the next
chapter.

The technological argument for similarities between styles of East and
West is made by Boas, who shows that interlocking S curves are inversions
of two symmetrical halves, so that what is above to the right is below to the
left in a rotation symmetry learned from pattern weaving and from making pots out of clay strips (*PA,* 37–39). (My Figure 1 in the Introduction is
redrawn from his diagram delineating these alternate inversions.) Watson
also claims the "influence of technique" to account for "the origin of the
vertical ridge which runs down the middle of the nose of most *t'ao t'ieh*"
(*SAC,* 36–38). Early Shang dynasty casters carved both positive models as
well as negative piece molds to make relief designs on the bronzes, and in
the late Shang apparently the "potentialities of joint seams were seized on
for positive effect" (*SAC,* 37). Watson charts a series of progressive turns
from origin to eventual use, from necessary technicality to decorated and
disguised technique to independent motif (38). Like Ruskin and Langer, he
also distinguishes between the technical construction of a design and its intended meaning. That is only reasonable because the amphibious components of any sign are the natural order of the physical vehicle that carries
the sign, whether breath, pigment, stone, or bronze seam—plus the signified context, the particular semiotic system, and the larger culture, which
stipulates its meaning. This is a basic category shift of signs, a first principle, if not a cosmic duality.

The amphibious sign, which combines this category shift, is under dual
control of signifier and signified. By reason of its odd construction,
whereby one cannot read it simultaneously as both full-face and two profiles, a *t'ao t'ieh* in split representation always draws attention to its turn
back and forth from its meaning to its own principles of construction,
when it is represented as two halves of a whole thing. If the stylized demon
mask is "really" two asymmetrical halves, then its unity is a fiction. And
though it may be a composite of more than one beast, it is still composed
of two opposing one-sided beings masquerading as full-faced unity. No
wonder that demons, monsters, and witches desire to know the exact order
of things, since much of their own nature is the half-nature of dragons
seeking their true complement, whether yin or yang. The semiotic point of
convergence in East and West is that theme and structure are twisted complementarities, because all turn back and forth in space and time to use the

dualistic character of signs from nature to create basic designs. A more radical thesis is that archaic sign/designs themselves are structured as they are because their structure reenacts methetically the complementarity of twisted natural forces, at a remove to be sure, and that makes all the difference for the one-sidedness of dragons. I shall discuss this asymmetrical twist as a variant of symmetry in later chapters.

S-shaped curves belong to a class of a few primal natural forms, from which many designs are derived. How they are deployed to "mean" various messages of duality is a different question. If I imagine myself an archaic artisan, for whom all of nature is coded from a few primal consonant forms, I implicitly assume that the structure of signs is of a kind with designs of natural force in curvilinear consonance. Note that if one imagines the yin-yang emblem as two dragons chasing one another in mutual creation, one easily falls into the assumption that logos and cosmos are the same, for there is only a shifting point of view between the first sign and the first design.

For many archaic artisans and story tellers, a tropical S is the curvilinear form that incorporates a rhyme about creation. To test this hypothesis of consonantal sign/design in the shape of a torqued S curve, using less dispersed examples than I have assembled so far, I next review the most famous dragon motif in the Western Hemisphere, the image of the Plumed Serpent, later called *Quetzalcoatl* in Mexican stories. Notice throughout the ways in which the primal form of the S as a pictograph carries a message about balancing creation by thermal frequencies.

The Plumed Serpent

Different cultures and tribes, and presumably also clans, interpreted the meaning of the image in different ways according to their own specific contextual needs—whether they were Olmec, Toltec, Mayan, Aztec, or the Anasazi, Hopi, or Navajo. Perhaps few would quarrel with this popular interpretation:

> The enemy of the serpent [previously described as a rain/fertility god] was the eagle. This giant bird dominated the heavens and was the only creature brave enough to attack the snake. The conflict between the representatives of the two primary natural forces could only be resolved if they became one god of earth and sky. So the quetzal (bird) and the coatl (snake) became one, Quetzalcoatl, the feathered serpent, the most revered and impressive of the ancient gods.[37]

That passage reintroduces Ruskin's compounding of bird and snake, as old as Athena and Inanna. Here is a more precise definition, one that shows the ways in which the iconic S as design-vehicle carries the possibilities of the narrative:

> American Indian religion in one of its many phases has tended to conceptualize the universe in terms of halves—the familiar Mother Earth or Father Sky, or Mother Earth and Father Sun—whose mystical product is the hero-child known to the Navajo as Naye'nezgani or to the ancient Mexicans as Quetzalcoatl. (To the earth-sky pair belongs the power of being, to the offspring is given the power of doing. The former, in other words, enjoys eternal or at least prior existence, serving as a procreative and salutary energy upon which the hero draws to finish the work of the world.) What is noteworthy, though hardly unique, about the deity Quetzalcoatl is that he represents not only the heroic "product" but the generative "halves" as well. He is synonyomous with the Spirit of Duality, the sophisticated Mexican concept incorporating earth and sky.[38]

Notice that these generative halves are the bases for any metaphorical move toward the concept of a "generative grammar." Here there is rhyme between the generating of beings and the generating of language, art, and knowledge. The principles of a common syntax compose all things.

Bierhorst briefly summarizes the tradition, beginning with Prescott's *Conquest of Mexico,* that personifies the Plumed Serpent in terms of a real but prehistoric hero who at some time took on messianic proportions. As he translates one of the fragments:

> Truly with him it began,
> truly from it flowed out,
> from Quetzalcoatl
> all art and knowledge.

So, in addition to the idea that the ideograph depicts a union of sky and earth, it may be interpreted also as the "form of the archetypal serpent who imparts cultural knowledge"(*FM,* 4).

Others have also interpreted the Plumed Serpent as "the primary archetype in Mesoamerican iconography." For instance, in his recent book *Legends of the Plumed Serpent* Neil Baldwin has traced several avatars of the archetype as connecting links in a history of Mesoamerica.[39] Beginning with the Olmec culture, he uses copious illustrations of feathered serpents, taken from the ruins of the past—stone statuary, relief carvings, clay products, painted codices—in an elegant demonstration of the resurgent power of the primary icon. He assigns chapters to successive cultures, such as the Maya, Aztec, Spanish conquest, Creole, Revolutionary Period, through to depictions of Quetzalcoatl by master artists of the twentieth century— Diego Rivera, José Clement Orozco, David Alfaro Siquieros, and Rufino Tamayo. The next-to-last chapter features D. H. Lawrence's novel *The Plumed Serpent,* which is quoted below, while the last chapter studies the ghostly return of Quetzalcoatl as a possible unifier of contemporary Mexico's mestizo culture.

According to Bernard De Voto, furthermore, at the very beginning of the Spanish Conquest, some thought that the Indians' recognition of the cross as a sacred symbol made them "instinctive Christians":

> But it was a native symbol in many tribes. It stood for the four houses of the sky and the four winds that came out of them, or for the morning and the evening star, or for the path of the sun, or in the form of the swastika for the rain god who gave life, and his bird that was the thunder, and the snake that was sometimes his avatar.[40]

One can see that it was almost impossible to avoid an adventitious but ideologically powerful imposition of a socially constructed lexicon, like a cross or swastika, upon a primal pattern of Four Directions.

According to Laurette Séjourné, Quetzalcoatl is the duality unifying the sun and the earth, spirit and matter, and he/it is also associated with the sacred cross of the four quarters of the earth and the four quarters of the sky.[41] Séjourné also illustrates various versions of the hieroglyph for "movement," which can also suggest the movement of the sun around the four quarters of the earth from east, south, west, and north (*BW,* 97). Bierhorst states that this sunwise movement around the four quarters symbolizes a rejuvenation cycle of death and rebirth for the god-hero, Quetzalcoatl, and later for the Navajo; the child as prankster-hero rises in the east, moves south as an adult warrior, enters the portals of death in the west, travels through the underworld in the north, where he prepares for seasonal rejuvenation once again in the east (*FM,* 92). This visual code of the four directions in space (and in time) serves as a plastic form for narrating the primal story. In the next chapter I describe the Four Directions of the sun as a diagram for delineating body symmetry; that is, as a primary orienting syntax.

The later personification of Quetzalcoatl as an enactment of superhuman qualities allows the putative superhero to be the mediator between an original contract about creation and a final comedy about the utopian ends of the culture. But the S-shaped hero, Januslike, accomplishes his tasks only in time with the cosmic seasons, according to the space of the cosmic quarters. One notices in passing for now that by coordinating seasonal time with the sun's local motion, space and time are automatically unified in a twist; that is, picture and narrative are not divided according to separate points of view in space as opposed to time. In this argument from original design, the recognition by humans that the hero is the wise serpent of deliverance amounts to the recognition of epiphany in the plot, that appearance at the proper time in the proper place of the daemon who can turn things into other things. His sacrifice is the inverse design of epiphany, the ritualistic turn upon the hero by the perfected community once his accomplishments have been dismembered and distributed through the community, but only when they have understood and inherited the curvilinear means by which to carry out the utopian task. Rituals and stories ensure that the community understands the timing of their utopian tasks. Here then the S seems to serve as the sign/design of the first turner, the archetype that stands for the vehicle of deliverance in the seasonal rounds of time.

Quetzalcoatl's temple at Teotihuatecán was aligned so that its axes conform with the four quarters. This consonance of earth and sky, where the circle of the sun arcs around the four sacred directions, surely prompts the central symbolic diagram of the Old Pacific Style. Imagining a cross in the center of a circle, an Axis Mundi, provides the schematic diagram for fixing the undulant movement of the wise serpent into regularity. In the next chapter I shall derive some bilaterally symmetrical designs from the sun's circuit around an axis of four directions.

One can also describe the Plumed Serpent as a sign of the various hydraulic motions of air and water. Consider, for instance, Brundage's synthesis about Quetzalcoatl as a polymorphous sky god:

> The coilings and windings of a snake represent perfectly the nimbleness and sinuosity of water, the uncurling of smoke in the still air, the flailing of a waterspout, or the dust devil sweeping over the land—all of them functions of the air. Among the Huichols, whom we consider as modern-day representatives of the ancient Mesoamericans, rain clouds are thought of as winged snakes. Indeed almost everything in the air—smoke rising from burning fields, lightning—even fire itself, was thought to be serpents inhabiting the atmosphere.[42]

So serpentine motion of helicoid shapes in air and water was considered by some Mesoamericans to be the common pattern of thermodynamic phenomena. Like those of the Dogon, helicoid designs seem to have been the first signs of energy transformations.

Let me make the point that these helicoid signs were seen presciently by Mesoamericans as carrier waves of the transportation of energy, by briefly mentioning the work of Sadi Carnot. The inevitable transport of thermal heat loss was the great discovery of Carnot, the founder of thermodynamics. I quote the opening paragraphs of Carnot's *Reflections* because they list some of the physical phenomena that characterized the Plumed Serpent:

> Everyone knows that heat can produce motion. That it can produce vast motive-power no one can doubt, in these days [1830s] when the steam engine is everywhere so well known.
>
> To heat also are due the vast movements which take place on the earth. It causes the agitations of the atmosphere, the ascension of clouds, the fall of rain and meteors, the currents of water which channel the surface of the globe, and of which man has thus far employed but a small portion. Even earthquakes and volcanic eruptions are the result of heat.[43]

His important act of synthesis occurs in the next few pages: "The production of motive power is then due in steam-engines not to an actual consumption of caloric, but *to its transportation from a warm body to a cold body, that is, to its reestablishment of equilibrium* . . . [italics his]" (*RM*, 49). In both prescientific accounts and in scientific accounts, helicoid shapes are the signs of the "transportation" of heat from a warm area to a cold area. Helicoid shapes are seen as the carrier waves toward a state of equilibrium. Always to be noted, nature is construed as that part of the communication-transportation complex that carries the message in wave forms.

To conceive a gigantic serpent as the hydraulic model of the cosmos seems to answer the most basic bootstrap question about the underlying first principle. It whorls in patterns that twist into curvilinearity. A gigantic serpent undulates in the air, as does the Milky Way in its apparent movement across the night sky, or as a turtle or snake floats in the water. The serpentine line schematizes the rhythms that support the earth and its functions.[44] In another work called *El Universo de Quetzalcoatl*, Séjourné shows that the "caracol" is a constant motif at Teotihuacán.[45] He maintains that its conchoid pattern was a sign of generation or of birth, which for him coincides with the tradition that Quetzalcoatl was the creator of humans. He illustrates a number of plumed serpents depicted as a caracol. He notes further that in Mayan hieroglyphics, the caracol signifies finality or totality (*UQ*, 50). In all of these caracoles, sign and design turn back and forth—

and in and from— each other, as totality and universality are thematically indicated by the partial emergence of duality.

A dragonic S seems to be the most abstract metasign, the one that is always doubling and twisting back to check on its first principles of construction. As the carryall signifier out of which all was constructed, the curvilinear form also pivots forward the other way to seek out an accommodation with eventual meaning for which it is also the formal vehicle. As a design about sign-making, as a metasign, it belongs to a class of visual tropes that W. J. T. Mitchell has called *metapictures*; these include geometrical pictures that circle back upon themselves in paradoxical self reference.[46] The dragon-shaped S curve seems to be an apt tropical vehicle for delineating double patterns of beginnings and endings.

This kind of thinking about Unity and prime numbers also characterizes much ancient gnomic literature, here exemplified by Arthur Waley's translation of the *Tao Te Ching:*

> Tao gave birth to the One; the One gave birth successively to two things, three things, up to ten thousand [i.e., noted as "everything"]. These ten thousand creatures cannot turn their backs to the shade without having the sun on their bellies, and it is on this blending of the breaths [a thermodynamics of warm sun and the cold shade, i.e., yin-yang "atmosphere"] that their harmony depends. To be orphaned, needy, ill-provided is what men most hate; yet princes and dukes style themselves so. Truly, "things are often increased by seeking to diminish them and diminished by seeking to increase them." The maxims that others use in their teaching I will use in mine. Show me a man of violence that came to a good end, and I will take him for my teacher.[47]

The Way seems to run from hot to cold to make atmosphere. But by what archaic logic can it be that dragons, especially the Plumed Serpent, a doubled creature, is still only half-complete? Perhaps because the serpent, even though a unifier, is customarily recognized as being only half of a cosmic duality, to use Rowley's phrase. About Quetzalcoatl, most discussants agree that the enantiomorph of the benevolent Plumed Serpent was the morbid jaguar who assumed ascendancy once the messianic hero disappeared, and who demanded human sacrifice as the medium of devotion. The jaguar is a perverse double.

Consider D. H. Lawrence's description of an archaic duality that grounded his work, both fiction and nonfiction, throughout his life. Here he interprets "the continual repetition of lion against deer" painted in the tombs at Tarquinia:

As soon as the world was created, according to the ancient idea, it took on duality. All things became dual, not only in the duality of sex, but in the polarity of action.

The leopard and the deer, the lion and the bull, the cat and the dove, or the partridge, these are part of the great duality, or polarity of the animal kingdom. But they do not represent good action and evil action. On the contrary, they represent the polarized activity of the divine cosmos, in its animal creation.[48]

When he describes the leopard biting the deer in the neck or haunch, "where the great blood streams run," one recalls the S-curved haunches of the Scythian buckle (Figure 9), and concurs then with his statement, "It is very much the symbolism of all the ancient world" (*EP*, 101).

Lawrence also wrote about a principle of rhythmic likeness in different things that (what Rowley had translated as "consonance" as noted above):

If we remember that in the old world the center of all power was at the depths of the earth, and at the depths of the sea, while the sun was only a moving subsidiary body: and that the serpent represented the vivid powers of the inner earth, not only such powers as volcanic and earthquake, but the quick powers that run up the roots of plants and establish the great body of the tree, the tree of life, and run up the feet and legs of man, to establish the heart: while the fish was the symbol of the depths of the waters, whence even light is born: we shall see the ancient power these symbols had over the ancient Volterrans. They were a people faced with the sea, and living in a volcanic country.

Then the powers of the earth and the powers of the sea take life as they give life. They have their terrific as well as their prolific aspect.

Someone says the wings of water-deities represent evaporation towards the sun, and the curving tails of the dolphins represent torrents. This is part of the great and controlling ancient idea of the come-and-go of the life powers, the surging up in a flutter of leaves and a radiation of wings, and the surging back, in torrents and waves and the eternal downpour of death (*EP*, 186–87).

In most of these examples the S curves of helicoid shape are thermodynamic. And a resurgent life force is the contested power.

In his novel *The Plumed Serpent* Lawrence used the symbolism of bird and serpent to mix his own combination of attraction and repugnance for the ferment of the Mexican people. His hero, Kate, was made to think of Mexico in this way throughout, especially when she was alone:

And then the undertone was like the low angry, snarling purring of some jaguar spotted with night. There was a ponderous, down-pressing weight upon the spirit: the great folds of the dragon of the Aztecs, the dragon of the Toltecs winding around one and weighing down the soul. And on the bright sunshine was a dark stream of an angry, impotent blood, and the flowers seemed to have their roots in spilt blood. The spirit of place was cruel, down-dragging, destructive.[49]

That is what Kate thought. Baldwin quotes the main character who imagines himself to be the returned hero-god: "I am the living Quetzalcoatl! . . . The serpent sleeps in my bowels, the knower of the underearth. And the eagle sleeps in my heart, the strength of the skies. I am Lord of the Two Ways, star [Venus] between day and the earth. . . . I am the spiral of evolution" (*PS*, 227). The spiral of evolution is perhaps Lawrence's most sweeping evaluation of the generative and regenerative pattern of the conchoid icon.

Preparation Out of Season

Quetzalcoatl is often associated with lightning and rainbow in story and decoration in order to stress the consonance of those signs of water and heat with the beginnings and endings of a growing cycle that depends upon warmth and wetness, sun and water. The interpretation of weather patterns by way of these lines in nature was one of the most crucial kinds of divination, of primal hermeneutics. Out of season, one plans for the next season by anticipating from the signs in nature. According to Mundkur, since humans first lived in caves, when the rainy season came around and snakes were forced out of their lairs seeking dryness, cave dwellers must have seen snakes as a sign of the impending rainy season.[50] Hence the habitual association of snakes with periodic changes in climate implied the age-old confusion between the sign of a condition, as harbinger or advent, and the cause of a condition, as the maker or designer of an event. But notice that when natural signs are shaped like natural designs, as in the curved water drops of the yin-yang emblem, there is no wedge by which to separate sign from design. So snake ceremonies were designed either to summon or to offset rain, depending upon a rainy or dry season. Here one finds another reason for the prevalence of wavelike periodicity that suffuses archaic culture, expressed as and atmosphere of S curves, in the all-dominating necessities of sun and rain that constitute the thermal duality of the growing cycle. And the universal patterns of withdrawal and reappearance, the ouroboric myth of eternal return, the undulation of snake motion, and the alternation of attraction and repugnance that many people feel for snakes, all may be habitually associated with the series of alternating inverted triangles and zigzags that sometimes summon the image of the serpent. The prevalence of this pattern of triangles and diamonds in

pictographs and petroglyphs of inscriptions on canyon and kiva
gests to some anthropologists that the diamond-backed rattler w
cific sign of the snake clan.

And yet one would not construe all of the intricate black and white designs on pots of the Anasazi culture to be derivations from serpent undulations, even though snake dances are still a common part of the ceremonies of the Hopi people, some of whom are remotely kin to the Anasazi.
The undulations of rotated duality are more immediately of one's body
than the seasonality of serpent devotions. For with one's natural bilateral
symmetry privileging balance at all costs, with two arms and two legs privileging the alternation of push-pull, of impelling and of drawing forth,
then in that dance-like resilience, that readiness in rhythm, S curves and
other swirls will be seen as the recurrent formal causes that help to shape
the concepts of yin and yang, yielding and domination, and the interweav-
ing of over and under in series.

For an elegant example of related kinds of swirls, Figure 12 represents
an Anasazi ladle made sometime about 900–1100 A.D. If you recall that my
eventual aim is to demonstrate that syntax in language is inherently the generator of physical *shapes* of speech acts in three-dimensional space, then it is
useful to begin to think about crafted objects that are composed in the
round. As Olson said, speech is solid. Here, too, I want to think about a
rhyming likeness between the physical craft of making the container out of
plastic clay and the apparent use of the container. The most unusual part of
the ladle is the enjambment of three half-spheres of wet clay into one three-
bottomed container. The side-effect of the squeezing pressure was the expression of a central Y shape of three legs that is the boundary which separates and joins the three half-spheres. The Y is a sign/design of a basic
junction in any pressured convergence of physical forces, as in the convergence of streams into the river delta, which intimated the force of Tao. A
Y-shape is part of a natural syntax that results from elements squeezed into·
a symmetrical sign/design. This straight-edged Y in the ladle is the physical
result of plasticine cylinders formed under the close packing of some kind
of external force, as the hexagonal shapes of honeybee combs, or in the ge-
odesic morphologies of soap bubbles grouped together.[51] This natural pressure of crowding formal forces into a topological Y shape is a basic symmetry of least action. As D'Arcy Thompson described in an example, the small
hexagonal shape of each honey cell is not "intended" by each bee that makes
one. Instead, each bee makes a rough cylinder as large as its body can make,

Figure 12. Anasazi Ladle. Larry Harwood photo, University of Colorado Museum, # 9542.

but the uniform pressure from surrounding cells squeezes each cell into the unintentional symmetry of a hexagon.

In my "close reading," the Y sign seems to be part of a natural syntax which is inherently related by pressure with the neighboring strokes of I, II, III, X, O, and S. I am beginning to make the point that together these designs may be read as material signs of a common syntax of symmetrical shapes in nature. In her painted decorations the potter has taken advantage of the squeezed presence of straight parallel lines and triangles that seem to derive from the central Y-junction in the clay. For the painted straight lines are painted in parallel to the straight walls of the central Y of the pot shape. The painted shapes rhyme with the materially forced clay shapes. In the half-sphere at the upper right she has painted a series of triangles that group into squared-off wheels or hexagons. Now if you randomly connect the legs of any Y shape, you find triangles that continually *group* 'naturally' into hexagons. So a potter who experiments with the possibilities of the material will see that circles and triangles offset one another as variations of the same mutual force that seems to be "in" the material of the signifier. The experience of making regular motions under material pressure will naturally reveal a few generative forms of circle and sequence, straight lines and swirls, which can be combined into innumerable variants. A Y shape will be seen as part of the same natural pattern of form as an S shape, or a double-S shape: three legs, two legs, four legs. Furthermore, if one studies the black background of the S shape in the container at upper left, one sees a black hourglass, with four quartered extensions, which rhymes with the white hourglass shapes of the inverted triangles. As noted above, all such ideographs stand for movement. All are *vortex* or *whirl* shapes, as Gombrich calls them.[52] For they seem to radiate and whirlpool in liquids, or to move centripetally. All of these whirling shapes are types of radiant motion, whether of sun or water. When the legs of any of these shapes trail off into curves at their ends, they become S curves, as in the Anasazi ladle.

The important point here is that a material cause of physical pressure can generate a natural syntax of related symmetrical forms. To be provocative: this vortex or rotation group constitutes a generative grammar of related existential forms in nature. If there is an inherent syntax of just a few forms of composition, it is learned only through the experience of an artist's alertness to the physical limits of the medium. The syntax of painted strokes may be drawn as abstract Xs or Ys or Ss. One could read the group as an abstract code of a symbolic logic, but then the symbols would be seen as a transcendental grammar that would seem to be doing the generating, much like our alphabet has appropriated the signs as part of a learned generating system. But the abstract signs are not doing the generating; instead, the patterns are generated by rhythmic pressures of physical materials under conditions of crowding into symmetries. If we mouthed those visual signs, as if the were letters in English, they would accrue the adventitiously associated sounds of our own spoken alphabet. Nobody knows exactly how those patterns signified to the ancient potter or painter. I shall continuously stress that natural syntax is a function of the material signifier, and in later chapters I shall disregard the illusory idea that a symbolic logic, as if it were an abstract syntax, is doing the generating of periodic structures in language.

The painter has offset the straight edges and the swirling patterns. They seem to refract and crowd each other away from any dominant centrality, and away from the absolute completion of any of the patterns. The skewed perspectives and the incompleteness of any of the shapes make one fascinated with the possibility of connecting the shapes into some larger pattern off canvas, into which they might be resolved. This kind of off-angled deflection and incompleteness characterizes much Anasazi and Mimbres design.[53] There is a combination of hard-edged fractal geometrical forms with curvilinear flows.

But can one turn from interpretation of pattern to interpretation of any conceptual meaning that the patterns might signify? If interpretation always starts from a stipulated point of view, what slant can be taken? I have partially answered the question of the Signifier, what was the ladle made *from*? But what was it made *for*? What does it signify? One is a "whence" question that is backward looking; the other is a 'whither' question that is forward-looking or anticipatory. There does not seem to be any "break-spotter" when all the decoration is seen as a continuity of offset swirls. What then was the utensil good for? Since the container is clearly a dipper or ladle, its intent was to transfer something, presumably liquid. Since virtually all other Anasazi ladles are merely one-half-sphere (derived originally from pressure packing clay around a half-gourd), this ladle exceeds the economics of utility. However, one supposes that all such archaic implements were dedicated to the original meaning of their eventual use, just as ornaments were presumably not mere jewelry but also prayers in another medium.

That generalization is not meant to idealize a potter or culture: anyone, an individual artist or an elite class or a theorist, can misuse a tool or transform a myth adventitiously, from its designed purpose to another end that serves one's interest. Yet the Anasazi ladle was designed for some purpose that transferred a liquid from a storage container to another use, for example, a stew pot. Ladles are not just containers; they are conveyers in-between. They mediate amphibiously between two distinct but related categories of use. Perhaps this potter wanted to be reminded of the sacred power of water to convey life. This may have been a ladle that served an overt ritualistic purpose, or it may have served an everyday devotion. In any case whenever users swirled the liquid, they would see the swirl shapes reenacted in the container for the thing contained.

In their own altar ceremonies the Hopi use a "medicine water vessel" that is handled with a ladle.[54] It is four-sided, painted with clouds and fish, and it is placed at the center of the altar's ceremonial design. For the transfer of life-giving Water is at the center of the Hopi seasonal universe. Apparently, to bring sacred water to the altar within the kiva is also to bring the gods and to open the ceremony.

In the desert culture of the Anasazi, the conduiting of water supply for their crops was solved in several different ways. I am always interested in the physical nature of the transporting vehicle, that which carries the message, not just metaphorically, as in "the ship plows the waves," but literally, as the ladle is carried from place to place. My premise has been that nature enters into communication by the carrier wave. Since water was a constant preoccupation for the Anasazi, one that often resulted in shifting sites for their dwellings, and possibly even in the eventual desertion of their most complex dwellings, one can understand that the maintenance of the water supply was central to their rituals as well as their everyday lives.[55] Planning for the economy of water was perhaps the primal divination in those desert communities. Since there was always too much water or too little, depending on the season, the "ladling" of swirling water in counterbalance to nature's extremes of economy and prodigality must have been a central transformer in their search for thermal "equilibrium," to use Carnot's term. Perhaps too extreme a conjecture, but still worth keeping, is that water ceases to be swirling and plastic when it freezes, as well as when its muddy components are baked in a kiln, fixing the shapes into angular patterns and risking the threat of fracture through too-sudden heat. These experiences also lend themselves to the craftiness of working with material and chemical transformations of the clay into a few basic kinds of morphological distortions. So the ladle may also remind that S-curved water is the life giver between the opposites of curvilinear fire and crystalline ice, whose fractal shapes are diagonals, triangles, and hexagons. Perhaps the designs of all Anasazi pots reminded their users of the skewed imbalance between the maximum and minimum extremes of nature: the experiential risk of increasing the thinness of the walls of the pot, modified by the thick strength necessary for containment of stuffs (Euler's law of soap bubbles expanding in volume while their membranes thin till they pop), the coils of clay that pivot around a central point at the bottom—a minuscule emptiness that is progressively covered over, the inevitable lines of fracture that will eventually seep through the container's walls, the meditation of seasonal economy and of plenty, of storage and of retrieval, of sending and receiving, the rotational prayer of balance and counterbalance.

What does this countervailing have to do with the Plumed Serpent? Ray A. Williamson generalizes, "In the historic pueblos, the plumed serpent is deeply respected as having an essential part in the balance of the world."[56] That is my main thesis for the Plumed Serpent. So my use of Carnot's "equilibrium" is not anachronistic. Williamson also provides a good summary of the prevalence of horned or plumed serpents depicted throughout the Southwest from prehistoric times. But as for the precise meaning of Plumed Serpents there is no unanimity. For instance, in an inclusive study about American Indian rock art, Polly Schaafsma shows some of the connotations associated with the spiral in the Hohokam culture, a people who apparently migrated northward from Mexico to southern Arizona sometime before 300 B.C.E., and who brought various items and customs from Mesoamerica.[57] For the Mexican Indians the spiral apparently had associations with rain and clouds, corn and serpents (some plumed), as well as the heart, and Grandfather Fire (*IRA*, 90). Its meaning is also complicated by an association with peyote rituals, since the pilgrimage to obtain peyote coincided with the seasonal arrival of the rain (*IRA*, 91). She concludes, "The spiral in western Mexico is also believed to refer to the god Quetzalcoatl, who is associated in part with agriculture and water" (*IRA*, 91). Later she writes of Quetzalcoatl:

> This deity, with varied and complex attributes, was recognized as the god of life, the morning, fertility, and agriculture and the patron of twins and monsters. As the god of wisdom, calendars, and learning, he was the god of civilization. He also took the form of Ehecatl, the Wind God, and at other times, he may be seen combined with his twin brother Mictlantecuhti, the Death God, thus summarizing the duality of life and death (*IRA*, 238).

So this spiraling unit of composition could be applied to a family of related meanings, like the S in our alphabet. It seems to be a diagrammatic form for indicating spatiotemporal kinship, but in all cases it incorporates a balancing antithetical curvilinearity. If a spiral is a basic unit of composition that stands for the sign/design for the Plumed Serpent, then it is a good formulation for junctured balancing of hot and cold currents, opposing seasons of dry and wet weather, the tumbling and turning of first principles of force and form, the patronage of art and knowledge. The formulation "Plumed Serpent," in its own series of relations, is as satisfying as a contemporary thermodynamic formula in its way, for neither do we now know in fact ultimately what agent it is that keeps on sending those shapely pattterns of hot and cold currents whose message is their own pattern of forces, and whose significance is our reconciling response.

A Few Natural Forms of Composition

To summarize here, *The Mustard Seed Garden Manual* opens with a simple description of how to paint a tree, "beginning with the four main branches." Because the manual instructs one in the Tao as well as in technical strokes, one learns that the four branches rhyme with the four symmetrical limbs of one's body, the four directions, the four kinds of wisdom, and the pushing upward and downward, backwards and forwards, of the ritual patterns of yin and yang. So one begins to group the various Y forks and branches into a visual syntax that seems to order the "various 'trees' of knowledge":

> As branches fork out from the trunk, so a road branches off into other roads and paths. When one knows the way (the Four Directions), the main route is clear and its landmarks familiar, no matter how many byroads and paths. Put your whole mind into what you are doing and you will be on the right path. The thousands of rules and ten thousand details are all based on this fundamental principle.[58]

To feel one's body as a forked confluence (of rhythmic dragon veins), with the way extending out in four conjunctive directions, that is a strange morphological feeling, even when you put your mind to it. I want to recollect this syntax of bodily forks and conjunctions, this first method of wisdom literature, as a cosmically ordered path, when I review various body symmetries of four arms and legs in the next chapter.

It seems that the archaic way to construe the category of S shapes in nature, Y shapes, W shapes, X shapes, Z shapes, and the like, was as an interrelated syntax of branched crossing-over models. These few calligraphic strokes can generate many other abstract designs. This inherent syntax enabled one to enact a visual language of painting in order to reenact the observed junctures in the world. That is why in so many of these drawings on pots the abstract geometries can interact with the realistic portrayals of fish and birds and snakes and mountains and so forth. There is represented a continuous interaction of syntactic patterns and patterned representations of creatures.

Because my summary about junctions seems utterly pedestrian, I close with an example of a very quick poem, this time from Japan. Issa wrote:

Dry creek
glimpsed
by lightning.

The poet Robert Hass writes: "To get it, one needs to imagine the interstice in the universe that a near lightning bolt rips open. Issa is in the dark; lightning; a sudden glimpse of its fierce zigzag configuration and below it the zigzag configuration of the creek, all at once, in bright, brief, unnatural light; then darkness again. This poem does throw immense weight onto the observer, in the dark, having been given a vision of the raw energy of the world."[59] The two zigzag strokes, or Z shapes, one in the form of lightning and the other in the form of the creek, rhyme with the quickness of cognition, and finally with the staggered syntax of the words in their zigzag order. As Ernest Fenollosa said archetypally about the way a sentence carries its message, "The type of sentence in nature is a flash of lightning."[60] The archetype of lightning carries the message in a quick but jagged assertion. All the configurations rhyme, but they rhyme interstitially, in their jagged intervals. Again, my thesis is that the syntax is an existential act of the material world, and the Y shapes or Z shapes are symmetrical signs and designs and not themselves the generators.

In order to bracket Hass's insight about rhymed configurations, I cite three scholars about the essence of Chinese poetry, one early, one from the 1930s, and one contemporary. Consider first this summary by Fenollosa about the nature of Chinese and Japanese poetry:

> All real poetry is just this underground perception of organic relation, between which custom classifies as different. This principle lies at the very root of the enlargement of vocabulary in primitive languages. Nature was so plastic and so transparent to the eye of early man, that what we call metaphor flashed upon him as a spiritual identity to be embodied at once in language, in poetry, and in myth.[61]

In this reading of language and one's being in the world, Issa's joining of lightning and creek bed is not a metaphor but a flashed archetypal identity of physical shapes, a natural syntax that in turn leads to an enlarged "vocabulary" about patterns of force and patterns of thought in primal languages. Notice that the dryness of Issa's creek bed not only tells one something about the season of the year, but also it reinforces the idea that the creek bed is a duct or channel for the elemental fluidic lightning, as the grooved lines in the Sicilian slab served as a fluidic duct for tracing an abiding pattern of creation. Not a metaphor, this configuration is an archetype of an enduring order of things that prefigures, by its shaped syntax, an identity with rhythmic language. More carefully and more skeptically than Fenollosa the popularizer, Stephen Owen describes a certain kind of poet who responds to *shih*, the character of the primal uncreated world, of yin and yang, which are nevertheless shaping principles that make up the created world:

> With no creative deity to emulate, the poet of the *shih* does not think to make the world anew; he participates in the nature that is; and in being of this world, he lacks the "creative" poet's aura of isolated divinity. The "Great Preface" of the *Book of Songs* presents the process of composing *shih* as a universal human impulse, not as a mysterious and singular gift.[62]

Owen summarizes: ". . . the poet is concerned with the authentic presentation of 'what is,' either interior experience or exterior percept. The shih poet's function is to see the order in the world, to compose via the pattern behind its infinite division. Like Confucius, he 'transmits but does not create'" (*TCP*, 85). "Composing" from this existential pattern, which is seen behind the divisions of nature, is also my point. If Confucius meant "transmits" as a metaphor, I would take it literally: nature carries its figures in physical strokes or patterns, which poets use and augment, in various languages, to try better to understand one's inner experiences.

Lin Yutang, furthermore, describes *shih* in a way that lets one see how the idea unifies both the essential character of a being, what it is, but also its essential gesture, that is, its characterizing compositional stroke:

> *shih*: gesture, posture, social position, battle formation, that which gives advantage of position in any struggle. This notion is extremely important and is connected with every form of dynamic beauty, as against mere beauty of static balance. Thus a rock may have a "rock posture," an out-stretching branch has its own "branch posture" (which may be good or bad, elegant or ordinary); and there are "stroke posture," "character posture," and "brush posture" in writing and painting and "posture of a hill," "posture of a cloud,". . . A situation is conceived as static, while a *shih* denotes that which the situation is going to become, or "the way it looks": one speaks of the *shih* of wind, rain, flood or battle looks for the future, whether increasing or decreasing in force, stopping soon or continuing indefinitely, gaining or losing, in what direction, with what force, etc. (*IL*, 439).

To interpret these archetypes is to figure in which way their patterns lie. The latter part of the quotation, which suggests that the characteristic posture can let one predict which way a thing is going, is very much like Ruskin's injunction to painters of landscape to heed the "leading lines" of the physical topography. Because Ruskin's passage enhances the discussion of the way artists can be taught to see and to compose by way of the characteristic attitude of a topographical feature, and more importantly, how they can employ a group of kinds of lines that reenact what Ruskin thinks to be the "governing lines" of landscape, I quote it at some length:

> I call it a vital truth, because these chief lines are always expressive of the past history and the present action of the thing. They show in a mountain, first, how it was built or heaped up; and, secondly, how it is now being torn away, and from what quarter the wildest storms strike it. . . . In a wave or a cloud, these leading lines show the run of the tide and of the wind, and the sort of change which the water or vapour is at any moment enduring in its form, as it meets shore, or counter-wave, or melting sunshine. . . . Your dunce thinks [these things] are standing still; your wise man sees the change or changing in them, and draws them so,—the animal in its motion, the tree in its growth, the mountain in its wearing away. Try always, whenever you look at a form, to see the lines in it which have had power over its past fate and will have power over its futurity. These are its awful lines; see that you seize on those, whatever else you miss.[63]

Being Ruskin, and speaking to beginners, he speaks autocratically of certain laws of drawing things in nature: the Law of Repetition, under which he includes the "symmetry" of repetitive succession, the Law of Continuity, the Law of Curvature, the Law of Radiation (*ED*, 254–77). The last carrier wave, of radiation, might be called J. M. W. Turner's main compositional format in his later paintings in which one sees only vortices of haze and steam and fire and light.

These leading lines that Ruskin sees in nature are the various carrier waves that shape the geology of a mountain, the leafing of a tree, the meandering course of a river valley, which John Constable often used to compose his landscapes. These governing lines are signs of those forces that ordinarily are unattended, the ways, for instance, by which a mountain is composed and decomposed. Geological fault lines show, like the movement of a glacier or a mountain, the way in which things that are ordinarily seen as being static are slowly moving in a characterizing action. So these compositional lines govern the attitude or characteristic gesture, the shih of things. Furthermore, if you turn back to the passage quoted in "Art/Frequency," in which R. L. Stevenson describes the plastic posture that characterizes the timing of an action—Odysseus bending his bow, Robinson Crusoe recoiling from the footprint, Christian running with his hands clapped to his ears (so much like Munch's agonizing painting *The*

Scream)—you see that *shih* is a term that might be transposed to help describe that most difficult attitude or posture.

Here another contemporary writer finds a characteristic syntax of jagged configurational lines, this time in Chinese painting. C. S. Smith is a metallurgist who transferred his understanding of the fractures, fault lines, and bondings of metals, in their crystalline and liquid phases, to the structures of art. Particularly in Chinese art he sees for himself the physical ways in which tree forks as dragon veins joined in the art of one of the greatest of painters:

> Wang Wei, a painter of the late T'ang dynasty, delighted in emphasizing the junction. For reasons evidently both philosophic and aesthetic, he played on the similarity between the branching connectivity of fissures in nearby rocks, the wrinkles and river valleys in distant mountains, and the branches of trees, the last, of course the model for thcm all (as well as for modern computer programming!).
>
> When the underlying structure is being examined, the style of the whole is not visible, for this resides only in an external view of the whole. Bohr's famous principle of complementarity is, perhaps, nothing but a statement that things react on different levels.[64]

Here the natural syntax of Y shapes is the same for landscape painting and computer modeling. We are on the way toward a model of natural syntax of Z shapes and Y shapes and X shapes that can serve as algorithms for a theory of in-*form*-ation. Also, Smith's insight about complementarily, seen as different levels, seems to me exactly like the reactive principle of push-pull, in which one focuses first on the underlying and transmitting Signifier—and then perhaps on the Signified—but not simultaneously, but only in a curvilinear succession of over and under. Notice that the impossibility of simultaneous focus is also the insight of Watson about the suture and the snout of the bronzed beast.

Fenollosa also wrote about the connectivity of tree branches in a way that is remarkably contemporary with my thesis about information:

> But the primitive metaphors do not spring from arbitrary subjective processes. They are possible only because they follow objective relations in nature itself. Relations are more real and important than the things which they relate. The forces which produce the branch angle of an oak lay potent in the acorn. Similar lines of resistance, half curbing the outpressing vitalities, govern the branching of rivers and of nations. Thus a nerve, a wire, a roadway, and a clearing-house are only varying channels which communication forces for itself (*CWC*, 22–23).

This Taoist compounding of transportation theory with communication theory, and with the brush stroke of the Chinese character, became possible because Fenollosa saw that branching diagrams are not mere abstract relations but rather that the communicating branches occur as resistances in natural channels of force, like Frost's and Ruskin's counterwaves. Like the Sicilian diagram of circulating life force and the thermodynamic theory of Carnot about the transportation of force by way of the passage from hot to cold locales, Fenollosa's theory about resistant channels of transportation and communication is not a metaphor, but is an archetype of necessary flow and tacit resistances to it. So the physical stroke of the artist's brush or pencil is an archetypal gesture (*shih*) that reenacts the *leading lines* of natural flow, from the past history to a future divination of a thing's course.

Later Smith says, ". . . there are not many basic units of composition and as large things merge into small things, and vice versa, both nature and the eye favor much the same principles of assembly" (*SS*, 22). These few basic rhythmic units of composition have been my pursuit throughout. Could it be truly that there are just a few morphological units of composition across the arts and that, as Herbert Read said, they derive from a kinship with the basic composition of things? Is there to be seen an inherent syntax in natural languages? Smith's point is that art's way follows nature's way with just a few patterned units of composition. The various junctional strokes in the paintings of Wang Wei are consonant with the forkings in natural structure, like Issa's zigzag, or like the Y-shaped Anasazi ladle. I have also noted in other kinds of archaic art that ball-and-socket symbols, drawn on haunches, signify a propellant force at the joints or junctions. We see a natural syntax of articulated anatomies, not just bodies of animals but the earth's prototypical body, as enacted in the Sicilian diagram, or in the prototype of the Plumed Serpent. For Smith, the economy of forms in landscape painting served in their resemblance to one another as models for taking one's way through the world. So, too, these archaic images and stories, with their few patterned strokes and junctions, may have served their audiences, consonantly, as models for composed action. Even though the composition may be very fleeting in its conjunction, as was Issa's quick impression, or halting, as was Creeley's patterning of "breath" with the sounds of braid and breed and brethren, still its afterimage abides in the way of the writing. In the traditional arts and ethics of China and Japan and America, to follow the leading traces of natural junctures in river streams and mountain ridges seems to have been a model for composing oneself. My hypothesis has been that these traced forms are what we know of the transmitting characteristics of carrier waves, and that this archaic syntax of a few basic units of composition can be drawn forth and glimpsed. To follow these runnels to their junctured sources seems to be a good way to understand in these few generative forms of composition the rudiments of natural syntax.

CHAPTER 3

Symmetries of the Sun Cycle

Synopsis

If one of the functions of the brain's cortex is that of pattern transformer, then its ultimate source for transformation signals is light from the
sun. If the sun is for earthly beings the material transmitter of energy, which
is converted into bodily uses, then it was also seen and drawn among primal
cultures as the vehicle or carrier of light, heat, and form to animal bodies.
Thousands of representations still exist of circles, arcs, crosses, and grids that
chart the sun's position and rotation. Among many first cultures, diagrams
of the Four Directions were conceived as primary points of bodily orientation to the apparent motions of the sun's cyclic rotation. As they are today in
any primary map of orientation, the Four Directions were a constant scale
used to measure one's place in space and time. Since the Four Directions derive from the stations of the sun's apparent rotation, the sun was seen by
early cultures as the constant source of scale into which all things fit. So
many of these early drawings are in fact patterns or models that diagram the
common forms of anatomical symmetry, and these diagrams embody the
formal symmetries and transformations that most animals share.

As I discussed in the previous chapter, the comfort zone of life depends
upon agricultural conversion of the sun's energy at different times during its
annual cycle. So, as *The Mustard Seed Manual* taught, the Four Directions
were seen as an inclusive branching or quartering diagram. As a diagram or
model of symmetrical fitness, it became a periodic grid used to measure
one's place in both space and time.

Instead of space and time, however, which are much too abstract, here I speak of scale and pace. Scale is the fit of one's body symmetry, and pace is the gait of one's four-limbed motion. These drawn arcs, circles, straight lines, and spirals composed an elementary visual syntax for reenacting one's orientation on earth with regard to the sun. As protogeometric figures, these primal drawings served as elements of patterning for a concept of the sun's rotation, in which the rotation itself was a propelling transporter. The syntax of lines re-inscribed the pattern of energy being carried in transit. This visual syntax, seen as the elements of the "solar constant," was a commonly shared set of "leading lines," to use Ruskin's and Bahktin's phrase, which gauged a constantly moving proportion of units in a series. Since they could be transformed and converted into other images, the inscribed lines seemed to reproduce (and perhaps help induce) the physical transformations of the energy cycle.

The idea of "consonance," discussed in the previous chapter as a "rhyme" between the physical strokes of an artistic medium and the presumed forces of nature, is the very idea of syntactic ordering. The model of the Four Directions was seen as a prototype, an *Urform*, for describing the symmetries of most animals on earth. To the extent that all animals share this fourfold symmetry, their components rhyme in similitude with one another. So the diagrams of the Four Directions enacted an early theory of symmetry sharing.

If the sun's apparent motion is the periodic *Urform* for symmetrical transformations, then it may be seen as the archetype for the subsequent syntax that follows the forms of a wave in time. Sunlight, our constant measure of all things in space and time, comes to us in waves. In this chapter I set forth some elements of symmetry theory in order to explain these protogeometric lines as a natural syntax, for the existence of symmetrical animal bodies in motion on earth is shaped primarily by the sun's gravitational influence. Unnoticed but enormous, a gravitational downward push upon animals shaped the symmetries of their bodies, which could propel them through water, earth, and air with least resistance. Although the principle of gravity, as such, may have gone unnoticed by early cultures, the shapes of fins, feet, and wings were seen as gait propellants in arcs and angles. In other words, the rotational movements of animal bodies were seen as recurrent models, in small, of the rotation of the sun at large. Like the microcosm / macrocosm fit of Renaissance theory, these small models can be seen as an archetypal syntax of periodic measure. For my purposes, symmetry theory is a way of describing a rhyme between rhythmic bodily motion and the rotational transformations of the sun's energy, weight, and rotation.

I am still moving toward the idea that linguistic syntax is a set of transformations of physical units in a periodic series of curved wave speech acts. Here I set up bodily symmetries to introduce the concept of a few rotational groups. The orienting syntax of the sun's rotation around a model of the Four Directions is seen as a symmetry group of recurrent patterns that carries or transports the body in rhythmic gaits. The symmetrical strokes of X shapes, Y shapes, S shapes, O shapes, and W shapes eventually became part of a visual alphabet, but the letters were a visual syntax long before they became arbitrarily associated with learned speech sounds specific to different languages. Before that happened, however, they served as a protogeometry for coordinating the regular motions of animal bodies under the sun. For almost any skilled craftsman who drew a symbol of the Four Directions, on a pot for instance, the symbol stood for a constant code or syntax whose continual lines could represent orderly rotation, location, pace, and rhythm.

Because the brain's cortex transforms visual images into spoken words, it uses a common syntax of periodic order to transform visual scales into spoken orientations, described in the next chapter as *deixis*. In terms of language, natural syntax may be defined as the rhythmic enactment of sounds or phonemes that shape the physical order of transmission in a sentence. Here, then, I introduce an essential concept of poetry—that syntax is a way of reorienting grammars in syncopation, where the rhythmic gaits of symmetrically cadenced feet reorient the "rules" of grammatical placement of units in series. In other words, the physical patterns of syntax as regular shapes of spoken sounds in the mouth, throat, and lips propel linguistic grammars, not the other way round. You cannot make the sun rotate backward. The energy of a sentence's flow, like the direction of a wave signal, is only one way.

Since, then, we are children of the sun, and our bodies a product of its rays . . . it is a worthy problem to learn how things earthly depend upon this material ruler of our days.—Samuel Pierpoint Langley[1]

Orientation: Body Symmetry and Asymmetry

Our bodies are universal carryalls. The regular patterned motions of bodies are sublimated vehicles for carrying the messages of life. Among first peoples, the rotation of the sun served as a means to orient one's body and bodily motions in scale and pace. The sun's course was seen as the archetypal model for formal motion around the Four Directions of the earth. Its apparent movement was seen as a Great Wheel of periodic rotation marked by four directional points. Among many cultures, as we shall see, this formal motion of the sun became a kind of "compass rose" for orienting one's symmetrical body both in space and time. The directions could also diagram the shared symmetries of the animal body. The Four Directions of the earth and sky, which are often called the Four Quarters, are reflected in the four-branched directions of the body's outstretched arms and legs. For instance, the most famous diagram of this tradition is not from primal art at all, but it is perhaps the defining image of the Renaissance return to nature. It is Leonardo da Vinci's diagram of the "Vitruvian Man" with his arms and legs outstretched in a squared circle.[2] George Herbert discussed this symmetry in the third stanza from his poem "Man":

> Man is all symmetry,
> Full of proportions, one limb to another,
> And all to all of the world besides.
> Each part may call the farthest, brother,
> For head with foot hath private amitie,
> And both with moons and tides.[3]

Herbert's insight about the correspondent proportions between heads and limbs and moons and tides is most rare, because the gravitational effect upon bodies was not yet widely known. Isaac Newton had not yet written about tides and gravitation in his *Principia*, and he did not write at all about the relation between body symmetries and gravity. The image described by Vitruvius in his book on architecture was intended to represent the symmetry of the human body and its fitness to scale with the rest of the natural world. Although it can be interpreted as a visual model that measures the rest of the world by man's powerful body, it was not described by Vitruvius to celebrate a man-centered universe, but rather to feature a branching model of the fitness of bodily proportion into the arcs and quartering angles of apparently universal proportions. "Symmetry is a proper agreement between the members of the work itself, and relation between the different parts and the whole general scheme, in accordance with a certain part selected as standard. Thus in the human body there is a kind of symmetrical harmony between forearm, foot, palm, finger, small parts, and so it is with perfect buildings" (*TBA*, 14).

Vitruvius's approach coincides with mine in featuring the semiotics of Stoic philosophy. In his first chapter, "The Education of the Architect," Vitruvius says, "In all matters, but particularly in architecture, there are these two points: the thing signified, and that which gives it its significance. That which is signified is the subject of which we may be speaking; and that which gives significance is a demonstration on scientific principles" (*TBA*, 5). It is worth quoting the Latin to see the history of these two basic semiotic distinctions: "Cum in omnibus enim rebus, tum maxime etiam in architectura haec duo insunt, quod signaficatur et quod significat. Significatur propostia res, de qua dicitur; hanc autem signficat demonstratio rationibus doctrinarum explicata" (*V*, 6). Rational principles of symmetry, proportion, weight, scale, and rhythmic motion, the demonstrating principles of Vitruvius, became crucial to Galileo and to the heliocentrism—not just anthropocentrism—of the Scientific Revolution.

Focusing on the symmetries of the bodies and joints in primal cultures is also worthwhile because the primary anatomy lesson uses the same kind of spatial and temporal orientation as the syntactical locations of spoken language, studied in the next chapter. I focus on first cultures because the later concepts that surround these diagrams, such as the Renaissance microcosm-macrocosm, become invested with ideological burdens of meaning and signification that are better explored elsewhere. If meaning is an adventitious tack-on, as Langer said, then I want to feature the protogeometric patterns of lines as gestural strokes that carry the message. For example, in the Christian tradition, the Vitruvian diagram of the Four Quarters is called a *cross of Saint Andrew*.[4] But recall the pattern itself. The arms and legs of the cross are oriented northeast/southwest and southeast/northwest. The body symmetry of outstretched limbs is always off angle from the north/south and east/west stations of the sun. Rotating the cross's arms or legs a one-eighth turn in either direction will make them overlap. That turn of transformational symmetry is called *rotation*. Imagine Leonardo's famous drawing, standing vertically, front faced toward you. The limbs of all earthly animal bodies are symmetrical when seen from this ventral point of view. Now imagine rotating the Vitruvian body 180 degrees on its vertical axis to see its hind view. The body limbs also are symmetrical from that point of view. But because heads and tails are differently constructed for different purposes, there is an asymmetry between head and tail, front and back, or anterior and posterior. Most animals walk in gaits on four feet, and thus their directed motion favors the head, with its symmetrically located sensors—eyes, ears, and nostrils. (Because energy waves come to all creatures in off-angled oscillations, the bodily sensors—eyes, ears, nostrils, and fingers—are symmetrically doubled to achieve an optimal set of receivers.) A gait has the fundamental asymmetry of directed motion, but, as we see later, a gait is a constant pace. Its orientation has a bias toward headfirst movement. I stress this bias because it carries over to many disciplines. The fundamental asymmetry of directed motion is forward propulsion, headfirst in a direction. It is this orientation that supports a bias toward transforming units forward in a periodic series. In symmetry theory, the forward-backward transformation of units along a line in a sequence is called *translation*. This directional bias underlies the orientation of a periodic sentence; its sequence of units moves us forward, gets us somewhere, as Plato said. Also, this is the bias of the passing of the arrow of time, in which the orientation of the arrow repeats units forward in a physical space while recalling the units backward in a proposition space of memory. Because no one can turn tail and literally go back in time, all recollection is putative. This combination of moving headfirst in physical space, while moving hypothetically backward in recollection, is the fundamentally amphibious combination of prediction and memory. One's animal body always exists in a continuous physical environment, but the work of one's cortex, one's consciousness, is always Nowhere and Nobody, ranging around "in" the future and "in" the past. One's cortical consciousness, because it is Nobody, is always inventing hypothetical spaces and times into which it projects fallacious metaphors of embodiment "in" the mind.

I stress this conundrum now because later I want to suggest that the amphibious compound of retrieving and predicting is a rhyme. Always becoming and always perishing, always searching whence and whither, the cortical consciousness always searches for meaning. It finds immense numbers of patterns, but the conceptual meaning is always a transcendental displacement before or after the fact. A rhyme is always an amphibious combination of front-back and back-front. It is a basic combination of symmetry and asymmetry. This anatomy lesson will become, in later chapters, the essence of a mnemonic hypothesis about rhyme as algorithmic translation forward in a patterned series. By *rhyme* I mean again the definition given to *consonance* in the last chapter: the strokes of syntax are actual physical gestures in space and time that are presumed to reenact and carry the patterned lines of force in nature.

By way of the physiology of speech, one moves necessarily forward, riding the troughs and crests of speech waves while ranging forward and backward in the proposition space of grammar, matching and predicting words with intent. Predicting and recollecting, one waves one's way through a medium. The syntax of physical speech acts, which ride the waves of explosions and implosions, means that sentences move in one way; they follow the flow of a wave of speech acts. Grammars and morphemes shape the hypothetical order of the sentence, backward and forward in antitheses and Ciceronian parallels, but they always ride upon the suppressed carrier vehicle of a physical wave form. The sublimation of the carrier wave was the point of the amphibious section in Chapter 1, "Art / Frequency." Models of action in real space and time are always asymmetrical in propelling bodies one way. In symbolic logic, algebra, geometry, and postulates of physics, one can go backward and forward in hypothetical space and time. But these symbolic spaces are enlightened fictions, not to be confused with existential bodies. Some of these issues of symbolism are discussed in the next chapter.

From this point of view about the front-back directions of consciousness, one can better understand Michael Leyton's recent hypothesis, based upon symmetry theory, that "Shape is Time."[5] He says that although there are countless studies of objects shaped in relation to space and spatial perception, there are no mathematical treatises about shape as time (*SCM*, 6). Leyton argues that while we live continuously in the present, we reach an understanding of the shape of an object by determining how it was made in the past. He postulates, for instance, that "asymmetry is the memory that processes leave on objects," and that "symmetry is absence of process-memory" (*SCM*, 7). It is important to notice that symmetry here is considered to be central to everyday acts of consciousness and causation, and not merely applicable to the recondite disciplines of quantum physics and mathematics. Leyton proposes that "all cognitive activity proceeds via the recovery of objects of the past from objects in the present" (*SCM*, 2). In later chapters he discusses art, language, and even "political subjugation" by way of symmetry. Here I want to establish an elementary natural syntax of motion by way of symmetry studies from first cultures.

What purpose did these primary visual models serve? The importance of the sun's motion as a universal syntax was that its periodicity, its periodic cycles, could serve not so much to divine the future at large, but rather to predict from memory the timing of one's acts within the recurrent cycle of the seasons. Within the cycles of the sun, memory and prediction turned upon one another. Many of the visual models, as we note later, served as divining boards or forecasters.[6] A common syntax of natural language may be interesting in itself, but like all learning, it might better serve as a model to predict probable futures, or as Plato said, to get us somewhere. The periodic rotation of the sun cycle was not just a static symmetry in space but a rotation cycle that marked periodic recurrence in time.

The awareness in memory of periodicity, of "recurrence," as Whitehead was quoted in the Introduction, is the first step in making measures that are abstracted from the particulars of flux and flow. It is worth repeating:

> The general recurrences of things are very obvious in our ordinary experience. Days recur, lunar phases recur, the seasons of the year recur, rotating bodies recur to their old positions, beats of the heart recur, breathing recurs. On every side, we are met with recurrence. Apart from recurrence, knowledge would be impossible; for nothing could be referred to our past experience. Also, apart from some regularity of recurrence, measurement would be impossible. In our experience, as we gain the idea of exactness, recurrence is fundamental (*SMW*, 31).

In the Introduction I also mentioned Whitehead's conception of flow as a first principle of time passing. Here, it is measured pattern within flow that is central to the experience of learning. In his examples from nature there is a tacit assumption that the recognition of recurrence presupposes the a priori existence of something that is sending those signals. Although Whitehead does not say so here, the examples of sun and moon and heartbeats and seasons give a definition of recurrence that also depends upon the awareness of alternating absences of things that were presumed to exhibit themselves in intervals, which gives pattern its contrast with presence. On earth, the sun does not shine forever; the moon glows in darkness but not in daytime; summer goes, and heartbeats are measured by interstices of stops. Recurrence of a certainty depends upon intervals of absence that demark the appearances of things in a patterned cycle. Whitehead's superb definition of pattern brings sciences and aesthetics together:

> The pattern may be essentially one of aesthetic contrasts requiring a lapse of time [or interval] for its unfolding. A tune is an example of such a pattern. Thus the endurance of the pattern now means the reiteration of its succession of contrasts. This is obviously the most general notion of endurance on the organic theory, and "reiteration" is perhaps the word which expresses it with most directness. But when we translate this notion into the abstractions of physics, it at once becomes the technical notion of "vibration." This vibration is not the vibratory locomotion: it is the vibration of organic deformation (*SMW*, 133).

Reiteration is the conceptual word, but a tune is the prime example. In the next chapters I take up rhymes and tunes as natural algorithms of reiteration.

This periodic recording of things that recur in alternating patterns of presence and absence, in which memory and prediction are rotationally connected through successions of rhythmic contrasts, could also be the experienced symmetrical pattern for learning a lexicon and a grammar. Periodic rotational recurrences of rhyming may be one of the basic ways a child learns language. I think it is the innate code for an "old sequence" of curvilinear motion. I cannot prove it, but I discuss prosodic learning by singsong in later chapters. But if I carefully study Whitehead's statement about exact knowledge, I surmise that its credence depends upon fundamental symmetry groups. In order to understand Aristotle's mutations of NA, AN, and IH, one must recognize *patterns* of recurrence within differences. Symmetries are not to be seen as merely static arrangements in space. Even when they are represented statically, splayed on a page, they are also to be inferred as recurrent primal forms that mark time passing in asymmetrical waves of recollection and anticipation. Although iconic images of symmetry are represented as being static, like the picture of a honeycomb, the stasis is more truly seen as a stratified phase in an ongoing rhythmic process.

Definitions of Symmetry

Introducing the study of Hermann Weyl's symmetry, James R. Newman admits to a possible ridicule:

> Symmetry establishes a ridiculous and wonderful cousinship between objects and phenomena and theories otherwise unrelated: terrestrial magnetism, women's veils, polarized light, natural selection, the theory of groups, invariants and transformations, the work habits of bees in the hive, the structure of space, vase designs, quantum physics, scarabs, flower petals, X ray interference patterns, cell division in sea urchins, equilibrium positions of crystals, Romanesque cathedrals, snowflakes, music, the theory of relativity.[7]

The unruly syncretism of this lot, the virtually postmodern phantasmagoria of allusions, is made spare by Weyl's baring of the symmetries of each. What is symmetry, and what are its roots in common use? Nowadays, symmetry theory redescribes with new import what are known as age-old principles of balance, equilibrium, and stability over time. The symmetries comprise a few basic geometrical transformations, and the symmetries themselves hold physical things tightly clenched into stable shapes over time. As I describe it in the Conclusion, symmetry is the formal element that conserves the energy of elements and atoms. Symmetry is also a way of describing certain "laws of form" in the natural world, the pursuit of L. L. Whyte, Noam Chomsky, and Stephen Jay Gould, mentioned in the Introduction. Symmetry theory can also be used as a scientific version of a linguistic structuralism, featured by Jean Piaget and others, in which the geometrical structure of a figure remains unchanged or "invariant" after it has undergone some transformations.[8] In the twentieth century the age-old study of symmetry was revived when physicists began to understand that there was a close relation between the conservation of energy and certain symmetrical transformations that left particles of atoms unchanged in their structures.[9]

What are the basic transformations of symmetry theory, and how may they be applied to curvilinear forms that seem to propel animal bodies that have symmetries of four limbs? What, then, are some of the functions of symmetry in animal bodies? How did primal cultures represent symmetry as four-branched modules for bodies that make waves? In this chapter I demonstrate how the coded old sequence of propulsion was represented by means of a few formal motions—translation, rotation, and twist. So here we see how the symmetries of the sun's rotation unified the syncretism of all things by a common syntax whose transformations were linear, circular, and helical motions. These are the regular motions that are necessarily enacted by the angled joints of the body that describe quartered arcs and rotations of those limbs. This recurrent syntax describes the old sequence of neural instructions; it describes the code of alternating propulsion that was preserved when animals left the sea. In the Introduction we saw with Arnheim and Blake how angles and arcs are drawn by jointed limbs of wrists, fingers, and elbows. And Aristotle's examples of the letters N, A, and others revealed that writing itself, the actual construction of letters, is based upon these regular transformations. If drawing and writing are enacted by jointed symmetrical limbs, perhaps speech, too, is part of a covert kinship. After sketching some structures of symmetries derived from the sun's apparent motions, I can begin applying it to forms of natural syntax in poems.[10]

Here is my reasoning. Natural syntax may be defined as a principle of composing, of assembling physical units in directional sequences. If natural syntax is a common pattern of composing physical units into orders, positions, and shapes, it must include a tacit set of instructions for physically moving units, such as phonemes, into a periodic order. Why? If symmetries are universal, and if they are compositional principles in the three space of natural phenomena, then speech acts, which follow the forms of waves through the three dimensions of their transmission, can be seen in terms of symmetrical transformations. In Chapter 4 I review the enactment of speech in the three space of the mouth cavity. Because syntax in language is a set of instructions for composing grammatical units, it must be a structured form of motion, of moving proportion. Syntactical structures should incorporate principles of enacting forms of motion. In the Introduction, when I quoted Aristotle's examples of position, order, and shape of written letters of the alphabet, I showed that these definitions depended upon orderly principles of transformation, of acting in motion. These tacit principles of motion in space and time are *translation, rotation,* and *twist.* Together they describe the motions of moving proportion. They reenact the moving proportions of the wave, the old sequence. This is a necessary turn in my argument. If natural syntax is a tacit set of instructions for assembling the physical parts of words in moving proportions, then it should follow the regular principles of symmetry theory, set out by Hermann Weyl and others, for those forms of transformation in three space.

Three Transformations of Symmetry Theory

Like Vitruvius, Weyl begins his famous study with "everyday" definitions: good proportion, balance, concord of parts into a whole, and harmony. For instance, he takes the precise meaning of the term *bilateral symmetry* from artistic intuitions. The sun's gravity forces objects straight down to earth, so the verticality of up-down is the main divisor of bilateral symmetry into right-left identities, like the allegorical balanced scales held up by blind justice. In his admirable study, Martin Gardner says, "The enormous preference that nature shows for vertical axes of symmetry is due, of course, to the simple fact that gravity is a force that operates straight up and down. As a consequence, things tend to spread out equally in all horizontal directions."[11] Although it may be a simple premise, this verticality is a distinction of utmost importance. Horizontal motion is a compensatory move of least resistance. The identity of left and right, distinguished only by a vertical line, is perhaps the first distinction of symmetry theory. Primarily because of the sun's gravitational force, verticality along a North Pole/South Pole axis is also the most common kind of symmetry. Picture a fir tree from directly overhead, that is, from the sun's viewpoint at mid summer. The tree will look something like a lotus, with concentric arrangements of limbs around the highest tip. The up-down direction of gravity lets the tree distribute its branches to catch the sunlight in an all-around or rotational symmetry. If I were to draw the tree as a profile, seen from a 90-degree angle, it would appear bilaterally symmetric. Notice that the sun is the source of two of the most powerful physical constraints for symmetry in nature: gravity and light, one of which is up and down, and the other is cyclical in rotation.

The letter A is bilaterally symmetric left and right once you split it vertically. Aristotle's letters H and I are bilaterally symmetric if you split them vertically or horizontally. Or picture an inkblot test. It is an essential design for a Gestalt interpretation of meaningful structure. It achieves its design from folding a sheet of paper onto a random splotch of ink. Left and right splotches meet in exact identities that are grouped by the vertical distinction of the fold's crease. Notice that you fold the paper backward along the vertical axis of its crease, and then you open the paper forward to reveal the new bilateral figure. The physical act of rotating the sheet backward and forward makes a rhyme of the two physical ink splotches, which is now seen as a new pattern of thought. The patterned result of this bilateral identity, aside from whatever the image might come to mean, is called *reflection* symmetry. When left and right, front and back, or up and down are *congruent* forms in a mirror, the transformation that continues to reveal an identity or *isomorphism* between two parts is a bilateral symmetry. Here too begin the math and geometry of symmetry theory. A fold of a sheet of paper can be seen as a plane turned back on itself 180 degrees. Notice that a fold is part of a rotation. Rotation is another main transformation that reveals symmetry. The letters S and N are not bilaterally symmetric but if you wheel them along a line as if they were within a hoop, they would return to their identical image after a rotation of 180 degrees, or two quarter-turns of 90 degrees. Now by way of this inkblot example we can begin to test a hypothesis about the relations between signs and symmetrical forms: to achieve a design, rotate or reflect a sign. In Figure 15, you can see how an interlace is drawn from overlapping arcs of rotation. In a poem, as we shall see, the more that syntactical forms such as rhyme, meter, and parallel constructions are repeated in their motions of return, the more the poem becomes shaped verse, like a rondo.

The basic symmetric transformations in three space are translation along a line, rotation on a plane, and twist of a solid shape. In everyday life most symmetrical transformations are combined: "When you walk forward you are undergoing a translation. When you turn a corner, it is rotation; when you ascend a spiral staircase, a twist."[12] This prime example is based upon the most common model that everyone can use—our bodies' symmetrical limbs that carry us somewhere by the old sequence. An "everyday" definition is a commonplace for a universal definition. The word *everyday* is a faded archetype of one universally recognized periodic rotation of the sun. Put another way, all such syntax-based symmetries are necessarily tested by the instruments that extrapolate prosthetically from the shared symmetries of all our bodies in Four Directions.

But let us return to twist. Twist through three space is central for the actions of S-curved helical waves. Any twist requires an *uplifting* through the symmetries of linear and planar dimensions, through the in-betweenness of three space, that is, lived space. Things twist into rhythmic shape. The voluntary motions of everyday life take place in a three space whose cardinal points of reference are linguistic commonplace—up-down, back-forth, and right-left, the orientations of the sun. An elegant example of twist is the principle of spiral leaf arrangement. The helical model is called *phyllotaxis,* and its description by Goethe was the result of his pursuit of the helical form of an *Urpflanze* (Cook, *The Curves of Life,* 95). Martin Gardner provides a witty poetic example of twisting plants from Shakespeare's *A Midsummer Night's Dream.* Most plants twist in right-handed helices, he says, like the woodbine or morning glory, but the honeysuckle twists to the left. That picture of complementary twists gives new meaning to Titania's speech to Bottom the Weaver, "whose top," says Gardner carefully, "has been transformed by Puck into the head of an ass"(*AU,* 57). Titania says, "Sleep thou, and I will wind thee in my arms. . . . So doth the woodbine the sweet honeysuckle gently entwist" (Act 4, Scene 1). One needs to use imagination to reenact the transference as a rotation upside-down with two complementary twists. So symmetry is a way to use primal frames of reference to measure bodies up and down and right and left to see what changes happen to them in the material world after they undergo a metamorphosis, like Bottom's transformation. Symmetry transformations update the ancient preoccupation with the way things and beings— from Shakespeare's fools to Ovid's gods—undergo bodily metamorphoses.

"Translation" symmetry is performed along a line. When images, figures, or units whose structures remain unchanged are translated as sections of lines they are said to be identical, isometric, or congruent. If you walk on a sidewalk, the square blocks can be grouped one after another into equidistant translations. You can count them to determine how many squares equal the length of a city block. You can use them to explain directions: "Go four blocks west, and then turn right, going north for another five blocks." Those cardinal opposites of front-back, up-down, and right-left, are the locations of X-Y axes on the Cartesian grid.

"Rotation," the main trope of this chapter, is the motion of cyclic groups. A tripod shaped like the letter Y is, as Weyl says, the simplest figure with rotational symmetry (*S,* 714). A one-third rotation returns it to symmetry. This is the Y shape of the Anasazi ladle, discussed in the previous chapter.

But a "twist" is what Weyl calls a *screw.* This is the trickiest of the transformations, and it is the same form as Goethe's endless screw, mentioned in the Introduction. Therefore, let me quote Weyl's description of the difference between left and right:

> To the scientific mind there is no inner difference, no polarity between left and right, as there is in the contrast of male and female, or of the anterior and posterior ends of an animal. It requires an arbitrary act of choice to determine what is left and what is right. But after it is made for one body it

is determined for every body. I must try to make this a little clearer. In space the distinction of left and right concerns the orientation of a screw. If you speak of turning left you mean that the sense in which you turn combined with the upward direction from foot to head of your body forms a left screw. The daily rotation of the earth together with the direction of its axis from South to North Pole is a left screw; it is a right screw if you give the axis the opposite direction. There are certain crystalline surfaces called optically active which betray the inner asymmetry of their constitution by turning the polarization plane of polarized light sent through them either to the left or to the right; by this, of course, we mean the sense in which the plane rotates while the light travels in a definite direction, combined with that direction, forms a left screw (or a right one, as the case may be) [*S*, 679–80].

Weyl is trying to show here that in nature it is virtually impossible to determine whether the universe prefers left or right. Or whether one could describe to someone from outer space what we mean by left hand or right hand. This fascinating dilemma is Gardner's main topic; his subtitle is *Left, Right, and the Fall of Parity*. But in Weyl's description, notice that the prime example is everyone's archetypal body action. Turn your bilaterally symmetric body left or right. Its symmetry is the felt measure of all things. In fact, all regular motion in three space is really a screw, sometimes called a *twist*, or a three-dimensional helix, which is the product of a translation plus a rotation.[13] In other words, physical areas of space are locally curved into troughs and peaks, like waves. To put it formulaically:

Translation + Rotation = Twist

In symmetry theory there are only three elementary motions, and all of them are experientially intertwined. Think of a machine screw, for which you use a screwdriver, as a linear rod whose rotational circles have been stretched along the line of the rod and simultaneously twisted and threaded. *To thread* originally meant to twist a filament. This twist is the asymmetry of the S-shaped curve and the formal motion of the wave in three space. The words *twist, screw, torsion, helix,* and *thread* are synonyms for the same kind of torqued motion in three space, whose symmetrical components may be seen as rotational circles combined with linear sequences. As I mentioned in the Introduction, these transformations, which leave their figures unchanged or invariant, are the actions performed upon the components of any code to formulate them into a group of symbols that generate a message to a receiver.

Discussing the creativity of some artists and scientists as visual and geometrical thinkers, Pinker cites an exhaustive study by the cognitive psychologists Roger Shepherd and Lyn Cooper (*LI*, 71–72). Volunteers were

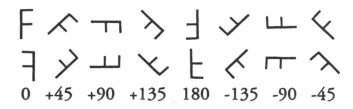

0   +45   +90   +135   180   -135   -90   -45

Figure 13. Diagram of the letter "F" on page 72, from *The Language Instinct* by Stephen Pinker. Copyright 1994 by Steven Pinker. Reprinted by permission of Harper-Collins Publishers, Inc.

shown thousands of slides of the letter F, each showing the letter from a different angle. Here is the diagram reproduced by Pinker (Figure 13).

The experiment was designed to test the reaction times of people who were thinking geometrically without the benefit of words:

> The subjects were asked to press one button if the letter was normal (that is, like one of the letters in the top row of the diagram), another if it was a mirror image (like one of the letters in the bottom row). To do the task, the subjects had to compare the letter in the slide against some memory record of what the normal version of the letter looks like right-side up. Obviously, the right-side-up slide (0 degrees) is the quickest, because it matches the letter in memory exactly, but for the other orientations, some mental transformation to the upright is necessary first. Many subjects reported that they, like the famous sculptors and scientists, "mentally rotated" an image of the letter to the upright. By looking at reaction times, Shepherd and Cooper showed that this introspection was accurate. The upright letters were fastest, followed by the 45 degree letters, the 90 degree letters, and the 135 degree letters, with the 180 degree (upside-down) letters the slowest. In other words, the farther the subjects had to mentally rotate the letter, the longer they took. From the data, Cooper and Shepherd estimated that letters revolve in the mind at a rate of 56 RPM (*LI*, 72).

Pinker concludes, "Many other experiments have corroborated the idea that visual thinking uses not language but a mental graphics system, with operations that rotate, scan, zoom, pan, displace, and fill in patterns of contours" (*LI*, 73). The experiment was designed around axes of symmetry, and the terminology is of *transformations*. Pinker accurately thinks of a visual syntax of strings, rotations, and shapes.

Although Pinker does not mention symmetry theory here, the examples and instructions tacitly depend upon these basic symmetrical transformations. Notice that in his diagram there are two different kinds of rotation that make the solution difficult. In the upper row, one rotates the letter in a mental great circle of four plus four directions, a compass rose. But in the lower row, one rotates each letter on its own axis, perpendicular to the plane space of the slide, in order to return the letter to its normal upright position. Although this is called *reflection symmetry,* its enactment requires a tacit visual thinking through a third dimension, in the round, as one pivots the letter on its axis to return it to the plane surface. When the letter pivots on its own axis, like a spinning ballerina, it achieves a fleeting image of a solid. Thinking in this visual three space, you see that the letter F must wobble and twist to return to normal equilibrium. This experiment is akin to Aristotle's thought experiment on the transformations of letters and their relationships to the description of atomic structure by Democritus and Leucippus. There seems to be a visual syntax of enactment symmetries, which we graphically use to locate the symmetries of atoms and other bodies in three space.

An Archetypal Body Plan

Consider Figure 14, a composite of several bilaterally symmetrical animal bodies from different cultures, drawn by Miguel Covarrubias.[14] These strange figures comprise his main examples for arguing a controversial thesis, called the *Old Pacific Style,* about the diffusion of cultures from China around the Pacific Rim to North and South America. These are but a few of the thousands of such quartered frog-like figures assembled by the anthropologist Carl Schuster, whose lifelong pursuit was a theory of symbols among primal cultures.[15] Covarrubias rendered these figures for his own book; Schuster, however, was less interested in the idea of diffusion than in the idea of universal symbolic forms. Schuster called these figures *hockers,* presumably because a *hock* is the joint in a quadruped's hind leg, which bends backward. The word *hock* is preserved for us mainly in the hock of a horse or in the culinary ham hock. So its verb form, *to hock,* is an action meaning to cut, to divide, or delineate at the four off-angled joints. Schuster thought this kind of symmetrical hocker could be seen as an archetype of the off-angled Four Directions of Four Quarters. These designs probably represent the principle of off-angled, bent-back limbs that propel bodies in rhythmic patterns of alternating gaits. Both Covarrubias and Schuster followed Franz Boas's early identification of dots and faces and eyes and circles as joint marks. Boas says:

> The most striking decorative form, which is used almost everywhere [on the North Pacific coast], consists of a round or oval field, the "eye design."

Figure 14. Drawings of Hockers from *The Eagle, the Jaguar and the Serpent* by Miguel Covarrubias. Copyright 1954 by Miguel Covarrubias. Reprinted by permission of Alfred A Knopf, a division of Random House Inc.

This pattern is commonly so placed that it corresponds to the location of a joint. In the present stage of the art, the oval is used particularly as shoulder, hip, wrist, and ankle joint, and as a joint at the base of the tail and of the dorsal fin of the whale. It is considered as a cross section of the ball and socket joint; the outer circle the socket; the inner the ball. Often the profile is developed in the form of a face: either as a full face or a profile.[16]

In the Tlingit example, for instance, faces are placed at the bilaterally symmetrical joints, presumably because the head swivels on the body as the eye swivels in its socket. In this sense, all the joints that angle off the body into hocks are workable because of the ball-and-socket connection. Schuster saw this kind of hocker connection and its numerous appearances in early cultures as an example of a historic survival of a characteristic "animal style" that could be traced from the late Paleolithic era. The Scythian buckle, illustrated in the previous chapter, is a side view of two-jointed hockers.[17] Schuster illustrates it as such.

Recall that Ogotomelli said the joints were the most important parts of a human body. And recall what Parmenides said about the relation between bent-back limbs and propelling thought. In this context, it seems probable that the representations of body symmetry were not just inherited designs but also were understood to be symbolic of a conceptual fittingness within natural orders of recurring proportions. There was a perceived relation between the ball-and-socket model of propelling symmetrical limbs and compositions of the four quarters, for if you draw and quarter the hocks, you get off-angled quarters of direction. Joints are pivots that transform the circular rotation of shoulders and arms into the propelling line of a leap, an arrow, or a javelin. In other words, these spirals, circles, and lines were not just doodles or tattoos. They also represented a set of archetypal body symmetries and asymmetries that repeated the elastic recurrences of earth's transformations. I illustrate this thesis with images that are not part of the Western canon of art history in order to demonstrate a commonly shared body syntax that I hope readers will come to think of as less strange as we go along.

For instance, there is strong evidence in some cultures that hockers were meant to combine into reproduction images. In his assemblage of Schuster's notes, lectures, and essays, Edmund Carpenter shows how primal cultures may have used hocks as visual representations of genealogical codes. If you join hockers continuously in weaves or drawings, they can represent the copulation of the mother and father into a third, fourth, and many figures. This is a fascinating and persuasive conjecture about these static diagrams as histories of generative actions. Genealogical histories, "First Kings, Second Kings, First Chronicles, Second Chronicles," are metonymic series of primal narratives. But the basic pattern is a visual and contiguous connection, like the shared symmetries of paper-doll chains, which one folds and cuts for children.

In *Language of the Goddess*, Gimbutas also illustrates many of these bi-laterally symmetrical figures as symbols of generation; only she sees them first as being "frog" images (*LG*, 251–56). In her gynocentric reading of these symbols, frogs come out in the spring, and their croaking often por-tends a woman's pregnancy. She illustrates many compound images of frog/women from prehistory. She also illustrates many Neolithic render-ings of these frogs or hockers as abstract stick figures (*LG*, 255). These styl-ized models are truly "matrix" diagrams, the Latin "womb," from which likely shaped creatures originate. That is, they diagram our shared shapes in unchanging symmetries of four off-angled limbs. That, too, defines "ma-trix": the essentially unchanging structure of a thing, embedded within variants, as in a "matrix sentence." The main point is that, once again, a uniform body is being structurally abstracted: the fourfold matrix is the primary carrier vehicle for the life of animals within their otherwise mul-tifold bodies—whether they be bird, fish, or hedgehog. Because this four-fold hocker or frog figure is composed of just a few syntactic lines, it is per-haps the very matrix of all "governing lines," as Ruskin called them, for it transports other bodies, and it rhymes with the larger pattern of the sun's rotation.

What these hocker or frog figures mean may vary adventitiously with the *langue* of different cultures. But I think that the fourfold structure of the ball-and-socket joint is a truly universal joint. It is the main connector of bones and ligaments that lets the four limbs angle off the body to get us somewhere. The structure turns our eyes in sidelong glances. A ball-and-socket joint is necessarily part of the common syntax of articulation that enables propulsion among four-legged animals. Returning to Figure 11, you see that the t'ao t'ieh is a hocker, with its eyes drawn as conventional ball-and-socket joints. Joints are the connectors that allow bilaterally symmet-rical animals to swivel, bend, undulate, twist, and thread. They let us lift and dance. They allow Blake's Creator to divide the world into geometries.

The four-branched hockers show how there is a necessary relation be-tween them and Arnheim's principle of simplicity in the use of angles and arcs. Whatever the figures may adventitiously "mean," the hockers must enact in their formal structures a syntactical pattern for the "animal style." They show us how the joints are a universal template with common syntac-tic structures, which interrelate with other physiological symmetries as rhymes. Because most complicated animals share these ball-and-socket symmetries at the four quarters of the body, the hocker depicts a reticulated natural syntax. There is nothing more commonplace. So instead of using the term *Universal Grammar* or even *universal syntax,* I like to use *common syntax* because it better expresses the commonly shared properties of our moving proportions. This is not a natural law; it is Read's aesthetic awareness of kinship.

The Measures of Scale and Pace

A hocker now can be seen as a static model that portrays the relative scale of an animal's body as it moves within its environment. In his singular study of the environmental fittingness of bodies within their local media, Thompson said, "The effect of scale depends not on a thing in itself, but in relation to its whole environmental milieu; it is in conformity with the thing's 'place in Nature,' its field of action and reaction in the Universe. Everywhere Nature works true to scale, and everything has its proper size accordingly" (*GF,* 17). He called the animal's retention of its relative dimensions, while it grows and forms within the physical limits of its locale, "a 'Principle of Similitude,' or of dynamic similarity" (*GF,* 18). In other words, its symmetries and relative magnitudes stay constant, even as it grows. So the measure of moving proportions must include pace as well as scale. This dynamic similarity is essentially a thesis about consonance, discussed in the previous chapter, in which the physical elements of the artistic composition rhyme with the forces in the local environs. For Thompson, the mathematics of an animal's relative growth and form comprise a principle of composition within its "field of action and reaction in the Universe." Every unit of measure assumes this principle of correlative similitude, in which an organic body grows constantly in form according to the physical limits of its medium. If a small hocker fits into a large pattern of the sun's motion around the four quadrants, then it is a cosmic model in microcosm. Put it the other way around: any presumed cosmology is a dilation of a constant magnitude from a small, known structural model to a large unknown conjecture. A cosmic model is a projection of a huge hypothetical space that uses the elements of composition that best fit into a small model. All the known conservation laws and symmetry laws, the laws of gravity, the laws of quantum physics, the motions of stellar bodies and black holes—all these must fit together to achieve a good cosmic model of everything nowadays. As little cosmic models, the hockers reinforce a primal supposition about one's place in a presumed order of things. They presume constancy between large and small measures of composition. *Dilation,* for example, as defined by Stewart and Golubitsky, is a principle of symmetry theory in which the form of an object retains its relative dimensions invariantly, even while its volume and surface expand progressively (*FS,* 264–65). A soap bubble or nautilus shell retains its essential form even while it dilates under constant transformations. So the abstractly drawn crosses, angles, circles, and spirals can be construed to constitute a primal syntax for measuring one's place in the regular order of rotating things. Today we use scales such as acres, miles, and light years to extrapolate from local shapes and scales to larger maps. Similarly, a hocker may be discerned as a template of physiological measure whose components—points, lines, arcs, angles, and spirals—can be broken down into an articulated syntax for delineating the scale model: "Everywhere Nature works true to scale, and everything has its proper size accordingly" (*GF,* 17). These models of the Four Quarters are scaled because of the symmetries of the sun's rotation and gravitation.

Since most earthly bodies share this primal hocked structure, their regular paces must follow the alternating rhythm of these off-angled quarter joints. For instance, in their book, *Fearful Symmetry,* Stewart and Golubitsky devote a chapter, "The Pattern of Tiny Feet," to the symmetries of animal locomotion (chapter 8). Throughout the twentieth century, the gaits of animals have been much studied and photographed. Just as an automobile must shift gears to accelerate to higher speeds, so, too, must an animal such as a horse shift gaits—a walk, an amble, a trot, a canter, a gallop—to propel itself to higher paces. Pace is the temporal measure of symmetrical scale. These recurrent gaits are characterized by alternating symmetries of placing feet on the ground in different patterns at different speeds. Depending upon the speed, the gaits and symmetries shift. In order to illustrate the different gaits as phased time intervals, the authors show a series of stick figures of a four-legged animal. The drawing is like a hocker; the body is simply a rectangle with four lines sticking off-angled from each corner of it. By way of this stick figure, they show a series of patterns of feet placement according to gait (*FS,* 200). They define a *walk* this way: "The legs move a quarter of a cycle out of turn (a 'rotating figure-8' wave . . .)" (*FS,* 196). Here is the archetype of the form of a wave. One can see Gans's "old sequence," discussed in the Introduction, as a pattern of alternating contraction waves of formal motion, but now they are described as symmetries of propelling gaits. One can conceive the *recurrent* pattern as a periodic motion, expressed as a figure eight, that is, as a continuously doubled S-shaped wave. To see and to conceive are separate activities.

However, if you are the kind of person who would let a millipede run down your arm, as the neurobiologist Charles Fourtner once put it to me, you could *feel* its feet on your skin as periodic changes in the amplitudes of the wave forms.

Stewart and Golubitsky show that there is a necessary relation between the coordinated symmetries of body shape and the phased geometrical motions that propel bodies in rhythmic recurrence. If there are just a few minimal patterns for propelling bilaterally symmetric bodies through space, then is that conceptual pattern transferable to the structures of syntax, which are carried through air by wave forms? Can a physical protogeometry for speech acts be expressed? Meter is also a pattern of tiny "feet." Iambs, trochees, and dactyls are jointed gaits of those feet, which speed up or slow down the pace and measure of words spoken in a line of sequence. For instance, Anne Bradstreet plays on the consonance between bodily and poetic feet in "The Author to Her Book," which is also her child:

> I wash'd thy face, but more defects I saw,
> And rubbing off a spot, still made a flaw.
> I stretcht thy joints to make thee even feet,
> Yet still thou run'st more hobbling than is meet; [18]

The symmetry of the periodic line in the sentence is achieved only if its jointed parts are made even. Bradstreet's lines may owe something to Rosalind's critique of a love lyric in *As You Like It:* "Ay, but the feet were lame and could not bear themselves without the verse and therefore stood lamely in the verse" (Act 3, Scene 2, 163).

Work songs, chanteys, and marches use those tacit metrics to evoke a communal gait for all the arms and legs, which are pushing and pulling or marching or dancing in patterned gaits of the old sequence. Herbert Read said, "Poetry can never become a popular art until the poet gives himself wholly to 'the cadence of consenting feet.'"[19] Is this use of "feet" for certain kinds of self-conscious speech acts just metaphor, which has illustrative value but not demonstrable truth value? These are issues for subsequent chapters.

But because poets have often noticed this link between the gait of real feet and the gait of a poem's measure, I preview upcoming chapters by quoting this famous little "Poem":

> As the cat
> climbed over
> the top of

the jamcloset
first the right
forefoot

carefully
then the hind
stepped down

into the pit of
the empty
flowerpot.[20]

Hugh Kenner accurately says that this poem is a "syntactic undertaking, purely in a verbal field."[21] If you read the poem aloud, you hear the phonemic foot get stressed as the cat's paw steps, ever so nonchalantly, into the available space. And the spoken words lift and fall with the paws. The pace of the physical reading carries the meaning literally, as the sound waves of spoken breaths, so that sound and sense, truly and not metaphorically, syncopate. In this example, the symmetry of the four-footed alternating gait of the hocker moves ostensibly with the starts and pauses of the line's momentum as it is read aloud. Notice that momentum is not metaphor. The momentum of the spoken syntactical order, not just the grammar, determines the order and shape of the sound waves, and it rhymes with the pace and scale of the cat's gait. The repetitions forward, the rhymes of alliteration and consonance, repay study, too. Poets and song writers know that physical syntax is the prime shaper of the rhythmic line of the sentence and that the grammatical order of sentences is not the absolute measure, but can be manipulated through inversions. I need to draw the distinction between syntax and grammatical order because some people think that grammar and syntax are synonymous. I define these distinctions in the next chapter. But I quote the poem here because I want to reinforce Thompson's idea of correlative pace and scale in nature as being appropriate not just for biological anatomy, but also for an anatomy of poetic syntax.

If there is such a code as a common natural syntax, it should be always before our eyes but beneath our scrutiny, as we use it without attending to it, just as we commonly use our limbs without attending to their coordinating conjunctions. In this chapter I am not trying to force a connection between forms of motion and the forms of syntax in speech acts, which are also periodic motions in three space. Here I want to demonstrate a primal syntax of geometrical forms that delineate a recurring and reciprocal connection between animal symmetries and the earth's symmetries of forces.

The function of these S-curved waves is to show continuity between figure and ground, animal and environment. These pivotal transformations, as pivots, demonstrate the continuity between recurrences of the earth's cycles and recurrences of animal bodies, as in the *élan vital* of the animals portrayed on the Scythian buckle. Demonstrated here is that there is not a hermeneutic circle of undecidability for primal cultures, but that bodies interact with the world in recurrent curvilinear patterns such as the figure 8. Although twisting helices refer backward in self-reference and forward in repetition, they always have momentum that is inevitably forward and not circular.

Even though the hockers are static, without the semblance of motion, they represent a working model of how regular motion is forever repeated and fed back in the world of living things. A picture is worth a thousand words, but a thousand pictures of one word repeat the *logos* or universal formula or moving proportion for saying the rest. That inherent patterning is the back-and-forth alternate wiring of the old sequence that presumably allows Meno's slave boy to recognize the forms of geometry. He did not have preexistent forms in his mind, but he, too, was shaped by the anatomical syntax of a hocker's symmetries, which let him recognize pictures, by the ball-and-socket joint of his eye, by the focus of his retina, and by the transformations of patterns in his cortex. If Socrates had asked the boy to draw a picture of himself, presumably he would have drawn a hocker-like stick figure, not because he thought bodies looked liked sticks, but because he knew that dots and circles and striate lines could reenact anatomical principles of shapes and connections. Discussing whether language is a homology with any other structure in the animal kingdom, Pinker notes that the structure of the eye evolved independently among different species "some forty different times"; it is therefore a "module" similar to what might be called a *grammar module* (*LI*, 348–50). Since the eye is a kind of ball-and-socket joint in primal art, as Boas taught, and if one agrees that natural syntax follows the forms of a reciprocating wave, then one might not need to find a grammar module or a grammar gene. Reciprocating waves are already self-reflexive in their forms of feedback.

To review the discussion so far, I have tried to show that if there is such a code as a natural syntax, it should be a set of transformations, derived from natural pace and scale, and encoded in everyone's nervous system as a set of instructions that formulates principles of moving proportion. By way of the hockers, I have tried to show that there is a necessary connection between bilateral symmetries in animals and their limbs of propulsion. But what is their relation to language? Here is Pinker again: "The overall impression is that Universal Grammar is like an archetypal body plan found across vast numbers of animals in a phylum" (*LI*, 238). Is the comparison of Universal Grammar to language an informative simile, which uses "like" as its ligament? Or is the archetype truly a homology, which portrays some kind of abiding fittingness? The four-jointed limbs of the body plan necessarily describe a few regular motions that I have grouped in symmetry theory as translation, rotation, and twist. I suggest later that these regular motions and bodily symmetries were recognized as patterns that followed the apparent motions of the sun. The Four Directions of the outstretched limbs mirrored the four directions of the sun's circuit. I demonstrate the sun's role soon, but first I want to give some examples of the tricky transformation called *twist*.

Twist

In all of natural history there is perhaps no more fascinating description of a physiological oddity than Thompson's of the narwhal's twisty tusk. He notes that if the grain of the tusk runs straight and true, not twisting in its growth like the spiral shell of a snail, then how does the tusk get its curvilinear threads? He further notes that a screw is made by combining a translatory and a rotary motion (*GF*, 217). A translation plus a rotation yields a twist. This is the formal language of symmetry theory. As we did in the Introduction, he begins with the dynamics of a wave, for its form provides the first clue to a morphological distortion: "The progress of a whale or a dolphin through the water may be explained as the reaction to a wave which is caused to run from head to tail, the creature moving through the water somewhat slower than the wave travels" (*GF*, 218). Thompson explains that while a fish has bilateral fins to keep its undulatory motion steady on a vertical plane, a dolphin and a narwhal ram their bodies forward by tail fins that give their bodies a slight rotary torque, much like a screw propeller, which in turn "tends to rotate the body about its axis, and to screw the animal along its course" (*GF*, 218). In the Introduction, we saw that an oligochaete compensates for this wrenching turn in the direction of the propeller's rotation by reversing its helices in alternating cycles. A narwhal must push ahead of its body a heavy rod of ivory, which does not bend and which is narrowly "coupled" to the skull, at the place where it grows: "At each powerful stroke of the tail the creature not only darts forward, but twists or slews all of sudden to one side. . . . The horn does not twist round in perfect synchronism with the animal; but the animal (so to speak) goes slowly, slowly, little by little, round its own horn!" (*GF*, 219). I recount this tale of a twist, not for its fascination nor for its questionable accuracy, but because it shows that there is a necessary coupling of symmetry and compensatory bodily motion that follows the form of a wave in three space. Further, its coupling will reveal a similar variation in syntactic forms. Later I describe a necessary relation between the morphological distortions of asymmetric syntactic forms used in sentences and the physical structures of bodies to which those syntactic forms refer in sentences. If the common syntax is twisted off its standard symmetries of syntactic form, then the morphology of the body shape is similarly skewed, as we shall see.

Nature's Endless Helix: The Total Syntax of the Universe

What qualities of twisting motion make it central to questions of syntactical form in both the arts and the sciences? The abiding answer was discovered by Louis Pasteur, the French bacteriologist (1822–95). His story has often been told, but in terms of symmetry theory, I primarily follow Richard Feynman's curvilinear "thread" in his chapter "Symmetry in Physical Law": "It seems that in the living creatures there are many, many complicated molecules, and they all have a kind of thread to them. Some of the most characteristic molecules in living creatures are proteins. They have a corkscrew quality, and they go to the right. As far as we can tell, if we could make the things chemically, but to the left rather than to the right, they would not function biologically, because when they met the other proteins they would not fit in the same way. The left-hand thread will fit a left-hand thread, but left and right do not fit."[22]

Pasteur's discovery was that certain microorganisms are "asymmetric" and that they select nutrients that "fit" their own right-hand helical forms. He began his experiments because he wanted to find out how grape sugars ferment into wine. He had known from a predecessor that quartz could twist polarized light either right or left, depending upon the lattice structure of the piece of quartz. His predecessor, Jean Baptist Biot, found that organic compounds of sugar and tartaric acid can twist light because of their own inner molecular structures. By patiently devising a series of experiments with tartaric acid and the rotations of polarized light, Pasteur was able to demonstrate that the asymmetric living organism only selects the threaded form of nutrient sugar that fits its own right-handed *corkscrew* shape. Martin Gardner tells this portion of Pasteur's story in order to move to the later discovery that carbon is the building block of asymmetric tartaric acids, and thence to the nuclear structures of protein threads, and thence to helical structure of DNA, the root of the genetic code (Chapters 12–14). This threaded helix is the building block of organic form in three space. Its moving proportion is the primal syntax for building an abiding life against decline. (A biology major once told me that he liked my Utopias seminar because "it is my only non-carbon course.")

In "Dynamics of Dissymmetry" Roger Caillois also describes Pasteur's discovery of a right-handed helical chain of life that fits only with its right-handed nutrient.[23] Discussing this phenomenon as an "inverse dissymmetry" (*DS*, 80), he generalizes that this form constitutes the "total syntax of the universe" (*DS*, 90). I assert that Callois's use of "syntax" for the threaded corkscrew is not a metaphor but an archetype of a living linguistic form. Certainly the most sweeping summary of the form of the protein base is J. Z. Young's: all species of life forms are "one flesh" unified by a living language. "This is no emotional statement or metaphorical figure of speech. One of the most startling revelations of molecular biology is that the same code of triplets of bases is used to define the proteins of all organisms. . . . We are indeed one flesh and we depend upon the information in a set of books, written not in a babel of tongues, but in a *single common language* [his emphasis]" (*SM*, 115). The complete passage about evolution and self-perpetuating organisms is a wise exhortation on the essence of living as an awareness of the interdependence of all species through this language.

What characterizes the human species as a distinct species? Every human being commonly shares the same brain structure with the same neural-anatomical "wiring" coded for speaking about the single common language that unifes us. Yet if there is indeed a common natural language, upon which all life is built and replicated, then its common trans*form*ations should also be the common forms for the generation of human speech sounds through the space and time of the speech act. Feynman explains how the three dimensions in space, together with time, are symmetrically implicated with the rotations through four directions:

> Real space has, in a sense, the characteristic that its existence is independent of the particular point of view, and that looked at from different points of view a certain amount of "forward-backward" can get mixed up with "left-right." In an analogous way a certain amount of time "future-past" can get mixed up with a certain amount of space. Space and time must be completely interlocked; after this discovery Minkowsky said that "Space of itself and time of itself shall sink into mere shadows, and only a kind of union between them shall survive."

I bring this particular example up in such detail because it is really the beginning of the study of symmetries in physical laws. It was Poincaré's suggestion to make this analysis of what you can do to equations and leave them alone. It was Poincaré's attitude to pay attention to the symmetries of physical laws (*SPL*, 94).

Here symmetry theory begins to be implicated with physics. If one substitutes for Feynman's image of "interlocked" space and time an archetype of a curved space-time that is twisted or threaded or woven over and under into a wave form, then one can see how the symmetry operations of translation, rotation, and twist are transformations derived from the spatiotemporal form of a wave. The idea that space-time is twisted or warped is beloved by sci-fi writers, but the concept is difficult to envision.

The Irreversibility of Translation

The implication of symmetry theory with biology is now easier to see. Basic biology texts discuss the successive untwisting and re-twisting of the helical strands of proteins and DNA. Also, the rungs of the twisting ladder—adenine, thymine, cytosine, and guanine—have significant molecular shapes, which allow them to bond only in four possible combinations: AT, TA, CG, and GC. These shapely inversions constitute a mirror or reflection symmetry. In fact, it is the peculiar twisting shape of each molecular bond that gives each of the base structures a slight twist or torque, which in turn creates a series of successive twists in the vertical threads that form the larger helical shape. That fundamental discovery by James Watson and Francis Crick led them to entitle one of their early papers "The Complementary Structure of DNA" (1954). Later, Jacques Monod used nature's endless thread to further his own theory of evolutionary change.[24] In *Chance and Necessity* he used the formal symmetries of DNA, nature's twisted thread, to show that "translation" is the symmetrical transformation that necessarily repeats or carries forward, within the structure of the molecule, any chance error in its replication. Hence the combination of chance and necessity. Replication, for Monod, is always replication forward, not backward. It is the organic version of repetition with a difference, wherein life is lived forward, as Kierkegaard thought, but it is recollected backward. DNA is a front-back rhyme. For Monod, the "irreversibility of translation" (*CN*, 110) carries forward any random change, which in turn provides the necessary one-way sign in this packet of chance and necessity. Monod concludes that evolution itself is irreversible because of a one-way translation. Finally, ". . . it necessarily follows that chance alone is at the source of every innovation, of all creation in the biosphere" (*CN*, 112). This sweeping conjecture works only because of the application of symmetry theory to the helical structure of DNA. The operations of translation, rotation, and twist are symmetrical transformations that help him delineate a theory of innovation in the biosphere. I'll need this idea of the irreversibility of translation later in the book.

As Coleridge quoted Petronius Arbiter, "Precipitandus est *liber* spiritus." Translation seems to be the algorithmic motion that necessarily precipitates forward the syntactic units in an additive series, but the chance inclusion of error frees up change. For Coleridge this is the predicting and recollecting pattern of the old serpent, which is also his model for reading. Can we now begin to think of a certain amount of past-future and left-right as the point of view upon an alternating and twisty wave, the old sequence that is coded in every body's nervous system and that interweaves chance and necessity?

Within this context of genetic evolution, Monod refers to Chomsky's emphasis on "the basic form common to all human languages (*CN*, 136). Monod is not scandalized by the idea of an innate code for language acquisition. For him language is part of the human genome system. No one has yet found a language gene, but many are searching. He thinks that this innovation of language therefore contributes to human evolution. When Caillois speaks sweepingly of "the total syntax of the universe," he is thinking of a biological code, but Monod suggests that if we wish to think anew about "innovation," about a creative thinking forward that gets us somewhere, we need to learn the organic symmetries of the helical form. Notice that biological innovation derives from a change in replicative pattern; similarly, I argue that an innovative thought is usually rhythmic pattern of thought that depends upon a kind of rhyming backward and forward, as with inkblot tests, toward a new interpretation or inference. But to return to Monod's DNA and Coleridge's wise serpent: To the degree that minute chances are replicated forward, while retaining from backward the other units in the series of the genetic code, one is to that degree liberated from a physical model of necessary drives and forces. One can substitute an evolutionary model of propulsion and purpose, derived from a translation, forward from backward. Monod's theory of chance and necessity gets us somewhere, moves us toward a biological theory of language whose syntax may be modeled as a one-way helix, translating, rotating, and twisting forward in space-time. I am still pursuing an *Urform*, the old sequence of the flowing form of wave, but the wave is now replenished by a fitting syntax of life forms.

My point is that these few basic transformations—of translation, rotation, and twist—are seen as formal operations within the over-and-over template of DNA that delineates the generation of further biological forms. The last sentence of *The Origin of Species* is rightly famous, but I want to stress Charles Darwin's idea of "forms" therein: "There is grandeur in this view of life, with its several powers, having been originally breathed by the Creator into a few forms or into one; and that, whilst this planet has gone cycling on according to the fixed law of gravity, from so simple a beginning endless forms most beautiful and most wonderful have been, and are being evolved."

What is remarkable about his discussion of a few compositional forms is that it might just as easily have been written by his grandfather, Erasmus Darwin, whose abiding theme was the creation of lovely life from a single form: a point, then a living line, that bends and rotates into a tube. For instance, in his long poem called *The Temple of Nature*, Erasmus Darwin, like his grandson, begins with Newtonian gravity, with attraction and repulsion, and he says there is a "chemical affinity" of fluidic attraction that manifests itself in "heat, light, and electricity."[25] This attractive force is called *Love*, the creator of life in the sea. Here is his vision about the origin of life from a few forms:

> In earth, sea, air, around, below, above
> Life's subtle woof in Nature's loom is wove;
> Points glued to points, a living line extends,
> Touch'd by some goad approach the bending ends;
> Rings join to rings, and irritated tubes
> Clasp with young lips the nutrient globes or cubes;
> And urged by appetencies new select,
> Imbibe, retain, digest, secrete, eject.
> In branching cones the living web expands,
> Lymphatic ducts, and convoluted glands.
> (Canto 1, 251–60)

Points of life began with the weaving over and under of a single thread into rings and tubes and convoluted glands. (I hope that while one might snicker at the couplets, one doesn't lose sight of the protogeometry of life forms.) A transformation from a point to a sequential line to a bending line to a rotating tube; in Erasmus Darwin's description this is living geometry, not just Euclidean geometry. It is no wonder that symmetry is built into cognition as a formal generator. Both Darwins saw that the effect of gravity manifested itself in a geometrical form of rotation. Theirs was an awareness of the primal relation between a conservation principle and a symmetrical one. Its common consent is that, under the influence of gravity, organic things seek equilibrium, and the formal manifestation is a symmetrical balance, like the righting arms of the scales held up by blind justice. Before leaving this form of biological geometry, recall that a large issue is the innateness of a generative syntax. Monod speculated that Chomsky's theory of innate forms of language acquisition followed from the template of DNA. So at least from the Darwins' point of view, life begins with a few protogeometric forms of composition. And though the Darwins did not know it, DNA is the format that transmits those characteristics indefinitely forward. Like a hocker, it is the century's generative matrix. I turn to Charles Darwin's theory of the origin of language in Chapter 4.

The Great Wheel of Syntactic Orientation

What do these reflections, rotations, and twists have to do with the sun? Both in terms of sunlight and gravitation, the sun is the main power source for all such mirrorings, circlings, and spiralings. The huge gravitational mass of the sun bears down equally in all directions on earth, so forms on earth spread symmetrically left and right, horizontally in all directions without preference given to one direction on the plane. But the sun's light comes to organisms in off-angled waves, both because of daily rotation and the annual tilt of the solstices. Bilateral symmetries of eyes, ears, and nostrils, as well as the sensitivity of the skin, correlate the off-angled impressions of waves that touch the sensory surfaces.

There have been many studies of the sun as the archetype for symbols of rotation among early cultures.[26] The association of the sun's orderly motion with rotary symbols of the wheel is also commonplace. One of the most elegant summarizing icons of rotation (Figure 15) is a Pythagorean diagram, drawn by Isidore of Seville. It twists and interweaves up and down and right and left, with four elements, four directions, humors, four seasons in a design of circles and half-circles. It nicely summarizes what would become the Renaissance cliché of correspondences between the human and cosmic cycles. S. K. Heninger studies many examples of this design of four directions and their multiples, which would come to be called a *compass rose*.[27] His analyses show that the patterns have meanings that are not

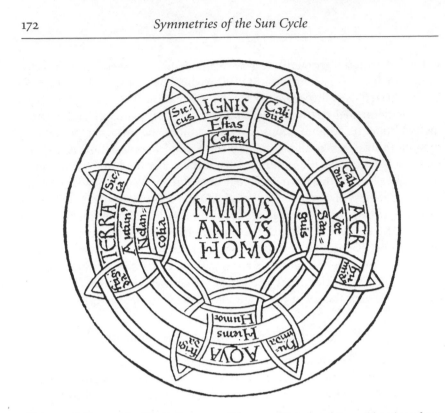

Figure 15. Isidore of Seville, Pythagorean Diagram of Interlaced Four Directions from *Liber de responsione mundi. . .* Augsburg, 1472. This item is reproduced by permission of The Huntington Library, San Marino, California.

clichés. Perhaps most important, the rotation model shows that opposite qualities cannot overlap each other directly in a dialectic of contraries. They must rotate and thread through neighborly qualities. For instance, at the top of Isidore's design, fire's qualities of dryness and hotness correspond to summer and choler. Water, drawn at bottom, has qualities of coldness and moistness, and corresponds to winter and phlegm. Heninger's important point is that those Pythagorean complements were interlinked through the adjacent qualities on their quarters, the half-circles that make the lace. In rotation symmetry through the periods of the world year, the annual year, the human year, opposite qualities are linked through their "mediate" adjacent qualities in the cyclical group. Complements do not oscillate directly into their opposites by dialectical inversion. Unlike one of the descriptions of the Chinese yin-yang mentioned in the previous chapter, there is no "bipolar complementarity" movement. In real space and time, bodies twist and torque and lace and thread adjacently through all media in waves. According to this model, therefore, there is no dialectic of two opposites in nature. The elements "weave," as Erasmus Darwin put it, in forms that take the long way around a rotational compass.

These designs were not seen as static models. They are models of periodically moving proportions. They all rotate through the four seasons in order to complete one cycle. The Stoics called this complete cycle of a day, a year, or a Great Year, its *periodos*. In the rotation cycle of the four seasons, the period is the least-time interval for which the cycle comes to completion. The mathematics of the four integers, in this case, is today called its *cyclic group*. *Periodicity* then can be defined in terms of least time in rotation, translation, or helix. For example, in physics the frequency of an S-shaped sine wave in simple harmonic motion is equal to the inverse of the time period (T): $f = 1/T$. I stress this physical equation to show how the awareness of periodicity is not just part of an outmoded history of ideas, and not just an arbitrary measure ticking in a metronome, but that it is grounded in our presuppositions about the physiological impressions of the Great Rounds of space and time.

Virtually all creation myths of the world's civilizations gather in their orientations of the Great Rounds a sacred grid or mandala of four tropical directions, whether explained as the homes of the four winds, four mythical creatures, or four mountains. For instance, as early as the late Chou dynasty, each of the four directions was assigned animals, colors, elements seasons, and phases in the yin-yang cycle. North was given associations of black, water, tortoise, winter, and major yang; east was blue, wood, dragon, spring, and minor yang; south was red, fire, phoenix, summer, and major yin; west was white, metal, tiger, autumn, and minor yin. In the center, at the hub, there was yellow, earth, and the human animal who could harmonize all the other directions and qualities. In other words, some ancient cultures indicated six directions: two from pointing straight up and down from one's own center, and four from pointing laterally. It is important to remember the up-down axis because gravity splits things into bilateral symmetry.

This axis of Four Directions was used to explain the past and divine the future. In the next dynasty, the Han, it was used to point the way through real space by a piece of mineral composed of magnetite. Primal hermeneutics—divination in time and *deixis* in space—could ritualize into sacred play. Some see the origins of all board games in this kind of quadratic compass rose: "The basis of the [arrow] divinatory system from which games have arisen is assumed to be the classification of all things according to the Four Directions.

This method of classification is practically universal among [early] peoples in Asia and America."[28]

By way of these examples of Four Directions I underscore the ubiquity of rotation symmetry in a model that could be used to divine one's place, not only in space, but also in the temporal narrative of the creation of kinship among all things. In literary theory this primal story is the seasonal theme of what Northrop Frye called *archetypal criticism*.[29] All creatures follow one ordained plot. The child as prankster-hero rises in the east, moves south as an adult warrior, enters the portals of death in the west, and travels through the underworld in the north, where s/he prepares for seasonal rejuvenation once again in the east. The heliotropic round of the Four Directions is explained by an ordained story, into which all things also fit.

To a rotation model of the Axis Mundi add the four ages of humans under each of the Four Directions: bright childhood to the east, hot adolescence in the south, melancholic maturity to the west, and phlegmatic old age up north. They follow the turns of the sun. Then the archetypes of the heroine's or hero's birth, adventures, accession, death, and eternal return could be seen as tropical turns. Each of these directions is an off-angled point of view, from which the narrative is told at that station of the cross. Point of view necessarily is associated with angles quartered off sunlight.

The primal story line must follow the rotation through successive and adjacent quarters: birth, childhood, maturity, underworld, and return of the main character. Each fourth direction is an off-angled point of view upon a stage in periodic rotation that is formulated as circle and sequence. There is no shortcut from one complement to another. The shortest way is 'round about. This is the meandering periodicity of Least Action. That is, one must traverse the rotation in a sequence through off-angled quarters. Isidore's model showed the off angles as half circles. But if you translate and rotate half circles over and under through three space, then you get an undulating series of S curves, as drawn in Figure 1 in the Introduction. In order to combine translation plus rotation through three space, one must twist through an oblique angle of a higher dimension. That is why the illusory perspective of drawing three space on a two-dimensional plane is twisted obliquely.

This fretful rotation is the archaic Way. The abundance of archaic ornaments (brooches, medals, pins) that represent two-, three-, or four-legged sun discs running sunwise, or clockwise, means that they are apotropaic; they turn away evil that runs countersunwise or "widder-

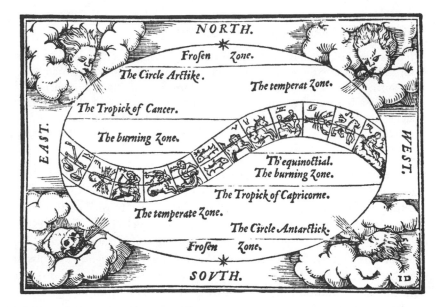

Figure 16. Renaissance Drawing of the Sun's Apparent Track. From William Cunningham, *Cosmographical Glass*, 1559, P. 64. This item is reproduced by permission of The Huntington Library, San Marino, California.

shins." In *The Curves of Life*, Cook says of the swastika as an ancient sun symbol:

> There is little doubt that it symbolizes the apparent movement of the sun through the seasons. The red eastern arm (for the earth) represents Spring and the morning; the southern arm is the gold of Summer noon; the blue arm of the west is Autumn sunset on the sea; Winter and white midnight are seen on the northern limb. Thus it symbolizes the life-giving sun, the origin of all things . . .(166–67).

In following the steps of the sun as heliotropes, people knew that there was no shortcut, no dialectic between opposites in nature. In the processional of the Sacred Round, the Way, one does not move directly between opposites, as for example, from east to west, or from north to south, or from youth to age, or spring to fall. Instead, on the Way, one turns obliquely through the mediation of the adjacent quarter. This is the apparent path of the sun, not only during its diurnal cycle, but also in its annual cycle. Figure 16 reproduces a Renaissance rendition of the apparent track of the sun as it meanders northward and southward, delineating and defining the respective Tropics (tropes) during the year.

The twisting S is the ecliptic line of the sun rendered as the zodiac, and Heninger provides a sixteenth-century explanation:

> It is named the Zodiaque either of this Greek worde *zoe*, which is as much to say as life, because the sunne being mooved under this Circle, giveth life to the inferiour bodies, or else of this Greeke word *Zodion*, which is as much as to say as a beast, because that 12. Images of stars, otherwise called the 12. signes, named by the names of certaine beasts, are formed in this Circle: and therefore the Latines doe call this circle *Signifier* because it beareth the 12. signes.[30]

The sun is the life-sustaining carryall. Its meandering curvilinear trace is zoomorphic. These Sacred Rounds, then, are not seen as a dialectic between opposites, but in the visual signs systems of traditional cultures, the way is with the Signifier of the sun's movement, which "carries" the rest of the zoomorphic signs. The S-curved track of the sun is the primordial vehicle for carrying the Signified. In *The Sound Shape of Language,* Jakobson says that Saussure, in his lectures, appropriated his basic theory of signs from this Latin tradition first taught by the Stoics, for whom a sign had a perceptible material known by the senses, as well as an intellectual element (*VIII*, 17). So the understanding of this amphibious carrier vehicle is very old. Nature enters ancient communication via the rotational circle of the sun, that is, as Signifier, which bears all signs.

The danger of these graphic diagrams of Four Directions is that they might sustain a false presupposition: the opposite categories, constructed by syntactical and grammatical distinctions, now seem to be "in" nature. But there are no opposites in nature, no poles, no positives or negatives, no dialectics of nature, no actual arcs or angles or helices, just felt tracks of frequencies, which we postulate as interactions of force and counterforce. That is why one needs a compass rose as a commonly shared model of moving form. Negatives and poles are the constructs of different lexicons and grammars. So are the words *translate, rotate,* and *twist,* but their operations are physical acts. Negatives have no place in a natural syntax. Unlike other animals, when humans formulate the concept of the negative, we are on the way to a self-conscious grammar. We feel the pressure of gravity, and we press backward and forward with our joints in order to walk upright and resistant on the way. But the oppositions are constructed with language as our own symmetrical frames of reference. These opposites that are constructed for diagrammatic inversion now seem not to be part of the grammatical model, but they can be projected as poles "in" nature. Not a projective allegory or a speaking picture, the polar concept seems to be the very thing itself.

For Wheelwright, the archetype of the sun-wheel-lotus is the main bearing vehicle of allegory. For Max Müller, such solar rounds are the origins of all myth.[31] My point is that most signs and designs of periodic motion, which later become the original elements of geometry and symmetry, are projections from the sun's apparent motions: rotation, translation, and twist. The more abstract point is that physical wave frequencies, seen as S-curved tropes of synesthetic inversion, whether from space to time or from time to space, were known as the design vehicles, the Signifiers of natural periodicity. They naturally twist us from the visual thinking of a circular plot in space to the sequential stations of spoken narratives in time.

The Sun as Carryall

The sun's journey through the sky was therefore seen as the elementary protogeometric model of periodic frequency in space and time. The sun was the vehicle, carryall, or omnibus for transporting the idea of periodic force that uplifts life (*zoe*). The type of container or vehicle might change in different cultures, but the sun's frequency was the formal transmitter carrying the message periodically as a source of energy. It might be Amon who navigated a riverboat with sun as his cargo, or it might be Apollo with his chariot and horsepower (Wheelwright, 124). They rode the great rounds of the sun as figures of destiny.

But this symbol of a solar vehicle as the form of a motion is not limited to primal cultures. For example, Arthur Eddington reminds us of our own astronomical ride: "With the earth as our vehicle we are traveling at 20 miles a second around the sun; the sun carries us at 12 miles a second through the galactic system; the galactic system carries us at 250 miles a second through the spiral nebulae; the spiral nebulae . . ." [his ellipses].[32]

Think about the subassemblies of Apollo's chariot in terms of the allegory, that is, as the diagrammatic parts of a solar powered vehicle:

1. Power Source: where does it come from, and what is it made of?

2. Chassis or Body: what is its biomorphic origin? Its bilateral symmetry? Its front-back asymmetry?

3. Transducer: what is the transformer, the converter from one kind of energy to another? An axle transforms rotary motion of wheels to linear motion. Perhaps this discussion is much too technologically reductive for some readers. But bear with me, because I need to show that the basic transformations

of symmetry theory also can be used to describe certain kinds of reversals in the directions of harnessed power. Recall that I described three elements of motion in the performance of Dennett's workaday crane. To return to the example of the bow and the lyre from the Introduction, consider the redesign of the bow from a weapon of war to a tool of domestic economy. Certainly one of the most useful tools in first cultures was a bow lathe for making fire. A bow lathe translates linear motion into rotary motion via a string wound in a twisting helix around a stick. Here again are the three forms of motion—translation, rotation, and twist—assembled together and put to work. Push and pull the bow back and forth; the rotary motion of the stick creates thermodynamic friction at the pressure point: motion, friction, heat, fire. Light, work, and fire from the sun are connected through the rotation of ball-and-socket joints. A basic transformer of knack technology, the bow lathe could make fire; it could also drill holes. In Hindu mythic art, the vertical axis of a bow lathe was sometimes drawn as a vertical Axis Mundi, or phallic Cosmic Pole, around which the rest of the creatures pivoted.[33] The name "Prometheus" meant "foresight." As fire-giver, he was often associated with the "fire drill," as it was also called. This machine tool, as Bronowski defined tools to make other tools, was the prime mythic model of work as deferred gratification, of anticipation of an end through arduous means. My point is that the myth of fire and the fire lathe or drill shows the primal syntax in the very composition of the machine: translation, rotation, and twist are the transformations that connect a body's ball-and-socket joints to the end of fire. Fire and the sun are related through a symmetrical syntax of body motions. So in this regard the Dogon storyteller Ogotomelli's assertion that the joints are the most important parts of a person is not laughable.

So what part of the assemblage, what coupling, what ball-and-socket transforms images of rotary space into those of linear and helical motions?

4. Instruction Manual: how does it steer with reins or tiller, turning right and left? The trace of right-left steerage is the elliptical S wave, whether in plowing the earth or tilling the sea.

Rotary, linear, and helical motions are combined into feedback loops as the vehicle travels through space. The feedback loop of a four-footed gait, one recalls, was drawn as a figure eight. These subassemblies are components of the transmitter. They are groups that subdivide the channels of force into geometric directions. This kind of heliocentrism meant that there was a version of cybernetic steering through the world. There is not just a vicious hermeneutic circle, but rather a reciprocating universal joint that pivots and connects and propels the body as vehicle through the elements in an old sequence through space and time. My reason for being perhaps too rigidly mechanical is to stress repeatedly that physical forms carry the message as syntactic vehicle. The physical signifier of the message, the form of the wave as carrier, is always referred with, but cannot be immediately referred to. When I refer to the physical world, I must always use mediating symbols that indirectly allude to other verbal concepts in their sign system. To circumvent this paradox, as I explain later, tropes are invented as physical gestures that seem to carry a physical meaning to a verbal concept.

The Labyrinth as a Solar Composition

In the previous chapter I cited Ruskin's technical description of a labyrinth. Perhaps the most pervasive figure for the representation of the concept of archetype itself is the labyrinth, the deep structure of the underworld, beneath the semiotic bar. When the sun sets under the horizon, it is destined to go to an underworld. What does that place look like? For many primal cultures, and later for some sophisticated cultures, the labyrinth was a hidden place of destining. As an archetype, perhaps the archetype for unconscious recurrence, it is the place where the hero turns to the underworld, in the dark night of the soul, to confront his monstrous double, but also his Ariadne.

Less widely known is that there are sets of technical instructions for assembling or drawing a labyrinth. I include this analysis under the title of "Instruction Manual" because a labyrinth can be constructed as a destining for the sun's regular motions. The labyrinth had its own "drawing manual." It had a syntax of composition. More intriguing for my thesis of a communal syntax, a labyrinth can be drawn from a set of instructions about the symmetries of the Four Directions. Schuster compiled hundreds of images of labyrinths from around the world, and he sought recipes for drawing them as well.[34] The most widespread instruction is Figure 17. The diagram sets the axes of four sacred quarters off from four symmetrical directions. But it also positions four arcs nested with angles and four dots or points nested within the arcs. These abstract arcs and angles, whose motor relations I discuss via Arnheim in the Introduction, can also be seen as abstractions of eye-and-socket, or ball-and-socket pivots. One can see the diagram as an abstract hocker whose joints are arranged in bilateral symmetries. Here is how the common syntax of static symmetries—position, order, and shape of the hocker—is associated with instructions about acting in motion. The instructions make up a recipe for transforming a set of symmetrical lines into an archetypal puzzle.

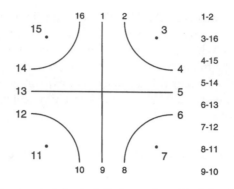

Figure 17. How to Construct a Labyrinth. Redrawn from *The Labyrinth* by Carl Schuster.

Figure 17 shows how and when to transform the static image of the Four Directions and the four off-angled and jointed limbs into rotational arcs that render another shape of the spiral labyrinth. In order to see the hocker as visual code, one must draw the arcs in a stipulated order; one must follow the periodic sequence of instructions, assembling the new concoction, as one follows a recipe, step by step. One gets a new shape, a new position, a new order, by drawing the motion of an arc, a semi-rotation. So this design of four quarters is a visual code for transcribing the periodic motions; that is, the bent-back limbs constitute a visual model for transforming the formal elements of composition into one of the great mysteries. If one recalls that Ariadne's thread through the labyrinth is the very archetype for narrative sequence and return, then to be able to construct a visual model of a labyrinth, perhaps by making the arcs with stones placed in arched sequences in landscape, one has a visual guide by which to script the narrative in its place. In all of these analyses, I still seek a few periodic forms of a motion, and how these formal motions were seen as commonplace acts of composition that derived from quartering the sun's rotation into an archetypal syntax. If you review Figure 8, the Sicilian tomb door, you see that it, too, is a bilaterally symmetrical hocker. If it does reenact the motions of the world's body, then those arcs and rotations and spirals can be seen as components of a labyrinthine circuit.

Because the Four Directions are associated with the sun's rotation, one can understand how the sun and the labyrinth might be associated. But there are closer ties. Coincidentally, in *Die Sonne als Symbol*, Rudolf Engler offers a drawing of solar motion as the origin of the spiral.[35] He thinks that the continuously rising sun may have been drawn as a spiral in order to

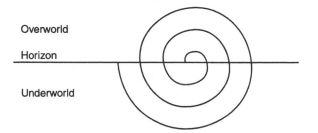

Overworld

Horizon

Underworld

Figure 18. Diagram of Solstices seen as Spiral. Redrawn from Engler, *Die Sonne als Symbol.* Displaced arcs of rotation, seen as a continuous spiral of sun risings and settings. The drawing can answer the questions, "Where does the sun go when it sets?" and "Why does the sun not rise and set in the same place each morning and evening?"

mark and explain the displacements of the solstices (Figure 18). Read the center of the spiral as if it were an arc drawn above the horizon line, as the limits of the sun's rotation on the shortest winter day of the year. The successively wider radiating arcs, from shortest to longest, mark the longer limits and moving locations of the summer solstices. But the spiral shape provides a reason for the displacements of the solstice by the regular displacement of intervals in the spiral form. A protogeometric form explains a physical motion. The continuous line of the spiral, drawn below the horizon, explains where the sun goes at night by continuing the rotation.

The spiral form and the labyrinth are closely related. But there is further evidence that the labyrinth can be derived from a solar hocker of four directions that marks the solstice. Schuster collected many photographs and drawings of labyrinths, drawn by early Southwestern tribes of America. Modern Pima and Hopi weave the labyrinth in their basketwork. But at least one archaeoastronomer reads this basket design of a maze, which the Pima call *I'itoi Ki,* or *The House of I'itoi,* as if it were a solstice marker.[36] My point is that these archaeogeometric forms are derived from the regular permutations of the sun's motions. These solar forms of symmetrical transformation provide recipes and blueprints for constructing one's zoomorphic form and destination under the sun. The sign/design of the four quarters is a visual code for orienting one's body to order, rotational position, and shape. What other transformations could be the basis for a common syntax of forms? If syntactic forms are physical forms, they should derive from this anatomical archetype of body symmetry.

To conclude this chapter, consider John Muir's study of the symmetries of the great pines of the Sierra Nevada Mountains. He describes what has been called the *phyllotaxis* of some plants, Goethe's *Urform* with which we began. Muir climbs one of the most perfectly shaped silver firs, which he describes as "the most charmingly symmetrical of all the giants of the Sierra woods," in order to get a feel for its structure: "How admirable the forest-work of Nature is then seen to be, as one makes his way up through the midst of the broad, fronded branches, all arranged in exquisite order around the trunk, like the whorled leaves of lilies, and each branch and each branchlet about as strictly pinnate as the most symmetrical fern-frond."[37] Muir's comparison with lilies and ferns is not just a simile, not just a figurative likeness, but an inherent structural symmetry perceived as being shared by other living things that he has seen and loved.

CHAPTER 4

An Anatomy of Natural Syntax

Synopsis

In the last chapter I described the sun's coordinates, together with a body's four-quartered fit into those orientations. It seems probable that there is a harmonic ratio between the neuromuscular motions by which we guide ourselves as symmetrical hockers and the patterned symmetries emanating from our solar environment. The fourfold anatomy of the archetypal animal body was seen as a compass rose for pointing the Four Directions of the sun. Furthermore, the generative model of DNA, with its irreversibility of translation, was described, by Callois, as "the total syntax of the [living] universe." So now I transpose what I have described about bodily symmetries to an anatomy of natural syntax.

In previous chapters I showed that carrier waves have, in themselves, significant configurations that shape in turn the visual or sound image. I spoke of the characterizing shapes and orders of the transmitting wave as a natural syntax. For instance, particular groups of light waves, reflecting off an object, send or transmit or carry the shapely signals that we translate into an image of a "tree." So a broad definition of natural syntax begins with the awareness of rhythmic groups within the carrier waves themselves. The abiding point from previous chapters is that the form of the propelling wave is doing the generating, while the particulars, so to speak, jostle up and down, back and forth, or twist over and under. The form of the wave is the form of natural syntax. That is not a Platonic notion, but it is a formal motion.

If a natural syntax exists, it has physical existence, in the form of a moving wave. The trick is to catch it in the processional act of generating, of transforming. If a syntax exists physically, it enacts a pattern of plosions in space and time—seen as moving proportions of pace and scale. If syntax exists physically in patterned portions of a space and time, as a natural syntax it will be subject to the same kinds of physical transformations that characterize other physical forms of "symmetry sharing" under the sun. So "natural syntax" in language may be defined as a conceptual term that stands for the rhythmic speech acts that modulate by shaping the spoken carrier wave. *Rhythm* here carries the standard dictionary definition: "an ordered recurrent alternation of strong and weak elements in the flow of sound and silence in speech." This fits well with Whitehead's definition of an iterative pattern of contrasting elements, quoted earlier.

As Pinker says, "Every speech sound is a *combination* of gestures, each exerting its own pattern of sculpting of the sound wave, all executed more or less simultaneously" (*LI*, 170). As we proceed, I explain how the trope of "sculpting" in the round is indeed an accurate way of describing the three-dimensional shapes of sound waves. Since there is not an adequate vocabulary in conventional language to describe this sculpting, I use figurative tropes; but I try to demonstrate that physical transformations of a symmetrical syntax can describe this kind of sculpting with precision.

Natural syntax can be broadly defined as a set of transformational acts, following the form of a wave, that arranges physical units into articulated sequences of rhythmically oriented three-dimensional shapes. I describe how these syntactical orientations or coordinates derive from a bodily oriented *deixis*. In previous chapters, furthermore, "consonance" was modified from a Chinese art term to be more widely construed as a "rhyme" between the physical strokes of the artist's body in a medium and the presumed patterns of nature that the strokes reenact. This is an idea of syntactical composition in any medium, and I define the term extensively in this chapter and the next. As distinct from the *concept* of natural syntax, a syntactic *act* is the body's actual modulation of physical strokes, or "gestures" upon the carrier wave, as vehicle. Thus the broad concept of a natural syntax is the array or group of possible transforming bodily gestures as acts of transformation. The acts of natural syntax are the actual transformations that orient speech acts in linear orders. Hence, for me, the anatomy of syntax in language is a group of actual physiological strokes, formed by the lips, mouth, and throat, which pattern the physical order of sounds in the sentence.

Here too I try to distinguish between grammar and syntax in language because the two terms are often used as synonyms. I suggest that syntactical

transformations in the actual speech acts of language are sequential orderings that are not necessarily determined by one's learned grammars, which
have their own conventional orders of syntax. In the construction of a sentence or paragraph, one manipulates certain grammatical conventions to
emphasize rhythmically the rhetorical effect. Yet in any artful crafting,
rhythms are also always physical acts, not just grammatical combinations.

Of course one certainly uses grammar to redirect the orders of sounds;
that is the traditional way of thinking about syntax as the ordering of parts
of speech within grammar. But one does not learn this art of rhythmic sequencing just by learning grammars. Each word in every sentence has its
rhythm, even a prose sentence, but the sentences and paragraphs of poems
self-consciously draw attention to the order, position, and location of word
sounds on the line. Part of my hypothesis is that the linguistic syntax of ar-
ranging the grammatical parts of speech into a rhythmic sentence is overlaid on the physical syntax of shaped carrier waves that come to us in orderly patterns. Recall, for example, the diagrammatic title "Art/Frequency,"
in Chapter 1, where the carrier wave underlies the superstructural Signified.
If the brain is in part a pattern transformer of carrier waves, and if the
brain transforms these significant shapes into metaphorical images, then
syntax in language, as it may have derived from the natural syntax of carrier
waves, is a common inheritance of the brain's development, to which grammatical skills are later appended.

An acoustic theory of symmetrical sound waves is not a new idea, for it
is found in Vitruvius's book on architecture. In a fascinating chapter about
the acoustics of Greek and Roman theaters, Vitruvius describes several ways
that builders designed the gradual rise of steps in the semicircular shape of
amphitheaters. The designs aided retention and amplification for the neces-
sary-to-be-understood parts of inflected Latin words. In Chapter 3, I said that
Vitruvius emphasized the importance of understanding the demonstrable
principles by which architecture is physically explained—how the Signified is
built or carried by the Signifier. Here is his principle of sound carriers:

> Voice is a flowing breath of air, perceptible to the hearing by contact. It moves
> in an endless number of circular rounds, like the innumerably increasing cir-
> cular waves which appear when a stone is thrown into still water, and which
> keep on spreading indefinitely from the center unless it is interrupted by nar-
> row limits, or by some obstruction which prevents such waves from reaching
> their end in due formation. When they are interrupted by obstructions, the
> first waves, flowing back, break up the formation of those which follow.

> In the same manner the voice executes its movements in concentric circles; but while in the case of water the circles move horizontally on a plane surface, the voice not only proceeds horizontally, but also ascends vertically by regular stages. Therefore, as in the case of the waves formed in water, so it is the case of the voice: the first wave, when there is no obstruction to interrupt it, does not break up the second or the following waves, but they all reach the ears of the lowest and highest spectators without an echo (*TBA*, 138–39).

Notice Vitruvius says that sound waves are not limited to a horizontal plane, but rise vertically as well. In Chapter 1 I quoted Trefil and Hazen about sound waves, and they use exactly the same analogy of the horizontality of water waves. According to Granger's edition of Vitruvius, the comparison of air to water waves is a Stoic concept, as is the idea that sounds undulate "spherically" through the air (*V*, 266–67n). For a Stoic example, here is Diogenes Laertius: "We hear when the air between the sonant body and the hearer is struck in spherical waves which impinge on the ears, just as the waves in a pool expand in circles when a stone is thrown into it."[1] Acccording to Vitruvius's account, the moving concentric circles expand and swell outward and upward in stages "by contact" with one another. Vitruvius's aim is to design half-circular rising steps that do not interrupt the flowing waves of speech. However, the acoustic waves do not rise spherically because of the semicircular shape of the steps; instead, the steps are built to allow the spherical waves to rise unimpeded as they carry the shaped parts of speech. Today one would say that the seats should be arranged so that, when filled with people, the spoken air waves would rise "in phase" with minimal interference; otherwise, the reflecting interruptions would make the waves jumble "out of phase" like slosh. One knows experientially that sound rises both vertically and horizontally because one can hear sounds on the top of a mountain shouted from below; one knows too that sounds reflect back in echoes. But here there is presented a Stoic theory of speech acts, which depends upon the three-dimensional shapes of sound waves moving in concentric circles outward and upward. It is based upon the rational understanding that sounds are carried, amplified, and broken up into shapely symmetrical patterns, by waves that move by contact. This ancient theory of speech acts is based upon the syntactical arrangements of sound waves projected in three-dimensional shapes. I suggest that this Stoic concept of spherical sound shapes supports the idea of a natural syntax that serves as a carrier vehicle for Latin grammar. Perhaps now one better understands what Needham meant when he was cited in Chapter 2 as saying that the Chinese developed a wave theory of the natural world that was similar to the Stoic conception.

So far as I can tell, my inquiries into the forms of the wave syntax of natural language most closely resemble the work of the linguist Kenneth L. Pike as expressed in his well-known essay "Language as Particle, Wave, and Field."[2] He wrote in the 1960s that the dominant view of linguistics was one in which linguists study units of speech as if they were static things, like still frames in a moving picture. Words, phonemes, and morphemes are the "particles" studied under this point of view. But from the point of view of actual speech acts, he says, language "is not made up of particles at all!" (*LP*, 135). There are not really sounds segmented into a sequence of particles. Instead, one confronts " a series of waves of movement or of a sound with the peak of a particular wave identifying the place in a sequence where a particular 'segment' was supposed to have occurred" (*LP*, 136).

Pike does not define the acoustics of waves further, so this "peak" might seem intuitive. But later he shows that sounds may not have such smooth peaks (and valleys)—which, I think, could stand for the compression and rarefaction of molecules, respectively. In actual speech acts, he says, not only do sounds fuse "where they bump into one another in the sequence, but the sounds which are 'due to appear' late in the sequence may actually be in part anticipated early in the sequence" (*LP*, 136). In other words, the pattern of syncopation allows the prediction of later sounds by the placement of earlier sounds in a line. Recalling the acoustics diagram in the Introduction (Figure 5), you can visualize the air molecules compressing as they bump and crowd in the longitudinal wave motion. In contrast to the static view of language, in which parts are seen "synchronically" (which is the primary point of view of structural linguistics), the wave sequence is seen as a dynamic study that includes timely phases of speech in rhythmic action. Pike's definition of *field* amplifies the wave view: "Instead of looking at language as a sequence of waves in a single flat wave train, language is viewed somehow in 'depth.' A word is seen as not a part of a sequence alone, but as part of a whole class of words which are not uttered at that particular moment" (*LP*, 140).

Notice that the hypothetical space of the field now requires the dimension of "depth," which shapes the field as a three-dimensional wavy landscape diagram. This "reservoir of potential speech acts" available to the community is for Pike the choice of a particular set of figures that have been featured over and against the ground of unchosen possibilities. His metaphor of a "reservoir" reinforces his earlier metaphor of "depth," lending unjustifiable credence to the idea of "wave" by way of a water reservoir. Nevertheless, the archetype of water held in reserve is a good figure of speech for the idea that Language (Saussure's concept of *la Langue*) is held in readiness as a potential act of speech. For Pike, the "field" includes the wider culture of possible human behavior. The range of language therefore is not just a "code" of instructions in a set, but it acknowledges the "holistic nature of phenomena" (*LP*, 141).

Instead of featuring the holistic term Universal Grammar, I want to study the sharing of a common generative syntax to suggest that its physical basis is more agreeably linked to the moving orders of physical waves from the physical world of transmitting signals. Unlike Pike's "depth" metaphor, I think of a three-dimensional wave as having an actual, not metaphorical, helical shape in the round. A natural syntax is a way of "sculpting" spherical shapes of speech waves into rhymes and rhythms. It is necessary to keep in mind that the speech acts of plosives are always three-dimensional wave shapes that move in a direction and sequential order that one can count as the fourth dimension of time. The physical world of four dimensions always transmits messages via a natural syntax of solid, shapely speech acts. The Dogon were right: speech acts are helicoid shapes. Why helicoid? Recall that in the physical world, there is no space that transcends three dimensions. So when dynamic shapes are cramped into tight three-dimensional spaces, they turn, twist, and fold over into helical shapes, like water hosed into a bucket. The symmetry transformation called "twist" is a physical turn that operates under the constraints of three-dimensional limits. A crowding or chunking in physical space and time will usually move through a series of torqued twists.

The terms *grammar* and *syntax* are often used loosely as if they were synonyms. But as we are seeing, there is a distinction worth preserving: syntax can also define the common orders of physical things as we arrange them in space and time. The physiological transformations of syntax seem to be a modification of natural language associated with the body's moving proportions in space. The actual transformations are innately given by the neural patterns of the previously described "old sequence" that seems to govern the rhythmic movement of all vertebrates. Because symmetry sharing occurs everywhere, the old sequence of ambulatory motion will also follow those rhythmic transformations. To the extent that waves, such as sound waves or light waves, travel in a direction, they have inherent momentum. Symmetry sharing also generates the syntactical shapes in the momentum of the moving sound waves that enact a spoken sentence. It has a physical rhythm, and it may have a patterned *gait*, as in Williams's catty "Poem," quoted in the previous chapter. Although any sentence may have a rhythm, according to the skill of the writer, the skill of rearranging these sounds into rare patterns is the art of the poet. So it is useful to separate the physical orders of syntax from the socially constructed conventions of grammatical composites in a given language.

Among linguists, cognitive psychologists, and other scientists, there is now sharp debate about the possibility of a Universal Grammar and the language instinct.[3] I am not trying to debate the issue, but I do think that if natural syntax follows the physiological form of a wave, and if it is part of the phenomenon called "symmetry sharing," then this kind of primal syntax is part of a neural coding that generates orders and shapes of physical units in a sentence. I do not think any other innate rules need to be posited for a Universal Grammar. I do think that grammars are locally learned sets of rules that arbitrarily fix manners of speech and etiquette, even as those rules change over time according to taste and fashion of prose styles. But the biologically existing shapes and patterns of neural-motor syntax have been with organisms since they responded to light and heat via the wave rhythms of the old sequence. I think that this more inclusive definition of syntax comes down on the *nature* side of the issue, while traditional grammar is *"nurtured"* as a set of conventions developed by generations of users.

However, in much literary discourse about European linguistics and hermeneutics, there still continues the normative concept that language is completely a social construction. So, this reasoning goes, terms such as *nature* and *identity* are only social constructions. I discussed this attitude in "Art/Frequency" when addressing issues of habit and prejudgment. In my usage, *nature* is the part of the wave spectrum that carries a message. By featuring a physical syntax, I suggest that there need not be postulated a Universal Grammar separate from a commonplace syntax of physical strokes within all languages, which follows a wave form of speech acts. I conjecture that these wave forms of syntax are embedded in sequential orders of grammars specific to different languages. It is a syntax of physically derived orders. Hence to fathom a natural syntax is to confront one of the more troubling epistemological questions of the century: how are our symbolic codes related to the physical world? My conjectural answer is that the syntactical transmission of waving forms of speech fits into the presumed order of frequencied things.

Bodily Etymologies of Syntax

The etymology of the word *syntax* privileges order at large, not just linguistic order, and in particular the orderly articulations of bodies. For instance, the first definition in *The Oxford English Dictionary,* hereafter cited as *OED,* is now obsolete: "Orderly or systematic arrangement of parts or elements; constitution (of body); a connected order or system of things." Two passages are quoted, the first from Bacon's *Advancement of Learning* in 1605: "Concerning the Syntax and disposition of studies, that men may know in what order or pursuit to reade." The second is from Edwards's *Demonst. Exist. God* (1696): "This single [argument] from the fabrick and syntax of man's body is sufficient to evince the truth of a Deity." A subdefinition is now also considered obsolete: "Physical connexion, junction." (In the previous chapters I spoke of the Chinese painterly stroke of "junction" as a syntax.) One passage is also quoted from Crooke's *Body of Man* (1615): "Their articulation doth not differ from the syntax or articulation of other parts." Throughout I stress this sense of physical articulation of things in bodily order. The concept of articulation still is used both in terms of verbal articulation and articulate body parts or joints. Within this definition of "junction," the insight of C. S. Smith about dragon veins as junctions can be understood as a natural syntax of anatomical strokes—which makes the strokes of a Chinese painting rhyme with the leading and connecting lines of the earth.

Over time, however, the syntax of a human body as physiological order was gradually lost in favor of a conventional linguistic order. Just as early as these bodily definitions, moreover, the word *syntax* achieved its abiding grammatical denotation: "The arrangement of words (in their appropriate forms) by which their connexion and relation in a sentence is shown"(*OED*). Orderly connection is the common denominator. This historical passage is from R. Cawdry's *Table Alph,* (1613): "*Syntaxe,* construction and order of words." An even earlier quotation from Ascham's *Scholem* also features sequential order: "In learninge farther his Syntaxis, by mine aduice, he shall not vse the common order."

According to the *OED*, the most general definition of syntax as a principle of putting things together in order is also the main feature of syntax as a mathematical term: "That branch of mathematics which deals with the various arrangements of a number of things, as permutations, combinations, and the like." This example is from Sylvester, *Collected Mathematical Papers* (1908): "The theory of groups . . . standing in the closest relation to the doctrine of combinatorial aggregations, or what for shortness may be called syntax." In physics, for example, the mathematical theory of groups is now used to describe the transformations of physical quantities, as in quantum physics, in which symmetries remain unchanged. I previously described the group of symmetrical transformations—*translation, rotation,* and *twist*—as a set of combinations that describe the syntax of a wave form.

In this array of definitions and quotations, the gist of syntax is a principle of physiological articulation not limited to linguistics—a broad compositional principle of orderly articulation, first based upon the body. In everyday conversation, however, the terms *grammar* and *syntax* are often used as synonyms. Even in painstakingly careful distinctions by linguists, the terms are almost interchangeable. For example, see John Lyons's definition of the traditional distinctions between syntactical structure and the inflectional forms of words: "Taken together, syntax and inflection are complementary and constitute the principal part, if not the whole, of what we are calling grammar."[4] For Lyons, syntax and inflection, plus syntax and morphology, are constituents of grammar, but not equivalent with it.

If the primary unit of linguistics is the sentence, then its primary characteristic is its structure of linear order. As Lyons said in an earlier book about the linearity of the sentence, it follows that

every sentence of the language could be satisfactorily described, from the grammatical point of view, as a string (or sequence) of constituents (which we conceive to be words).

As an abstract illustration of what is meant by the term "string" (which is a technical term used in mathematical treatments of the grammatical structure of language), we may consider the following instances:

$$a + b + c + d$$

The plus sign is employed here (other conventions of notation are also to be found in the literature) to indicate concatenation ("chaining" together). The string results from the combination of the constituents, or elements, in a particular order. [5]

In that passage, the root principle of these jumbled tropes of strings, chains, alphabetical letters, and pluses, is an interconnection by linear patterning. The quotation itself is an effective demonstration of mixing sign systems, comprised of different codes that one uses in expository writing to make up for the deficiencies of a language. But because of the very jumble of mixed metaphors, the idea becomes clear enough to proceed. This standard concept of sequential grammatical order lets one draw a distinction between the typical grammatical sequence of subject-verb-object in English as opposed to subject-object-verb in German use. Grammars prescribe the conventional order of words in a language, and syntax describes the sequence. But here I think some linguists take a wrong turn. For some, there seems to be a kind of primordial grammar that generates orders and hierarchies of sentence patterns. Because all sentences have these patterns and because all children articulate the patterns quickly, there seems to be some kind of *Ur*-grammar in every child's head.[6] Suppose, however, that a physical syntax is separable from grammar. Suppose that physical syntax is derived first from the *Ur*-sequence of the carrier wave that has been carrying neural messages in many forms of transmission through the body. Then suppose that the sound groups that govern speech must fit into the patterns of a carrier wave only in certain limited ways, described as symmetry sharing, like the groups of shapes in the Anasazi vessel. Then one could construe a physical order of a limited array of sound combinations that would govern the grammatical relations in their normative orders. A *natural* syntax then may be initially described as the group of physical patterns that shape the sounds of speech into limited symmetrical transformations. In later chapters I describe a Darwinian theory of the sounds of musical tones as a precedence for human speech.

Here I focus on the idea of natural syntax as a physical order of words in a sentence. I do not mean the physical objects to which the words in a sentence refer, such as *doorbell* in the sentence "The doorbell rang." This sets up the contorted issue of physical referentiality. To reiterate a point from "Art/Frequency," the physical world, or nature, in my sense of the term *natural language*, enters the picture not so much in reference *to* but in carriage *by*; that is, nature is the portion of the communication that carries the message by breath, light, or other wave form. By the physical increments I mean those sequences of spoken noises that constitute phonemes, spoken one after another, but still twisted together as the three phonemes that compose the word *tree*. To check my own usage in that sentence, why is the word *twisted* not a metaphor? Because the spoken sound "tree" is a physically shaped, helical wave that is twisted through three space and time; its sound shapes are literally twisted into the transformations that constitute the unity of three phonemes. A breath is not usually thought of as a thing, but it is material, wavy shape, like the Dogon's plosives. Being part of bodily articulation, syntax is—first and foremost—the physical order of the breaths that are spoken and transmitted as air-wave disturbances. Or in reading words on a page, one perceives the syntax as the sequence of light rays that shape the positions of icons constituting letters in an alphabetic code.

There is a fairly important issue at stake here. If there can be found a common "alphabetical" code of syntactical transformations in the physical orders of transmitting waves, then there may be a syntactic unity for the composition of strokes within drawing, music, and poetry. Perhaps there are only a few strokes of symmetrical transformation, a few orderly units of natural composition, as Herbert Read and C.S. Smith assert, such as the patterns of consonance, among those physical shapes that ride waves.

My conjecture about natural syntax assumes that human language is existentially connected to the physical world by physical waves as carrier vehicles. I think that most will agree that organic beings and inorganic media respond differently to periodic oscillations in the physical world. Consider how an inorganic medium such as air reverberates when a bell is struck. The waves carry over from metal to air by symmetry sharing, but the air does not respond expectantly to the clang as a sign. While the air molecules have been disturbed and reshaped into new waves, the new waves have been transformed by organic beings into the natural syntax of a sign.

This existential premise about natural syntax may conflict with the method of those who define *syntax* (or *grammar*)" as being a set of purely formal transformations, a symbolic logic with its own rules, without being encumbered by any relation to semantic considerations. Yet the range of definitions from the *Oxford English Dictionary* provides material for a more ample definition of syntax than the algebraic or logical one premised, for instance, by Pinker in this passage:

> Grammar is a protocol that has to interconnect the ear, the mouth, and the mind, three very different kinds of machine. It cannot be tailored to any of them but must have an abstract logic of its own.
>
> The idea that the human mind is designed to use abstract variables and data structures used to be, and in some circles still is, a shocking and revolutionary claim, because the structures have no direct counterpart in the child's experience. Some of the organization of grammar would have to be there from the start, part of the language-learning mechanism that allows children to make sense of the noises they hear from their parents. The details of syntax have figured in the history of psychology because they are a case where complexity in the mind is not caused by learning; learning is caused by complexity in the mind. And that was the real news (*LI*, 125).

I am not trying to undermine the syntactic method of those linguists who use symbolic logic to parse a sentence or demonstrate how the mind works. But I am trying to show the possibility of thinking about a wider concept of a physically ordered syntax, a complexity there from the start, which generates transformations in the three space of speech acts. For instance, algebra and symbolic logic are comparatively late developments in the history of abstract languages. Algebra is an example of a purely logical syntax. The advantage of algebra as a logical syntax is that the abstract patterns may be used irrespective of a particular number, and they may apply to any numbers, as well as to any physical quantity that may be assigned to numbers.

In the following passage Whitehead distinguishes the difference between the principle of abstract patterns of sequencing in an algebraic equation and arbitrary sequencing in a sentence. He explains that the history of algebra is a development of a technique for representing "finite patterns." First he says that algebra is part of the historical development of language itself. Whitehead shows that algebra was developed in order to be precise about the significance of linear patterns within its equations; whereas language in general is "casual" in its orders of linear meanings within a sentence: "It is true that language strives to embody some aspects of those meanings in its very structure. A deep sounding word embodies the deep solemnity of grief. In fact, the art of literature, vocal or written, is to adjust the language so that it embodies what it indicates."[7] In the next chapter I say more about this metaphor of embodiment in terms of poetics. Whitehead wants to show that, unlike a literary sequence of embodied sounds, and unlike algebra, the sequence of words in a standard sentence is usually composed so that its grammatical sequence is *not* set in parallel to the significant sequence of ideas to be represented:

> But the larger part of what language *physically* [my italics] presents is irrelevant to the meaning indicated. The sentence is a sequence of words. But this sequence is, in general, irrelevant to the meaning. For example, "Humpty-Dumpty sat on a wall" involves a sequence which is irrelevant to the meaning. The Wall is no sense subsequent to Humpty-Dumpty. Also the posture of sitting might have been realized simultaneously with the origination of the sitter and the wall. Thus the verbal order has the faintest reference to the idea conveyed (*MG*, 82).

But in writing an algebraic equation, Whitehead notes that sequential marks on the paper are "instances" of the inverse pattern to be transformed: $x + y = y + x$. I would say that the significant sequence of algebraic marks *reenacts* the principle of inverse patterning to be conveyed. It assumes that the transmutations occur without reference to physical quantity. Its symbolic reversals and inversions occur in a hypothetical space, within which time has no truck. One can translate back and forth endlessly in a conceptual space. It doesn't matter that the letters x and y are rotated into a new sequential order without considering that the latter half of the equation repeats the letters forward, but in reverse temporal order of delivery. The physicality of the syntax has been rarefied and sublimated.

What Whitehead doesn't say, of course, is that the sentence he quoted is part of a famous nursery rhyme; perhaps he assumes that his English-speaking readers know it. Whenever you recite the whole poem from memory or read it aloud, the lines of the sentence *do* take on a poetically significant, non-casual, *physical* momentum in a direction and a sequence. I discuss this idea of a momentum of reading poems in the next chapter. The poetic lines have a significant syntax of spoken syllables, the pattern of which is repeated forward in each subsequent line. The momentum of the sequential order of sounds is all one way, so the thematic impossibility of reconstructing Humpty-Dumpty's original structure is underscored by the inevitability of the physically moving syntactical speech acts. "All the king's horses . . ." The moral of an irreparable sequence is repeated by the sequence of speech acts.

So the nursery rhyme, from which Whitehead lifts the example, proves the exact reverse of what he was trying to demonstrate about the "casual" pattern of physical ordering in a standard sentence. In real space and time, you cannot reverse the order of things and make y revert to x. Algebra works only in a conceptual space. You cannot recompose Humpty-Dumpty's body parts into his original syntactical articulation. "All the king's horses and all the king's men couldn't put Humpty together again." Once you *translate* a physical entity forward, as Monod asserted, you cannot reconstitute it backward in exactly the same format—even if it rhymes forward. Why not? Because the moving forward of one algorithmic unit in time changes its relationships with the rest of its set. In the crowded mass of a wave form, every particle jostles, shifts, and twists. And of course, the onomatopoeia of the name "Humpty-Dumpty" *embodies* his plump, oval, egg-shaped body.

If you have known the nursery rhyme since childhood, you may have seen its accompanying illustration so often that the original intent of the rhyme has been forgotten. You immediately recall the iconic image of an egg sitting on the wall. But the rhyme was originally couched as a riddle, without the illustration, so children were supposed to guess what object might be referred *to*, by way of the verbal description.[8] (The deep structure of the egg is of course the archetype of the primal generative matrix, Eurynome's Universal Egg, the total syntax of the universe in the ancient world.) Probably too, Whitehead recalled that Lewis Carroll, who as Charles Dodgson had taught Euclid at Cambridge, gave Humpty-Dumpty the character of a nonsensical lexicographer. He says to Alice: ". . . *my* name means the shape I am—and a good handsome shape it is, too."[9] It is nonsense to confuse shapes and patterns with words and grammars, but perhaps it is not at all insignificant to rely upon a pattern for the underlying physical syntax of bodies and the sequential momentum of sound waves in speeches.

It is one thing to use symbolic logic to parse a sentence in a hypothetical space and time. It is something else again to assert that a child's brain, which exists in real space and time as part of the irreversible flow of physical events, uses an abstract logical syntax to generate sentence structures. I do not think that I need to know how to resolve that issue within linguistics. Rather, I am trying to substitute a broader definition of syntax for poetics, which describes how the mind generates sentences physically, just as one learns a nursery rhyme or a song, wave after wave, rhyme after rhyme.

But since most of us in literary studies pass quickly from the parsing of sentences—and even more quickly from the physical study of light and sound waves—toward questions of signification, let me stress the a priori nature of the perception that carries the syntactic signal. If I see a tree and wish to identify it as a type of tree to my partner in a dialog, then I translate the patterned light waves that slant off all points of that tree onto my retina, and I also assume that my partner will receive the waves in a like-minded way. We share those symmetries in common. Maybe I'll say, "Is that a sycamore?" Maybe she'll respond, "No, it's a buttonwood." Because we share a learned lexicon, we can give the tree different names as we discern the patterns of the leaves; the locations of the branches; the texture of the trunk; the shapely contours of the whole. The important point is that we do not see the light waves as such; instead my partner and I commonly sense the light waves and perceive them as if they constituted an object out there in a perception space. As John Muir assumed when he described for us the symmetries of a silver fir, we expect that we are seeing and commonly sharing the same symmetries of wave shapes. The sensed waves whose shapes carry the patterned message are suppressed in the commutation into meaning, but they are nonetheless shared—"shared" in the sense that the light waves break up and reshape in similar ways for the two perceivers.

The physical impulse of the sense impression is usually suppressed in the move to signification. Meaning seems to be an adventitious sublimation of the patterned forms of acts and gestures, as imposed by learned lexicons and grammars. Just as we "see" through the picture plane of an oil canvas, seeing the oil-paint strokes as images of a person in an oak grove, so too we see through light waves, perceiving them not as rays but as word images. To infer waves as the sign of an object, like a tree, is to receive the physicality of the sense perception and transform it into a lexicon and grammar. When we watch a movie, we watch real moving light waves projected onto a screen. But we see those real optical waves, not as if they were contours and patterns of light, shadow, and color on a two-dimensional screen, but as if they were slanting or refracting through the screen into a perception space of illusory objects and people. As with the movie screen, so it is with all physical impressions—we see through slanting physics to the implied meaning. All patterned thinking, by way of a syntax of orderly signs, seems to be refractive or slanted thinking—by means of the physicality of the patterned waves that sequentially carry the shapely message.

As noted in the Introduction, these light waves are not illusions. They are real patterned wave forms of transmission. As Bertrand Russell said years ago:

> In the first place, images are not "imaginary." When you see an image, perfectly real light-waves reach your eye. . . . We have here an example of a general principle of the greatest importance. Most of the events in the world are not isolated occurrences, but members of groups of more or less similar events, which are such that each group is connected in an assignable manner with a small region of space-time.[10]

For Russell, the theory of groups, symmetry, and the gravity of space-time were linked by Einstein's new conservation principles. If a mathematical group can define a syntax, then a natural syntax can be defined by conservation principles. In the Conclusion, I say a bit more about the connection between symmetry and conservation. When a formidable philosopher like Russell speaks of an "event of the greatest importance," one thinks again. I suppose that Russell is here contesting the vexed question of Hume's skepticism about identity. How, Hume asked, do we infer from one event to another? Only by way of habit or custom, he said, for he averred that there is no necessary causal connection to be found between isolated events A and B and C. I ventured a brief excursus about habit and prejudgments in Chapter 1. Now, however, from the point of view of the new physics of Einstein, the rhythmic groups of light waves that constitute an image are *really patterned* in a small region of space-time. The group is not just socially constructed by conventions. The neighborhood is a spatiotemporal composition.

If light imagery is structured in groups, patterns, or chunks of shaped waves, then the matter of greatest importance is that we now know that very fact about physical *patterning* of groups in locales. The identity of any physical composition is not mere custom or habit or significant fiction; the linkage between events A and B and C is not just a customary order of symbols. Our brains know how to transform those shapes of light into our own recognizable groups of images. If natural syntax is constituted from the shapes of light waves, then the rhythmic patterns of language reenact patterned groups that move beyond skepticism toward probability.[11] Again, in the next chapter, I suggest how the concept of a grouped pattern of thought leads to an important idea about inference in reading poems and patterned inference in thinking at large.

So the customary suppression of the physical syntax, which carries the message as light wave or sound wave, and the favoring of the arbitrary meaning in a lexicon or grammar is a kind of sublimation. As I noted in "Art/Frequency," Karl Marx favored the inversion of the Hegelian *Aufhebung* in a similar way. Marx asserted that light waves were real. If, however, the physicality of the sense impression is usually suppressed in the move to adventitious signification, then one can understand why syntax as the measure of physical order is habitually discounted in favor of grammar. So does Hume go by the boards? By no means. Custom and habit and training—all the social constructions of reality—come into play in the *grammar* of the question of communication, in the learned associations between seeing the image of a light-reflected object and saying its name. These customs are the rules of various grammars. This transformation of seeing into saying is the arbitrary but necessary habituation of lexicons and grammars. But recall that *habits* are in themselves electrochemical patterns of neural circuits. When I have learned something, the learning circuits shut down or inhibit other possible neural circuitry in favor of the sublimated habitual circuit.

This sublimation of physical syntax in favor of a transcending grammar can transform meaning into an indeterminate mystery. Without an accompanying physical series of sense impressions about a transmitting order of thing-like waves, the social constructions of grammar remain the only order of the day. For instance, in an essay instructively titled "The Common Language of Science," hereafter cited as CLS, Albert Einstein spoke briefly about the beginning of language: "The first step towards language was to link acoustically or otherwise commutable signs to sense impressions. Most likely all social animals have arrived at this primitive kind of communication—at least to a certain degree."[12] The crux is to understand how sense impressions are linked to commutable signs. As I suggest above, by way of Russell's passage, a *patterned physical* syntax is a demonstrable commutation group.

Otherwise, Einstein writes, confusion ensues in the further development of language: By "the frequent use of the so-called abstract concepts, language becomes an instrument of reasoning in the true sense of the word. But it is also this development which turns language into a dangerous source of error and deception. Everything depends on the degree to which words and word-combinations correspond to the world of impressions" (*CLS*, 108).

If natural syntax is the articulate order of shaped patterns in the helical transmission of breaths or light images, then it participates in the world of sense impressions. In keeping with Einstein's use of "common language," I am attempting to reconstruct the etymological definition of syntax as a common physical order of shaped sense impressions that carry the message for a grammar. In this usage, everyone is neurally equipped with a *common syntax* of shared symmetries for transforming sense impressions into semiotic patterns. It is commonly shared to the degree that we use the same parts of our anatomy to transform wave patterns into communicative images, and it is commonly shared in the sense that wherever one looks in nature, one finds symmetry sharing.

The point so far is that the physical world enters language use, not in reference to an object, such as *to* the doorbell, but in the transference of the signal, the physical medium that transmits the sentence over sound waves or light waves. *Doorbell* is a word chosen from a normatively constructed lexicon, *la Langue*. Most people would agree that the word itself is an arbitrary construction that refers only symbolically to a set of other words in the English lexicon and grammar. Its meaning is an indefinite set in a conceptual space. Even the word *I* in a sentence such as "I rang the doorbell," is socially constructed by the grammatical prejudgments of the English language. Let us grant that "I" is a sign of the times. But let us not discount the physical body that carries the sublimated message. The articulated syntax of physical waves is transformed and socially sublimated by means of a lexicon and grammar, but to say that is not the same as saying that some transmitting wave is not impulsively operating as force, energy, dynamics, power, or whatever word might be used to signify the physical order of the syntax.

The Solar Constant

Radio waves and light waves are periodic frequencies upon the electromagnetic spectrum. Some names of familiar portions of waves on the spectrum are microwaves, X rays, and radar. (Sound waves are not usually grouped as part of the electromagnetic spectrum because they move slower than the speed of light.) All of these impulses have frequencies that the human body cannot discern without the aid of instruments to receive them. And yet they pass through our bodies as if through porous media. Although they make up our unwary environment, like radioactive dumps, our bodies still respond to their presence. Their periods of impulses affect organs even though we are unaware of them as physical transmitters of impulses. Most creatures on earth are attuned to the slanting rays of the sun. Only those in the depths of the earth and the sea respond to different heat frequencies.

The anatomy of eyes in sockets has evolved to focus these angular rays. The linguist Steven Pinker says that the separate evolution of about forty kinds of eye forms among different species suggests that there is a basic "eye module" (*LI*, 349). In the previous chapter I discussed eyes, together with other ball-and-socket joints, as basic models of symmetrical transformation in primal art forms. I suggested that, among primal cultures, they were drawn as part of a body syntax that included other groups of joints, considered as junctions of the Four Quarters. If you premise light waves as having character as they bounce off objects, then perhaps you do not need an a priori module that nature will follow. It is true that once a model of a circuitry like the old sequence gets set up within a vertebrate species or among related species, the pattern will be passed on, for natural compositions seem to be old models that work over time through chance adaptations.

The point is that different kinds of eyes have evolved over species far removed from one another, not just among vertebrates. If brains are in part pattern transformers, then they will evolve to receive wave frequencies in just a few forms, which are symmetrically composed by the solar constant of light and immense gravity. The epigeneticist C. H. Waddington says that one does not need to premise fixed modules for biology.[13] He thinks that in biology there are no basic units from which other proportional relations are derived. I mention this difference of opinion about premises because literary theorists might wish to remain skeptical about thinking in terms of absolute modules of perception. Similarly when I use the word *archetype*, I do not mean that there is some sort of Platonic preplan that makes everything evolve according to an ideal formula. Just because there are inverse-square laws that describe many parts of the physical world, that doesn't mean there is a primordial inverse-square law that is doing the governance. That kind of assumption retrofits a symbolic logic back upon the tendency of things in the physical world to balance out symmetrically in three space.

Think of the object that we identified as a tree, perhaps a buttonwood. The round shape of the eye's lens bends or refracts most of the rays leaving the tree in our direction. Many of the rays converge at one point on the retina to shape the image; "the eye's lens is said to *focus* the incoming rays."[14] The lines of the light are bent by the arc of the lens to focus the rays upon the retina. The study of the anatomy of the eye, as well as the arcs and angles that structure the light rays, are combined in the discipline called "geometric optics." These are the protogeometries that I have also been describing in terms of symmetry theory. The angles and arcs of this geometry, drawn sublimely by Blake in "The Ancient of Days," derived in part from the circular shape of the eye in its rotating ball-and-socket joint, plus the motion of light in a straight but slanting line, lend credence to the premise that anatomical perception is based upon the implied presence of something to be perceived.

This is not naive realism, but evolved realism. Norman Kemp-Smith coined the term "natural belief" for the inescapable assumption that the act of perceiving implies a tacit awareness of some "mind-independent reality."[15] As Martin Gardner said in another context, "Electromagnetic waves are ideal for giving the brain an accurate 'map' of the world" (*AU*, 62). His carefully constructed sentence might be taken as my basic premise about the carrier waves that map the world with a physical syntax. Suppose I write the sentence "The doorbell rang three times: two longs and a short." This is a sentence about a group of sound waves whose pattern is one of alternating repetitions and contrasts translated forward in sequence. Although the sentence is about the physical transmission of sequences of rings by *sound* waves through the air, the syntax that carries the message would be the sequences of *light* waves that the reader commutes into images of letters as they were written in their orders. Each letter has its characteristic light wave, but we see them as images. *Natural syntax then may be broadly defined as the conception of physical order derived from the sequential pattern of shaped waves being transmitted on the part of the electromagnetic spectrum chosen to transmit a message.*

Our anatomies have evolved as self-tuning systems that are primally attuned to light waves from the sun. Perhaps the previously quoted passage by Langley states the matter too reductively: "Since, then, we are children of the sun, and our bodies a product of its rays . . . it is a worthy problem to learn how things earthly depend upon this material ruler of all our days." The constant measure of all our days is the sun's periodicity. That periodic measure which makes it a constant is the speed of light at 186,000 miles per second. This speed is the constant of the electromagnetic spectrum. No wave on the spectrum travels faster than this constant speed of light. Even though we are not explicitly aware of it as speed, much less a constant speed, the light of the sun is a stable measure of the physical order of sequential patterns. Natural syntax, in this sense, can ultimately be understood as a conception of physical order derived from the constant sequential patterns of light.

Deixis

I stress the associated definitions of orderly syntax and bodily articulation because many people have studied the roots of language in anatomy. In fact, in the early years of linguistics as a discipline, those who studied the structure of the language sometimes transferred morphological terms from early chordate anatomy.[16] For example, Goethe as poet and natural scientist studied the forms of both language and plants and animals. As I mentioned in the Introduction, Ernst Cassirer showed in *Philosophy of Symbolic Forms* that, for Goethe, form per se is the medium of biological metamorphosis over time. To form is to *trans*form in time. Goethe apparently thought that poetry and natural science were related through the "comparative method" of form. I have been discussing a syntax of comparative forms, a few forms of composition seemingly shared among several arts. In another work, Cassirer wrote, "This peculiar intermingling of being and becoming, of permanence and change, was comprehended in the concept of form, which became for Goethe the fundamental biological concept."[17]

Also in *Philosophy of Symbolic Forms*, Cassirer proposed a body-based theory of language that is worth recalling. In an early chapter called "Language in the Phase of Intuitive Expression—Space and Spatial Relations," he discussed a concept of space and time that developed from a self-awareness of the body's location in space: "Once [a human being] has formed a distinct representation of his own body, once he has apprehended it as a self-enclosed intrinsically articulated organism, it becomes, as it were, a model according to which he constructs the world as a whole. In the perception of his body, he possesses an original set of coordinates, to which in the course of development he continually returns and refers—and from which accordingly he draws the terms which serve to designate this development." (*PSF*, vol. 1, 206–207). In the last chapter, I showed the body's coordinate fit with the four cardinal directions of the sun. Later, I show how "cardinal" coordinates can be seen as a model of body coordinates for designating patterns of vowel sounds as parts of syntax.

Some of the standard body-based coordinates are inside-outside, before-behind, above, and below. Skin surface, head and tail, and back and belly are coordinates that can become increasingly abstract. For instance, head-tail can become front-back, and then anterior-posterior, or before and after. They retain a basic asymmetry between head and tail that, when transferred to anterior-posterior, can be applied to the asymmetry of time, for in time, there is no exact repetition of an element from an anterior to a posterior place in a group or set. Translation is always translation forward with a slight difference. But these body-based locations remain coordinates, and, according to Cassirer, they articulate locales in space that eventually became parts of grammar. Here, for example, he describes the development of grammatical cases:

> In general, we find that Indo-Germanic case forms served originally to express spatial, temporal, and other outward intuitions, and only later acquired an "abstract" sense. Thus the instrumental was first the "with" case; when the intuition of spatial *togetherness* passed into that of accompanying and modifying circumstances, the case came to indicate the means or basis of an action. From the spatial "whence" the causal "whereby" develops, from "whither" the general idea of aim and purpose (Cassirer, vol. 1, 208).

Teleology, purpose, and ends themselves, can be traced to the case of whither. Cassirer says that this bodily assumption about spatial and temporal orientations is called "the localist theory of cases" (Cassirer, vol. 1, 208). In the previous section, I labored over Russell's important principle of local groups of patterns.

Cassirer summarizes the priority of this localist theory about spatial relations:

> And this would make it probable a priori that language can proceed to the expression of purely "intellectual" relations only after it has detached and as it were "abstracted" them from their involvement with spatial relations. The nominative represents the agent of action, the accusative or genitive designates its object, accordingly as it is entirely or partly affected by the action. . . . Indeed the more we consider those languages which have shown the greatest fertility in the formation of "case forms," the more we become convinced of the priority of the spatial over the grammaticological signification (Cassirer, vol. 1, 209).

By way of this localist priority, I am also suggesting that a body-based syntax in real space takes priority over the developments of grammar, which operates according to abstract rules that have developed over time.

In modern linguistics, this localist grammatology of space and time has been subsumed into a more inclusive term that linguists call *deixis*, from the Greek for "pointing" or indicating. According to Lyons, it is a technical term used to "handle the 'orientational' features of language which are relative to the time and place of utterance" (*ITL*, 275). Lyons says unequivocally, "Every language utterance is made in a particular place and at a particular time: it occurs in a certain spatiotemporal situation" (*ITL*, 275). Let us agree that the grammars of languages developed culturally within a history of language groups, the effect of which was to refine what Cassirer called the "coordinates" of space and time until they were formalized in the abstract rules of cases and tenses.

If one supposes that every utterance is indeed peculiar to a place and time, then it is crucial to be clear about what kind of pointing or indicating is being discussed. Because I want to separate a physical syntactical order from grammatical order for the nonce, I want to use P. F. Strawson's concept of a deictic space in order to describe the idea that the space and time used grammatically in the subject and verb of a sentence, the combination of which is called "predication," is not to be confused with physical space and physical time.[18]

The sentence "Socrates was healthy" asserts something via a subject and predicate that is spatially factored according to Socrates' existence in the space of Athens, and it asserts something about time in the use of the past tense. This predication exists in a proposition space and time maintained by the grammar of the sentence. Recall Whitehead's idea about the casual/not causal order of units in a standard sentence, and then notice the sequence of grammatical units of my example. The word order of the words, which convenes the word *Socrates* to be anterior to the word *healthy*, has nothing to do with the space and time of the assertion about the man's condition in Athens at a point in space and time. For Strawson, the overlap of the conceptual space of a logical proposition upon physical space is a "metaphor":

> Thus one may say: every general concept occupies a position in *logical* space (or in *a* logical space), a position which it can wholly share with no other. This is not to say that one general concept cannot be wholly contained *within* the logical space occupied by another or wholly contain another within its logical space. And indeed both these relations can hold. The metaphor has some merit. Mimicking a symmetry, it proclaims an asymmetry. Attempting a parallel, it reveals a divergence. For a logical space is not a space [his italics] (*SP*, 17).

Put in terms of my working distinction between syntax and grammar, the grammar of the sentence imitates a symmetry between its proposition space and the long-gone space of Athens, which is only metaphorical because its subject-verb combination proclaims an asymmetry. Furthermore, the grammatical combination of a spatial noun and a temporal verb imitates the actual space-time of curved-wave speech acts.

According to this effort to redefine syntax as an order of physical events, the perception of the real world enters the sentence not in reference to Athens or a healthy body, but in the act of reading the order of the slanted light rays that carry the images as if they were patterned letters, which we interpret by way of their Aristotelian positions, orders, and shapes: AN, NA, H, I. Or if you read the sentence aloud, the real space is the order of the "sculpted" phonemes, transformed into morphemes, words, and grammatical orders. In a standard sentence, as Whitehead noted, this physical order is casual or trivial. But in a line of a poem, it is a crucial pattern. That crucial pattern of shaped thought in poetry is the thesis of the next chapter. Like the leading lines of Ruskin's landscape painting, the lines of a poem are governed by the rhythmic shapes of the natural syntax.

So far you may be assuming that the physical order of the syntax is all one way, but I always mean to imply a listener in a dialog. In a dialog between two people—speaker and listener, writer and reader, sender and receiver—one is always in an inverse relation of mutual transformations: the speaker or writer translates from logical concepts into chunked groups of physical units along light waves or sound waves of the carrier frequency in one way, while the listener or reader transforms the real spatiotemporal waves into *spatial* images and concepts in an opposite order. For the sender, it is first the Signified and then the Signifier; for the receiver, it is first the Signifier and then the Signified.

In that context, it is especially useful to recall Mikhail Bakhtin's idea that a real utterance in speech is never isolate; it always occurs in a reciprocal dialog, either explicitly or implicitly.[19] For instance, the sentence about Socrates may bring to mind the tribunal's order that he poison himself. For another example, "The doorbell rang" implies that someone rang it, as an announcement of an impending dialog. It can also take on literary portent as an imminent plot. Doorways are always liminal. If a large part of literature is about foreshadowing, then this sentence announces a liminal moment. The grammatical *deixis* is widening to be part of the hypothetical space and time of a literary plot. For example, in the 1960s, Rex Stout used the sentence *The Doorbell Rang* as title for his warning about the impending threat of J. Edgar Hoover's witch hunting and ominous invasions of privacy.[20] The idea of utterance in a dialog is not necessarily that of an ideal community of inquiry. One group's ideal community may be interpreted as a dystopia by another group or an outsider. As with Socrates' inquisitors and Hoover's FBI, the utterance in dialog may be part of a vindictive communal plot in space and time. Community is an ideal concept in the abstract, toward which we may work in a real locale, but its ethics are measured existentially in physical action upon the bodies, minds, and speeches of others who may or may not wish to be part of the communal dialog. Stout used the sentence as an introduction to a conspiracy theory, arguably one of the main literary and film plots of the decades since the 1960s. So a sentence about a doorbell may be the beginning of a literary narrative, or it may introduce a theme that scripts the ideological history of a political period.

Spatiotemporal Coordinates for Semantics and Syntax

If the defining presupposition of spoken syntax is the physical order of plosives, and if it can be separated from the proposition space of grammar, then syntax and grammar can be brought together again by poets into splendid rhythms, as we see in the next chapter. (In the last chapter, I quoted Williams's poem about the adept convergence of physical syntax with his own singular use of an American grammar, not English.) This broad topic I defer until we think through the anatomical coordinates of front-back, inside-outside, and above-below, as body anatomies that develop into concepts: front-back distinguishes the concept of sequential order; inside-outside defines the concept of a solid shape distinct from the surround; and up-down distinguishes the coordinate of vertical position or location. As I mentioned in the Introduction, *order*, *position*, and *shape* are coordinates used by Democritus for first describing any atom.

Consider the verticality of up-down, the first measure of bilateral symmetry. As soon as these categories are used in language, of course, they become semantic and grammatical indicators. Melissa Bowerman has extensively compared and contrasted language acquisition among children who have different primary languages:[21] Much of her work involves the spatial aspects of pointing words, because, as she says, space itself is a paradigm for matching words with non-linguistic concepts such as the topologies of handling or manipulating, as well as the visual thinking of pictures. For example, in studying the spatial preposition *on*, she lists ways that the expression is used in English, and she represents the expressions with line drawings of objects. Here is her list of expressions that are scaled in an implicit order from objects *on* something to objects *in* something: cup on table, bandage on leg, picture on wall, handle on door, apple on twig, apple in bowl. These phrases exemplify certain kinds of physical morphologies, and she arranges them so as to match the preceding list of expressions: "support from below" for cup on table, "'clingy' attachment" for bandage on leg, "hanging over/against" for picture on wall, "fixed attachment" for handle on door, "point-to-point attachment" for apple on twig, and "inclusion" for apple in bowl. She says there is a gradient of physical support exemplified in the objects linked by the prepositions in the list; all the objects are supported in some way. In all the listed objects drawn in a topological space and spoken in semantic space there is a tacit unity, and it is gravity. The objects and the prepositions tacitly presume a support against the physical pull of gravitation, even as gravitation tacitly defines the symmetrical axis of a vertical pull up-down.

We always protect against Humpty-Dumpty's fate of syntactic disintegration. Because we all sit metaphorically "on" a wall, we try to conserve our symmetries for as long as possible. In other words, even though languages may differ in the ways they handle such prepositions, they all seem to account for a common experience of space in nature—things fall down, like Newton's apple from a twig, and they fall down and break, unless they are supported or symmetrically stabilized, like Humpty-Dumpty. But the principle often remains tacit—not unconscious but tacit—until one studies the physicality of a thing or a wave. Gravity makes for the first distinction of up-down verticality and bilateral symmetry.

If semantic *deixis* depends upon the body's orientation in space, as it aligns with the effect of gravity *on* objects, how might these coordinates help to bracket the physical waves of the transmitting medium? It is easy to see that letters on a page are characterized by order, position, and shape. In any dialog based upon the alphabet, we perceive essentially the same shapely patterns of light rays in order to distinguish the letter A. As a community of readers in an implicit dialog, we perceive essentially the same patterns of light rays that let us distinguish between the letters A and N. Although the commutations are part of a socially constructed code that we learned as children, the shapes themselves are physically induced each and every time we perceive them anatomically. We tacitly assume that anatomical commonality in every dialog.

It is easy to see how written letters could be defined physically as light waves being shaped, positioned, and ordered into sequential forms that would be grammatically induced by the conventions of Greek and English languages. But what about spoken words? The phonemes that we project through the air via sound waves are patterned three-dimensional shapes of phonemes, which we read as morphemes within the conventions of English grammar. They take their shape, their position, and their order in a sequence by way of the grammar, but the concepts of sequence, order, position, shape, and motion are not conditioned by the grammar. The breaths in vocalized order are carried by the direction of the sound waves.

The Oral Round of Four Directions

Is there a tacit structure of shapes, positions, and orders that surfaces only in speech acts? Consider that the structural linguist Roman Jakobson diagrams vowels neatly in a pair of dyads that localize sounds in the mouth cavity (vol. 8, 1988: 86):

	front	back
high	i	o
low	e	a

But I want to avoid the use of binding dyads. Instead of being caught in a dialectic between opposites, think of rotation through adjacent areas, as with the model of Isidore of Seville (Figure 15). Think of these locations of vowel sounds as points in a vocal round or rotation group. Sound the vowels and heed their locales so their cardinal points can be located as physical signs in the mouth. John Lyons lists three phonetic "dimensions" for the "cardinal vowels" that define "the total configuration of the oral cavity." He means *real* physical dimensions of three space, and the vowels are really *cardinal*, as they are located in the cardinal directions on a compass rose:

1. Close (or high), like *i* and *u*, where the jaws are held close together, in opposition to open (or low), as *a*.

2. A Front vowel, where the tongue touches in the front of the mouth, like *i* and *u* in opposition to a Back vowel like *a*.

3. A Rounded vowel is made by rounded lips, like *u, o*, versus Unrounded vowels, like *i, e*.[22]

These vowel sounds are confined to real locales in the three-dimensional space of a rounded body cavity. Speech for Lyons is truly a *poly*-dimensional action. They are true phonemic commonalties, for everyone's mouth cavity is shaped with a similar anatomy. Vowel sounds are commonly structured physical carryalls in all languages because every human has this mouth structure. The sounds are projected; they are *ex*-plosively impelled by breaths. The rhythmic motions of sound waves, reshaped by the formal turns of syntax, are projected through the air by the physiology of throat, mouth, and tongue. They must follow the form of the wave of air waves, so that the vocalized morphemes are literally superimposed upon the air waves of the speech act. The speech act and the syntax are usually modulated into the meaning of the wave by a locally applied grammar. As we learned in the Introduction by way of Sobel's illustration (Figure 5), this kind of *interference theory* of meaning comes about by the modulation of three waves: the carrier wave on the electromagnetic spectrum (such as the transmitting radio wave), the coded wave signal, and the waves of sounds with varying amplitudes that shape meaning.

Can Jakobson's diagram of the location of vowels in the mouth cavity be combined with Lyons's orientation of cardinal vowels? Can they be considered as a rotation group of symmetrical invariants? What if we diagrammed these cardinal vowel sounds as if they were four cardinal points within the compass rose of the mouth? Imagine these cardinal vowels as rotation points on the Great Round of Four Directions, studied in the previous chapter. To check your orientation, chant or project aloud, "Fee, Fi, Fo, Fum—I smell the blood of an Englishman." The first four syllables—*e, i, o, a*—box the aural compass counter-sunwise, widdershins, for it is a bad incantation, said by the man-eating giant in "Jack and the Beanstalk."[23]

These cardinal vowels are the invariants that remain at their locations in the oral round of the mouth while the lips purse and the tongue twists them with consonants into a syntactical series. The opposites—up-down, back-front, round-un-round—which diagram the aural sphere, are not reflectively symmetrical. The symmetry of this model is rotational through three-dimensional space of the mouth cavity. My point is that the aural round is a spatial, three-dimensional concavity, but the speech act is primarily temporal, being a before-and-after pattern of an ordered wave series, composed with explosive and in-taken breaths. Because the speech act modulates and shapes the sound wave, the rhythmic vehicle is a spatial-temporal *twist* of a helical wave, derived from the rotation of sounds and the translation of the series, sent forth by the physical momentum of the curvilinear waves. Because the amalgam is a spatiotemporal weave of sound and sense, linguists such as Jakobson speak of the syllables as if they were wave forms that "peak or crest" or "slope and slant" (Jakobson, *SSL* vol.8, 1988: 87). Speech sounds and oceanic images are identical because of the serried formal motion of waves in frequencies. This is the old sequence, for waves run through all archetypal bodies.

Next, try to align this set of cardinal vowels on a compass rose with the rest of the associations that have been made with the Four Sacred Directions. Where I want to be first on the compass rose is at the station of the Bright East, on the right-hand side, facing the rising sun, that is, at the beginning of an allegorical round. Strange to say, sound and sense line up as in the vowels in "bright east."

In the previous chapter, I noted that colors have traditional associations with the Four Directions on a compass rose. In the Chinese diagram, yellow is located at the center, as a particularly reconciling human quality. But in most traditional calendrical roses, yellow is located with east and the choler of the rising sun. For instance, following Jakobson, Lévi-Strauss mentions the synaesthetic associations, among primal cultures, of phonemes, words, and "calendar terminology."[24]

Synaesthesia, the interconnections of the different senses, has been extensively studied by linguists since Wolfgang Köhler first outlined the fundamental role of "light-dark" oppositions in both vowel and consonantal patterns (Jakobson, vol. 8: 194). Note that I am still following a sun rotation of light to dark, from dawn to dusk. Also, a light-heavy opposition is a kinesthetic sense, similarly based upon the sun, the up-down direction of gravitation on earth.

Although I cannot take up that topic systematically here, I can at least suggest a physical reason for the association of bright and east. Jakobson notes that the great art theorist E.H. Gombrich played synaesthesia as a structuralist parlor game, with the dyads "ping" and "pong" (*SSL*, vol. 8: 192). Is ice cream a ping word? Soup? Watteau? Rembrandt? So *i* is brighter than *u*, he thinks, for reasons of binary opposition. Jakobson summarizes this kind of structural linguistics:

> The hypothesis that light-dark is a universal attribute of all senses is constantly being tested in new domains. More and more the continuing inquiries into the inner organization and grouping of colors reveal a concrete coherence between speech sounds and colors and give rise to the thesis that sensation should be described in terms of polar oppositions (*SSL*, vol. 8: 195).

What is the universality here, if any? If one recalls that the polar oppositions are points of view, there for convenience, like the North Pole or the tropic of Capricorn, and that the vowel sounds really are located by the mouth and the tongue in four cardinal directions around the three space of the mouth cavity, then one can move beyond the polar reifications of structural linguistics to a commonplace, symmetry-based syntax that seems to reenact rounds as tropes. These aural locations represent a kind of prescience of physical symmetries in three space. For example, one might say that "scarlet" is a pong word. But what if scarlet were associated with the westward setting sun, with imperial maturity? "Scarlet" is pronounced so that its location is low and back on the phonetic round, and it is associated on the dark turn of the Great Round.

The color scarlet is a clue to what must be the physical reason why light and dark, sound and sense, are worldwide synaesthetic associations. Again, the reason is based upon wave frequencies as the universal carryall. Light and sound are two extremes of frequencies that bracket the electromagnetic spectrum. The reason for this kind of synaesthesia may be the Doppler effect. It is a relative motion of both light and sound. Remember the poetics of a train whistle? As it whistles and comes closer to you, the pitch of its sound is raised, to an *i* and an *e* of "shriek," because the train's speed makes the frequency impulses reach your ear at a quicker rate than they would if the train were standing still. When the train speeds away from your vantage point, its whistle pitch is lower because the frequency reaches your ears at a slower rate, as in the long nostalgic *o* of "hoot." As Hank Williams sang, "Hear that lonesome whistle blow." The Doppler effect works relativistically for light waves as well: If the train's searchlight were moving toward you at a high rate of speed, it would appear white-blueviolet. If it were receding from your vantage point, and if there were a searchlight on the caboose, it would appear to be scarlet red. So the synaesthetic pairings of light and sound line up with the relativity of the Doppler effect. In their sonar systems some bats apparently use a kind of Doppler effect to measure their relative distance from moving objects.

Tongue Twisters

These shapes of sounds in the oral round of the mouth are nicely detected in tongue twisters. "Peter Piper picked a peck of pickled peppers," and other silly tricks, were collected in anthologies of nursery rhymes. But they also demonstrate how a slip of the tongue in the phonetic order of syntax, which carries the message, can change the lexical and grammatical meaning. "How many sheets could a sheet slitter slit, if a sheet slitter could slit sheets?" A rotation of the syntactic phoneme from *sl* to *sh* causes embarrassment, especially for a child, as the grammatical order remains invariant but the meaning changes. This is of course trivial stuff. But knowing how to do a simple act is often confused with trivial knowledge. Here is a less trivial example. The most torturous tongue twister that I have found is self-referring, as it asks readers to parse the differences between the order of syntax, the order of grammar, by way of the underlying order of twists—the over-and-under topology of twining a string, and the necessary combination of four-dimensional shapes in phonemes:

> When a twister a-twisting will twist him a twist,
> For the twisting of his twist, he three times doth intwist;
> But if one of the twines of the twist do untwist,
> The twine that untwisteth, untwisteth the twist.

> Untwirling the twine that untwisteth between,
> He twirls, with his twister, the two in a twine;
> Then, twice having twisted the twines of the twine,
> He twitcheth, the twice he twined, in twain.

> The twain that, intwining, before in the twine,
> As twines were intwisted, he now doth untwine;
> Twist the twain inter-twisting a twine more between,
> He, twirling his twister, makes a twist of the twine.[25]

This group of instructions about the consequences of twisting and untwisting twine in various sequences was written by John Wallis, the mathematician whom Newton made famous by giving Wallis credit for teaching him how to begin thinking about a new form of symbolic notation, the calculus. Wallis was also a grammarian. Wallis was thinking through the conundrum that the sounds and sense of speech acts seem intertwined with the physicality of twisting twine. Twist is the simplest act of symmetrical commutation that makes the shape of a wave in three space, and the twist of phonemes is the most basic speech act, as in the lyric twist of the patterned sounds in "Lauralee." Any modulation of a carrier wave by the superimposition of a phoneme is in reality a reshaping of a twisting helical wave. Whether this weaving or twining is just metaphor, or is in fact an archetype of transformation, is an issue to be tested in the next chapters.

Order, position, and shape are *static* descriptions, but the stasis is just an assumed quiet; in reality everything is in moving proportion. Translate, rotate, and twist are *process* transformations that answer Aristotle's call for motion, deictic actions derived from body coordinates. *Translate* is a front-back repetition forward. *Rotate* is an up-down inversion. *Twist* is an inside-out inversion, or topological fold, or wrap, or any action that turns back upon itself through another dimension. These terms are grammatical verbs, but their vocal commutations are based upon anatomy. They depend upon a group of body-based coordinates that orient us in a proposition space.

They are also processes that can govern the motions of groups of related sounds in an articulated series. The terms *translate, rotate,* and *twist* are lexical of course, as are their appearances in a grammatical sequence: "Translate and rotate one long ring and then two short rings into two shorts and then a long." That sentence is in the injunctive mood. In a dialog it expects a responsive action. So the nervous system understands and rotates the relative positions of ring duration in the injunction. In response to the directive, one's dialog partner actually rings the doorbell three times, first by ringing shortly two times, and then by extending the signal into one ring of longer duration. But the action of ringing the doorbell is of course not the same kind of action as directing someone to ring the doorbell. The act of ringing ratifies, via Einstein's commutation of sense impressions, the articulate sentence that is spoken in a dialog. The act in real space confirms the meaning of the sentence spoken in proposition space. Both the grammatical speech in proposition space and time, as well as the responsive act of ringing the doorbell, depend upon tacitly understood physical sequences. As I noted about the physicality of sentences, we usually read through the physics to the meaning. Unless it is self-referring, like a tongue twister, the sentence does not heed the physically exploded phonemes from the mouth and throat. And to heed the phonemes in that sentence, you would need to repeat the sentence aloud several times in order to heed the patterns of phonemes. More about the patterns of phonemes later.

Suppose that the act of ringing three times, with two shorts and a long, is a secret code. A presumed but hidden receiver would not be listening exactly for the duration of each ring, but rather one would be heeding the way that they were grouped together. In that case, the group pattern might be coded as a signature. Maybe the group of rings stands for "Joe." In that instance, the three rings, of two shorts and then a long in front-back order, would be a syntax. The listener would read through the syntax for the lexical equation of "Joe." But what if the true code were not rings, but knocks:

> Knock three times and whisper low
> That you and I were sent by Joe.

Then the secret listener, who guards the door in the song "Hernando's Hideaway," would know that the syntax of physical sequence was correct, but the physical action of ringing instead of knocking would mean that those who desired to enter were sending a false code. The natural syntax is the physical series whose pattern is common to both ringing and knocking. The grammar stands for the arbitrary agreement that three knocks is the correct code for "Joe." So the body actions of ringing and knocking and whispering in physical sequence constitute a syntax of sounds that in real space and time are prior to the grammatical overlap of proposition space.

Because I have discussed how nature enters into communication by way of a transmitting wave, I can define the word *action*, which has been featured in this section, in a way that is unconventional, but still in keeping with the principles of communication by wave forms. Any act, such as the actual opening of a door, is a transmission of a physical wave, in which the transmission speaks for itself.

The Orderly Actions of Things

In order to focus some of these syntactical issues about *deixis* and action in sentences, I end with a poem by Emily Dickinson about the unknown design of trees:

> Four Trees—upon a solitary Acre—
> Without Design
> Or Order, or Apparent Action—
> Maintain—
>
> The Sun—upon a Morning meets them—
> The Wind—
> No nearer Neighbor—have they—
> But God—
>
> The Acre gives them—Place—
> They—Him—Attention of Passer by—
> Of Shadow, or of Squirrel, haply—
> Or Boy—
>
> What Deed is Theirs unto the General Nature—
> What Plan
> They severally—retard—or further—
> Unknown—[26]

About the design of these trees there is no symmetry of Four Directions; there is no orientation of the inner life with a metaphorical

compass rose. There is no lexical identification of sycamore or alder or perceived order of phyllotaxis. The only branching diagrams to be found
would be those that might be imposed upon the odd grammatical order—
that kind of linguistic syntax—in order to get the semantics right. But just
in order to review the ambiguous meanings that come about by
Dickinson's oblique grammatical use, I rearrange the first verse (and only
the first verse) into the conventional grammatical order of a prose sentence: "Upon a solitary acre, four trees maintain [their place or themselves]
without design, or order, or apparent action." Her improper use of English
grammar is the transitive verb *maintain,* which requires some noun to
complete the assertiveness of the predicate. But that impropriety is what
she wants, because "maintain" is placed all by itself on the fourth line, just
before the pause and progression of the stanza break. Its *position* anticipates a sequential *order* of a single stressed foot that constitutes the last line
of the next three stanzas. "Maintain" is to be grouped with "But God," "Or
Boy," "Unknown." Notice that the vowels of each iambic foot rhyme with
one another. So "maintain" anticipates a following pattern of similar
sounds. They are flat rhymes: the long *a* of "maintain," and the *u* and *o* variants in the next three lines constitute a syntactic group or pattern of oddly
unemphatic rhymes. And, of course, because there is no object to complete
the predication of the verb "maintain," the grammatical nonsupply forces
one to fall back upon the intransitive act of maintenance. Of maintaining
(one or four) in ignorance of "Plan."

There is no apparent design, order, or apparent action to these things,
these trees. They are alone, for the most part; their neighbors are the sun
and the wind and God. The trees get only the most casual attention. If they
are attended to, it is only by chance—by shadow, by squirrel or boy. The
haply attending boy ends the line, and the phrase rhymes with "Or God."
So chance and design get juxtaposed by way of the conjunction *or,* which
has been already used in the phrase "Or Order, or apparent Action." Loosely
paratactic, the repeating of these coordinating conjunctions reenacts the
apparent absence of a master plan. The abbreviated grammar of the line
"They—Him—Attention of Passerby" reenacts the apparently casual absence of design, for I must substitute an assumed but absent parallel construction to fill in the attenuated grammatical scheme: the Acre gives to
trees a place, but they give "Him," the Acre, the casual attention of passerby.
There is just a glimpse of figure-ground relation, but one where trees may
measure an acre. In other words, the casualness of the grammatical construction enacts Whitehead's assertion about the casualness of the order of

speech parts in sentences. Only Dickinson has made a rhyme between the casualness of the grammatical sequence and the larger lack of place or plan.

The words are also flat and commonplace. There is no exceptional glitter of cochineal to make the landscape extraordinary. Only a faint personification animates it sympathetically, by way of "Acre . . . gives." These trees not only lack a compass rose of orientation upon a featureless and solitary acre in space, but they also seem to have no "Apparent Action" in time. This issue of action anticipates the issue in the last stanza of making inferences from temporal patterns. Whatever their deed might be, what plan they retard or further, is an unapparent action, unknown. The trees may have place, by virtue of the acre—and they point glancing attention to the acre in turn.

And yet, because of the flatness of the words, I detect no nostalgia, no poignancy, no tone of elegy for their being "Without Design." There is apparently no semantics of transference. If that acre of trees stands as an objective correlative for the poet's own lack of orientation in space and time, her apparent place or action in a presumed order of things, there seems to be only grammatical indication or semantic innuendo that would transform them into symbols of her own lot. Because one's actual choice of each word in a language is indeed an enactment of self, the choices of words that compose an assertion are always, at least tacitly, self-referring. So Dickinson's word choices in a certain physical order are her primary orderings. What I do hear is a syntactic continuum of orderly sounds as I read the four stanzas. There apparently is no spatiotemporal order among the four trees, but there is a halting and retarding sequential pattern to the momentum of reading aloud the four stanzas. That is, the primary order in the poem seems to be the usually suppressed order of the transmitting frequency. While there may be no apparent orienting Plan or Design in the placement of the Four Trees upon the acre, there is a faintly discernible pattern of physical order among the sound shapes, the metrics and off-rhymes that compose the order of the transmitting syntax. The halting pattern of the first stanza *maintains,* just as the ending iambic foot repeats, through the next three. But I do not detect that this is a poem about romantic subjectivity, in which the poet or artist subjectively creates her own order through an ardent pursuit of some object of desire. Nor is it like the aesthetic inquiry of Wallace Stevens's "Anecdote of the Jar," in which he questions the imposition of artistic roundness upon the slovenly wilderness in Tennessee.[27] Instead of comparing this poem to another, I think it better stands alone. The trees stand so. But the poem can serve as introduction to the halting syntax of sculpted breaths and indeterminate orders of things that characterize the wresting of an American grammar and syntax from English rules of composition. Perhaps nobody, woman or man, poet or reader, likes to be made the instance of somebody else's measure of comparison. Dickinson's stand-alone writing introduces a new form of syntactical wrestling with the syntactical order of things. Like the word *maintain,* which anticipates the end stops of three stanzas, this kind of writing anticipates the halting syntactic patterning of W.C. Williams, Denise Levertov, Robert Creeley, Susan Howe, and others.[28] So let me conclude with the idea that the poem is composed in a syntactic pattern that may or may not be a reenactment of an order of things. It may be agnostic about "Design," but it is insistent about the sound and sense patterning of "maintain." So it does not insist, as do Ruskin and Langer, that life is full of pattern, while meaning is an adventitious add-on. The "Four Trees" has a remoteness from both assertions, even while the pace of the syntactic feet repeats forward, retarding and yet furthering the "Unknown."

CHAPTER 5

Rhymes and Rhythms of Natural Language

Synopsis

In this chapter I discuss the "old sequence" of wave forms as if they were the moving proportions that propel sentences by means of rhythmic speech acts. I use the group transformations of symmetry theory—classified in previous chapters as translation, rotation, and twist—to try to catch a common syntax "in the act" of speech-acting. Because the physical elements of syntax carry the message in rhythmic waves, they themselves are the recognizable elements of rhythm. Because poems depend upon spoken cadences, a few lyrics that exemplify the physical shapes of sounds will be studied.

Because the physical shapes of poetic sounds repeat in cadenced intervals, they may be said to *rhyme* in unconventional ways. Here I feature unconventional rhymes as certain kinds of patterned inference, in which the syntactic movements of speech acts are the moving proportions that shift the thought inferentially from the memory of rhymes to the composition of new thought patterns.

Darwin Makes Music Precede Speech

If the rhythmic elements of poetry and song are a form of physical syntax, then their putative origins are worth reviewing. Among several competing theories about the origins of language, one of the most time-honored is that speech began in dance and song. By reviewing this old argument, I continue to foreground the idea that language usually reenacts rhythmic speech acts. By means of this review, furthermore, I suggest that the musical cadences of dance and song, as transmitted by air waves, may have the same rhythmic syntax as speech. In other words, that which evolution carries over from animal communication, once speech emerges, is an "old sequence" that coordinates certain rhythmic breaths when walking and talking. Finally, by way of Darwin's musical theory about the evolution of language, I privilege the importance of poetry and song in any putative theory of the way that language is generated. As one conjectures about these originating matters, I suppose skeptically that there may not be just one original cause for language coming into being, if only because among the elements of nature there are few other occasions when there has been just a single reason for the inception of anything. Even Charles Darwin hedged slightly in his conjectures about origins of language in species.

But within this context of the rhyme and rhythm of natural language, I want to recall Darwin's fundamental argument in *The Descent of Man* that speech was preceded by musical tones.[1] As Donald Fleming says, Darwin "had supplied a perfectly plausible account of the emergence of poetry, singing, dancing, and ornamentation in sexual selection. He had even assigned to music in the distant past the tremendous cosmic function of generating language."[2] I think this idea is exactly right: once upon a time musical tones helped to generate language, and now the sounds of musical cadences help to drive new inferential patterns among children and adults.

The issue in any theory of "generative grammar" is what kind of inherent patterning is doing the generating, so I here support the idea that a musical syntax of patterned tones probably preceded grammar in the evolution of language. Darwin on this disagreed with Herbert Spencer, who wrote in "The Origin and Function of Music" that music followed speech in "the idealized language of emotion." In order to support his own argument Darwin cited the pioneering anthropologist Lord Monboddo (James Burnett) in a footnote: In *Of the Origin and Progress of Language*, Monboddo argued that "the first language among men was music, and that before our ideas were expressed by articulate sounds, they were communicated by tones varied according to different degrees of gravity and acuteness."[3]

Darwin was not so much interested in the history of ideas, or else he might have also cited social philosophers like Rousseau, Herder, Adam Smith, and Percy Shelley, all of whom, in France, Germany, and England, conjectured that the first speech was song.[4] In fact, in a chapter called "False Analogy Between Color and Sound," Rousseau, preceding Monboddo, speculated about what he thought was a false analogy between the angle of light wave refraction and the angles of sound vibrations in French usage.[5] He made his argument by way of acute and grave accents. Darwin, however, began with the physics of vibrations and their effects upon animals with very simple physiologies. For example, he cited Hermann von Helmholtz's *Physical Theory of Music* to say that "crustaceans are provided with auditory hairs of different lengths, which have been seen to vibrate when the proper musical notes are struck" (*DM*, 568). Physical vibrations have been my starting point, too. Darwin proceeds "up" the scale of species—from the antennae of gnats, which respond to music, to bird songs and courtship, to gibbons who apparently sing in pure musical cadences during their seasons of "courtship"—to argue that song and speech are interrelated through the allure of sexual selection.

In one of his shortest and most dramatic sentences, Darwin says, "Love is still the commonest theme of our songs" (*DM*, 570). The language of love helps to replenish species. The courtship rituals of song and dance are designed to persuade humans to sexual action, but in acceptable rituals. For many species, Darwin says, the songs of animals help to reproduce the species by carrying the message, so reproduction is anticipated by the vibrating repetitions of song:

> All these facts with respect to music and impassioned speech become intelligible to a certain extent, if we may assume that musical tones and rhythm were used by our half-human ancestors, during the season of courtship, when animals of all kinds are excited not only by love, but by the strong passions, of jealousy, rivalry, and triumph. From the deeply laid principle of inherited associations, musical tones in this case would be likely to call up vaguely and indefinitely the strong emotions of a long-past age. As we have every reason to suppose that articulate speech is one of the latest, and certainly the highest, of the arts acquired by man, and as the instinctive power of producing musical notes and rhythms is developed low among the animal series, it would be altogether opposed to the principle of evolution, if we were to admit that man's musical capacity has been developed from the tones used in impassioned speech. We must suppose that the rhythms and cadences of oratory are derived from previously developed musical powers. We can thus understand how it is that music, dancing, song, and poetry are such very ancient arts. We may go even further than this, and, as remarked in a former chapter, believe that musical sounds afforded one of the bases for the development of language (*DM*, 570–71).

In this remarkable passage, and others like it, Darwin presumed to intimate how, via courtship rites of dance and song, instinctive drives were transformed into normative rituals. For Darwin the allegory of love was the music that helped promote sexual selection. The courtship rituals of love and jealousy, rivalry and triumph, reenact the early songs of love. Within the developing "series" of several species, Darwin assumes, somewhere along the line of development from the musical sensitivity of vibrating antennae to the rhythmic courtship of humans there occurred a ritualized shift from physical transportation of sounds by what I have called the old sequence of alternating forms of motional tones to the symbolic vehicle of rhythmic love songs. According to Darwin, musical sounds and rhythms are "one of the bases."

The best contemporary summary of this kind of song-and-dance argument that I have read is Robin Dunbar's chapter called "First Words."[6] Perhaps, Dunbar suggests, dance and song organize humans into group actions through euphoria (*GG*, 146). In this skeptical context of many causes, it is important to mention that Dunbar suggests plausibly that perhaps language did not evolve only among male rituals of exchange and coercion, but also that females bonded within their own kin groups by way of "maiden speech." Still more verbal than males, she speculates, women perhaps spoke first.

Here I should recall that many years ago, the mythologist Max Müller dubbed different theories about the origin of language with three tag lines. The one about concerted communal action through organized dance and songs and work chants he dubbed the "yo-he-ho" theory. The "bow-wow" tag characterizes those who believe that language originated in imitation of animal sounds, and the "pooh-pooh" tag represents a theory that language began with cries of emotion.[7] Although humor may not have originated language, it may be, as Müller implies, the best way to avoid the linguistic wars over the origins of languages.

The Cadence of Consenting Feet

In quickly reviewing these song-and-dance theories, I am not so much concerned with trying to isolate a cause for the origin of language as I am interested in featuring the early relations between song and dance and their mutual dependence upon a rhythmic physical syntax. Just as music theorists speak commonly about the syntax of music, regardless of language, so we may think about certain rhythmic properties in the order of speech acts that are not governed by grammars. I briefly discuss musical syntax in a later section. Rhythm is first and foremost the coordination of bodily motions of arms and legs through symmetries of gaits that propel animals through different media. While I speak of the rhythm in sentences, for instance, I assume now that rhythm in language implies a carry-over from body motions of hands, legs, and feet. In primal music and poetry, for example, the prevalence of two-four time, as well as the four-beat line in the standard poetic line, depends upon the "two-foot stride" of human pace. [8]

The intimate linkage between poetry and dance may be introduced by Herbert Read's assertion, cited in Chapter 3, that poetry can never become a popular art until it incorporates "the cadence of consenting feet." Over the course of a long career, Read often wrote about rhythm as an aesthetics of one's fitness into the natural world. For instance, Read, following George Saintsbury's classic *English Prose Rhythm*, constantly featured rhythm in his *English Prose Style*. His biology-based thesis concerns rhythm in prose as a knowing combination of emotion and order:

> There is an intimate biological connection between sensation and rhythm. Pain and sorrow are often expressed in rhythmical swaying movements; joy is expressed in rhythmic dances; religious emotions in ritual—there is no need to expatiate on such a commonplace of social psychology. The voice has its visceral controls, and though it would be rash to assume that the rhythmical reactions of the viscera and larynx to a strong emotion are the rhythms of the accompanying speech, yet these physical connections should be remembered since they are the basis of those refinements of expression which art introduces. What else is art, or conscience and intelligence for that matter, but a subtle extenuation and spiritualization of the gross physical responses of the body to its environment?[9]

To suggest just how Darwinian is Read's environmental theory of rhythmic speech, I cite a passage from Darwin concerning human emotional development, as it is regulated by the cerebral system:

> He who admits the principle of sexual selection will be led to the remarkable conclusion that the cerebral system not only regulates most of the existing functions of the body, but has indirectly influenced the progressive development of various bodily structures and of certain mental qualities. Courage, pugnacity, perseverance, strength and size of body, weapons of all kinds, musical organs, both vocal and instrumental, bright colors, stripes and marks, and ornamental appendages have all been indirectly gained by the one sex or the other, through the influence of love and jealousy, through the appreciation of the beautiful in sound, color or form, and through the exercise of a choice; and these powers of the mind manifestly depend upon the development of the cerebral system.[10]

Although it may be difficult to imagine some bodily function more reductive than sexual selection for the coordination of song and dance and song, I must briefly describe some contemporary research into rhythmic patterns of body movement. The preconditions of speech patterns may have been, in actuality, the neural coordination of movements of the feet with the rhythmic movements of breathing and chanting sounds. The stresses in poetry are still the stresses that are inflected when a morpheme is syncopated with a musical foot beat.

Today, for example, neurobiologists speak of central pattern generators when describing the cerebral coordination of rhythmic body functions, such as breathing, swimming, and walking, among many species of animals. Sten Grillner recently described the sets of neural networks that govern repetitive motions such as breathing, crawling, walking, and running.[11] He calls these governing groups of neural patterns "central pattern generators." He describes separate neural generators for the motor control of each leg of a horse; these neural patterns interact with different limbs in different ways when a horse increases its gait from a walk to a gallop (*NN*, 66).

Although Grillner does not mention the symmetry theory of gaits described in Chapter 3, biologists certainly do routinely study these and other neural rhythms and patterns. Grillner's own work focuses upon the isolation of the neural networks that govern propulsion among lamprey eels. His description of the lamprey's locomotion through water is similar to Gans's description of most fish: "In response to signals emitted from the brain, wave after wave of muscle contraction and expansion pass from head to tail down the body of the fish, propelling it forward through the water" (*NN*, 66). Beginning with Ruskin, D'Arcy Thompson, and with Gans, I have been pursuing a wave theory of rhythmic motion, wave after wave. Most important for Gans's hypothesis about an old sequence of neural propulsion, which stayed with animals when their ancestors left the seas for the ground and air, is Grillner's summary of his own work and that of his colleagues in research:

> Because the earliest invertebrates used only undulatory swimming for locomotion, the networks that later evolved to control fins, legs, and wings may not be all that different from what my colleagues and I have already discovered. Evolution rarely throws out a good design but instead modifies and embellishes on whatever already exists. It would be most surprising to discover that there were few similarities between lampreys and humans in the organization of control systems for locomotion (*NN*, 69).

Because biologists have recently found central pattern generators that govern breathing, it is not surprising for me to learn that other biologists have also discovered connective neural pathways for the songs of birds.[12] In humans, the cerebral cortex transforms the inputs received by organs such as the retina and ear drum (which receive and begin to convert wave signals), and it also controls voluntary movements, as well as speech acts that direct other kinds of emotional activity. By way of this brief review, I stress the primacy of physical wave patterns in the acts of breathing and singing. Perhaps most important, if one wants to pursue the idea of a generative grammar, which would be the base for acquiring speech, one might study the rhythmic patterns of song. For my purposes, it suffices here to feature the musical syntax of a few poems.

The Two Languages

By using the phrase "The Two Languages," Kenneth Koch means to describe the words within poems as being part of a "second language," constituting their "physical nature," which is the sound that makes up the musical component.[13] He starts with the basic idea of inflection that every spoken word has a rhythm: the word *father*, or *FATHer*, gets a stress different from the word *before*, or *beFORE*. He says that one is pronounced "*DUM* da"; the other is pronounced "da *DUM*." Long before, Saintsbury said that a trochaic word like *father* has a "falling" rhythm, while the iambic *before* has a "rising" character (*EPR*, 478). These metaphors of rising and falling metaphorically describe rhythmic motions of speech acts from the oral round of the mouth, discussed in the previous chapter.

Much of Saintsbury's study of English prose rhythms involves a foot called the "amphibrach," which he avers to be common in English prose but rare in poems. Its foot, composed of a long syllable between two short syllables, like the word *romantic*, is the main unit that makes for the "undulating or rocking movement" of English prose (*EPR*, 478, 481). Undulating is of course the felt motion of rising and falling waves. But look more closely at the Greek word *amphibrachys*, which means literally "short at both ends." Its rhythmic syntax of da *DUM* da can be described by a symmetry operation: in the sequential *order* of three syllables, *locate* two unstressed or unaccented syllables on either side of a stressed syllable. These quantities and stresses have almost nothing to do with grammar and everything to do with syntax. Unlike the double lives of linguistic amphibians described earlier, those who use a language composed of both a sensible and an intellectual component, an amphibrach is a fundamental measure of prose rhythm, which is, in its rhythmic measure, pure syntax. For those who find difficulty in separating grammar from syntax, the fundamental beats of prosody are good practice.

In his essay on prosody, "Listening and Making," Robert Hass begins with a premise about pattern that I have followed throughout: "we are pattern-discerning animals, for whatever reason in our evolutionary history."[14] Listening for the "keener sounds," Hass says, we attend to the rhythm itself, which he defines as a principle of "recurrence and variation" (*LM*, 113). Koch also says that the second language is governed by patterned rhythm: "The poet is led in unnecessary directions by the musically related language, and readers are led there in their turn" (*MD*, 21). The "physical nature" of this second language leads poet and reader in unexpected directions. As a physical frequency carries a message, or as the lohan is sustained by the curvilinear forms of the way, the poet and reader are led by the physical rhythms into unexpected patterns of thought.

When asked "what does it mean, poet?" Robert Browning replied, "Your brains beat into a rhythm."[15] So far, this is pretty basic prosody, but consider more pointedly the idea that the physical nature of spoken language, rhythmically carried, leads to unexpected patterns of thought. Consider, for example, the basic artifice of conventional rhymes at the ends of lines as a way of moving toward new patterns of ideas. The mathematician Stanislaw Ulam reported :

> When I was a boy, I felt that the role of rhyme in poetry was to compel one to find the unobvious because of the necessity of finding a word which rhymes. This forces novel associations and almost guarantees deviations from routine chains or trains of thought. It becomes paradoxically a sort of automatic mechanism of originality. . . . And what we call talent or perhaps genius itself depends to a large extent on the ability to use one's memory properly to find the analogies essential to the development of new ideas.[16]

Dennett also quotes this passage as part of a fascinating discussion, which suggests that within evolutionary changes there are not so much "*laws of form as rules for designing*" (his italics). Nature's rules of assemblage are like designing the variables of a conventional sonnet (*DDI*, 222–23). In composing a sonnet, he says, if you make a change somewhere in one line, you must correct the variants of formal constraints in most of the other lines because the sonnet's format is so compact. In that context, Ulam's idea about rhyme and originality seems worth a closer look because the physical character of the repeated sound waves, which comprises the phonemes, leads the poet and then the reader toward a new semantic understanding. When two words rhyme, the same phonemic sounds are repeated forward, making the brain remember "backward." I qualify "backward" because the physical element of sound is really repeated in the reading, but the brain does not truly go backward in the same physical way. In time, nothing is ever repeated exactly. That is why, as Jacques Monod was cited in Chapter 3, the unit of translation in DNA is never an exact repetition. Monod's phrase was "the irreversibility of translation." In terms of the recurrence and variation of rhythmic syntactic units, repetition is always forward with a slight variation. In other words, the slight variation in repetition is both the patterned rhythm as well as the variant pattern of thought, as when *bird* rhymes with *word*. Nevertheless, rhyming is an act that repeats a sound almost perfectly, and in symmetry theory the repetition of a physical unit along a line is a Translation. Translation is transformation along a line that repeats the structure of a physical unit invariantly, without variation, even though its syntactic position has changed in the order of the line. *If the second language of poetry is the physical rhythm of patterned sounds, repeated forward and recollected backward, then in poetry the basic unit of Translation is rhyme.* It is the basic unit of composition in poetic syntax.

So Ulam's idea provides a good way to rehabilitate rhyme in poetics, should one need to do so. The physical syntax of artfully rhymed sounds is guiding the thought in unexpected directions, by means of the articulations of juxtaposed sounds that are nonetheless heard as morphemes. Not just the grammar, but also the physical nature of the second language is guiding the unobvious thought. The syntactic sounds are repeated forward along the lines, while the memory of the repetition nevertheless leads to a new and variant idea. The semantic meaning of the whole sentence is presumably a thought that has never before been repeated, until this occasion. Even a verbatim quotation of someone else's sentence is presumably repeated exactly within a context of new ideas.

Recollecting a rhyme from memory is at the same time the inference of a new idea. "Humpty Dumpty had a great fall." Put another way, rhymes are memorable because they repeat a sound, but they chunk a thought into a *patterned* composition that is "unobvious." So the physical syntax of aural articulation in spoken and heard sequences of orders, composed of phonemes but heard as morphemes, leads us to new patterns of thought. Rhyme is the age-old pattern generator of thought. What is the relation between thought itself and the idea of pattern generation?

This notion of rhyming physical elements leads to an unobvious idea about the nature of thought itself. Recall the apt subtitle of William Calvin's book, which is *Seashore Reflections of the Structure of Consciousness*. If almost everything is communicated to us in physically moving proportions of wave patterns, and if the brain is in part a pattern transformer of wave patterns into images, then what is called *thought* itself may be a sort of emergent rhyme. Perhaps thought itself is always a pattern of thought, the pattern of which allows one to infer forward. By way of poems and other art forms, un-obvious meanings come into our brains in patterned alignments and shifty beats from the physical world.

Unlike a word such as *thought*, which is a stand-alone concept only when it is made to stand alone as an isolated word, the act of inferring seems to be an impending pattern that is repeated forward and yet trans-formed by that which went just before. Thinking is carried forward by the physical patterns of the physical world. According to Calvin, who makes a similar point about thinking as a "cerebral symphony," Percy Shelley put the concept of thought more assertively. What happens to thought after death? According to Shelley, most people in his era believed that thought and sensibility remained unchanged after death: "However, it is probable that what we call thought is not an actual being, but no more than the re-lation between certain parts of that infinitely varied mass, of which the rest of the universe is composed, and which ceases to exist as soon as those parts change their position with respect to each other."[17]

Every time one becomes aware that the kaleidoscope of the physical world is shifting its patterns by means of the physical signal that we are hearing, a new pattern of thought is "composed." In this regard a physical syntax is the set of transformations that constitutes the second language of poetry. Just as music and poetry may have been the primal forms of lan-guage, in an evolutionary sense, so too is poetry the most abiding physical form of language, but seen as patterned thought.

Unconventional Rhymes

When a poet's brain beats into a "cerebral symphony," not only is the cortex attending to the massive rhythms from the rest of the world, but also it is transforming those beats into new rhyming patterns of the poet's own design. For an example of an uncustomary kind of inferred rhyming, I quote one of Robert Creeley's more recent poems:

> Someone told me to stand
> up to whoever pushed me

down when talking walking
hand on friend's simple
pleasures thus abound when
one has fun with one
another said surrogate
God and planted lettuce
asparagus had horses cows
the farm down the road
the ground I grew up
on unwon unending.[18]

Albert Cook says of this poem, "Here the enjambments, coupled with a constantly ruffled accentual pattern, keep the voice pushing along and at the same time provide openings of sound large enough to accommodate God and lettuce in the same line, but in a movement too fast to allow pauses for further inference." Cook attends to the physical aspect of what Koch called the *second language* of the moving sound pattern, which in fact keeps the voice pushing along, while the mind also *simultaneously* tries to catch up with the momentum of sound and make an inference about what is being predicated. Notice first that Cook speaks of the reader's voice that it is being pushed along by the poet's syntactic formulations of physical quantities of sound; but if one speaks of the poet's voice who composed the rhythms, then the voice is in a sense composing the syntax and grammar. In other words, from the poet's point of view his voice is doing the pushing, and from the reader's point of view, the composed syntax is doing the pushing of the reader's voice.

What Cook means by "inference" cannot be reached until the poetic sentence has been completed in the final word of the last line. Perhaps one might infer that the thought of the poet's sentence could be crudely summarized by linking the first lines with the last lines, "Stand your ground, though you can't win it." This conjecture doesn't seem quite right, though. Perhaps the uncustomary rhymes might help one understand the patterns of thought, the principles of composition that make up the interiorized patterns of consciousness that is this poem. For example, one might notice that the last line links with the first line through rhymed repetition of the word *up*, which begins the second and ends the eleventh line.

This exact repetition forward of the fused phonemes that constitute the sound "up" is called in my lexicon of symmetries a *translation*. An exact translation is the memory of a perfect recurrence. I notice, too, that the last line also links with the first by the distantly repeated rhyme of "someone" with "unwon." Then I see that "one has fun with one" is an example of the funning rhymes that pervade this poem. Note these pairs: talking / walking, abound / ground, unwon / unending. All these are translation symmetries whose transformations of sound into sense move the voice along, as the poem is read aloud. The poem would have had a rotation symmetry if it had included the words *one's own* with the group of words *someone* and *unwon* because "one's own" is made up of a set of phonemes in which the "one own" reverses or rotates "unwon."

The accentual patterns are studied by reading it aloud often; then its strange enjambments of sound and sense are revealed. I mention only one such instance, the pause upon "another," as one realizes that the word *another* serves a twofold purpose as two kinds of grammatical shift. That is, the word *another* ends the idiom "fun with one/another" even as it be- gins the subject of another clause: "another said. . . ." But as Cook says, in reading the poem aloud, one doesn't pause to make an inference about that enjambment. If the word *another* serves as both grammatical begin- ning and ending, then there might be a hint about the meaning of the word *unending*. *Positioned* to end the poem, it could mean "begin again." I then realize that there is a parallel grammatical construction between "Someone told me. . . and "another said." They are parallel subjects that share a conflated predicate; one reads through to the last line to complete the assertion of the sentence. The enjambments at the beginnings and ends of lines, together with enjambments in the middle (talking walking, asparagus had horses) when combined with the absence of standard predication and the absence of standard punctuation, render the poem oddly groundless in terms of customary *position, order,* and *shape* of the elements being referred to.

Although standard punctuation usually marks the periodicity of a phrase, clause, or complete sentence, in this poem the whole period would be called, in a basic writing course of English composition, a *run-on sentence* because it is punctuated or pointed only by a period at the end, which ends the period of the assertion. Its run-on character of "talking walking" is reinforced by the enjambments, and its stops and starts and swings, which constitute its pace, are only marked by the variable accents of the words strung together in idiomatic groups. Locational pointers, mentioned in the previous chapter, such as "down" and "up" and "upon" and "when" here rhyme with "one" and "won" and "un" and "fun" in mon- otones, the effect of which is to confuse the speaker's location with his self- reference. One is grateful for the particularities of the friend's surrogate

God—lettuce, asparagus, horses, and cows—but one suspects that they are just there as deflecting substitutes for the poet's unending groundless pace.

Like Dickinson's local poem about four trees upon an undistinguished acre, this poem allows the inference that the physical aspects of nature enter communication, *not* by way of reference to objects, but by way of the overriding syntax of the physically accented sounds, the second language, pacing the sense. The words such as *cow, horse, hay, asparagus, lettuce,* and *the farm down the road* are deictic symbols which do not refer to the real world but refer in memory to the rest of the words that constitute the group word *farm.* Notice, too, that this poem is not behaviorist, although it seems to be about walking and talking. It enacts the inner speech of a mind grappling with its next changes. The natural world is not referred to; nature is the tacit wave frequency that carries the symbolic message by means of the physical syntax of quantitative groups of sounds, Koch's second language. By means of English grammar, the words associate with other words in the poem and thereby summon Strawson's hypothetical space and time of the sentence's subject and predicate.

If recurrence of similar sounds is the aural source of memory, then recurrence of similar words that rhyme by recollecting backward and anticipating forward is a combination of memory and prediction central to poetics. As Whitehead said, there is no measure whatsoever without recurrence. In Creeley's poem, however, the recollection of recurrence is troublesome. A new pattern of enjambed words is inflected up and down in my reading it aloud, while the words *up* and *down,* used in the poem, receive a harmonic sound stress of rising and falling, as in Williams's syntactic poem about the stressed foot of the cat's variable paw. The unexpected fusion of sound and sense play off the grammatical grouping of expected idioms, such as "stand up" and "down the road," to create new meanings through unexpected rhymes. In other words, the poem exemplifies the kind of discovery featured in Ulam's insight about rhyme as discovery. A rhyme is a sound-memory of a word remembered backward but that discovers forward. *Recollection and anticipation are linked by the pattern of the rhymed words, and the pattern is literally a series of phonemic waves. Thought is carried and grouped by the physical patterns of S-curved wave forms.* I think now that I have demonstrated the rightness of Coleridge's statement, quoted in the Introduction, about the movement of patterned reading as the reenacted motion of the "wise serpent" in its old sequence. Also I have demonstrated the hypothesis that these four-dimensional waves are the shapes of a physical syntax, and they are transmitted in groups. As Bertrand Russell was quoted in "The Anatomy of Syntax," these wave frequencies are the essence of Einstein's revolution about a curved space-time, and they are patterned in warped, folded, or twisted "groups." In this sense,

Cook's word *inference* is a conjectural pause that, Janus-like, combines the memory pattern of what one has already heard with the impending discovery of what one has not yet known.

So in fact there is a lot of forward-backward in this poem, as well as up-down. Then, too, Creeley's odd coupling of "unwon unending" makes me remember Feynman's "past-future," quoted in Chapter 3. One infers that the cerebral symphony of this poem is its unending pattern of the dissonant masses of off-rhymes and halting paces. The unwon-unending meaning of the poem does not stand alone as an independent thought or concept. Thought, as Shelley speculated, is not an independent, surrogate, or transcendental concept. Poetic thought at least has no locatable ground, upon which to stand our ground, other than as an inference about the uncustomary rhythmic pattern of the syntax and grammar that went into its elements of composition and that reoccur when one reads it aloud.

What if the only "ground" to stand upon were indeed Creeley's strange pattern of spoken thought, "up/on unwon unending," which, as Shelley said, stops existing, "as soon as the parts change their position with respect to each other." Then the locus of Shelley's "infinitely varied mass" would be understood by reenacting through reading aloud the poet's own peculiar composition of physically recurring sound waves. For here the words of *deixis*, such as *up, down, on*, and *upon* (Bowerman's locational words), do not orient one in any space in the world or on a farm; instead they meaningfully endure in a poetic time only so long as they are read as patterned linkages in the syntactic order of sounds.[19] This sort of meditation about duration seems to be part of Creeley's poem, if only because certain words, by virtue of their syntactic position in the line, undermine their lexical and grammatical meaning. For instance, "unending" ends the poem. The group of words "up/on unwon unending" has its own proportionate syntactic order but very little grammatical order. Is it fair to say that grammar and syntax are mutually twisted? Reading it aloud, one reenacts the poet's composition of a physical syntax of sound rhythms, and one heeds this syntactic order more carefully as one puzzles over the semantics of "the ground I grew up on" being linked to the last line. What is that ground? Is it a farm? A set of adages about standing up straight and tall? Or living with simple pleasures?

The Push and Pull of Musical Syntax

As the speaker in the poem said, "when talking walking," the pace of the talk and the walk go together, as do Williams's cat's paws. This rhyme between walking and talking, singing and dancing, is accomplished by the physiological foundations of musical syntax. Leonard B. Meyer asks once again the age-old question whether there is a bodily explanation for musical pitch systems and syntax. He believes so: "The central nervous system, acting in conjunction with motor systems, predisposes us to perceive certain pitch relationships, temporal proportions, and melodic structures, as well shaped and stable."[20] He says, furthermore, that one's bilateral symmetry, the on-off signaling of the nerves' synapses, and the systolic beat of the heart all contribute to a preference for a few simple and memorable durational proportions: 1:1, 1:2, 1:3, or 2:3. Meyer believes these to be norms of musical syntax because, as simple recurring "perceptual patterns," these proportions lend themselves to easy remembering (*MAI*, 289). This idea supports my thesis that body symmetries and characteristic wave patterns of thought are intimately correlated through the transformations of transmitting waves into rhythmic tones of memorable perception. For Meyer, these norms of musical syntax are moving proportions of intervals that aid perception and memory. He qualifies this assertion about "natural" modes or perception by agreeing that these patterns must also be learned and that they are conventionally modified into certain musical styles. But he says wisely, "The fact that something is conventional and learned, however, does not mean that it is arbitrary, anymore than showing that something is 'natural' is to assert that it is necessary" (*MAI*, 288).

A poetic system of orderly beats that counts intervals as moving proportions may be customary in terms of metrics, as with iambic pentameter, or it may be natural in terms of body symmetries and rhythmic asymmetries. Presumably one finds pleasure in a poetic or musical line not just because it has a naturally rhythmic beat, but also because the poet's voice slightly strokes the rhythm with dissonant starts and pauses that vary the dominant tempo. In conventional poetry, standards of metrics and grammar do not impede one another. But consider the way that the word *another* in Creeley's poem is rendered ambiguous by its *position* at the front of the line. I have already spoken about my uncertainty about reading it as the object in a conventional grammatical phrase such as "with one/another," or as the subject in "another said." That grammatical ambiguity is accomplished by its syntactic position where the word is voiced at the head of the line's *order*. Because of the ambiguities of grammar and syntax, sound and sense are rendered slightly dissonant.

Doesn't this imply that the syntax and the grammar cannot be focused upon simultaneously? In my definition, syntax is the construction of the physical articulations and patterns, while grammar is the assemblage of learned conventions. I have already discussed a "push-pull" effect in the plastic arts. According to gestalt psychology, a phenomenon of perception is the impossibility of focusing simultaneously upon both the physical gesture and the illusion of meaning. In "Art / Frequency" I cited C.H. Waddington's discussion of Hans Hoffman, and in the next chapter I mentioned Watson's observation that one can't focus simultaneously upon the suture and the snout of the bronzed beast, which calls attention to its own composition as the bilaterally symmetrical fusion, at the snout, of two profiles. This slight but necessary switch between stroke and image is presumably a function of the cortical switching from wave frequency as transmitter to the meaningful image. While reading a poem like Creeley's, one discerns that this slightly staggered switch between syntactic stroke and its attendant meaning is a kind of dissonant disclosure, which accounts for the reason why the words move slightly too fast for the inferences to catch up with the pace of the musical syntax. The reader's voice follows the poet's voice as it reshapes the expected grammatical inferences with unexpected reformulations and unexpected rhymes, as one hears, for instance, the unexpected rhyme of "hand" at the beginning of the fourth line with "stand" at the end of the first line.

The meaningful sound of rhyme is an essential algorithm of thought, for in hearing and attending forward to the meaning, one also remembers backward when the sound is repeated.[21] If to remember is to match or rhyme one thing by way of another, as in a mnemonic, then the modernity of a "mnemon hypothesis," as Young called it, may be linked with the oldest figure of memory, with Mnemosyne, the Mother of the Muses, whose invocation summons the power of harmonic remembering across the arts.

This featuring of patterned syntactic sounds over grammatical expectation makes for a novel kind of mnemonic inference. Perhaps poetic inference is "unwon unending." Maybe that kind of inference about sound-alike words is a simple pleasure. In the phrase *unwon unending* the sound is not Pope's echo of the sense. Instead *the natural language of the patterned sounds drives the sense.* By admitting the physical world into the poem by way of the physical patterns of a transmitting syntax, the poet reorients the question of the century: how are our thoughts related to the physical world? As with the *shih* poets, much of Creeley's poetry allows the inference that natural language reenacts syntactically the shapes of the transmitting wave. The translation forward of certain sound-alike words means that there is a consonance or rhyme between the sound and the sense. But when the rhymes are off-rhymes, and when the syntax drives the voice to violate cues of conventional grammatical order, then that awareness is pleasantly dissonant, like a voice or instrument twanging just slightly off key. This alternating stress upon sound and sense implies a staggered kind of poetic inference or dissonant disclosure of thought by way of patterned intervals.

Tropics of Discourse

A trope is a figurative "turn" that shares the same Greek root of *tropos* with the tropic of Capricorn or the tropic of Cancer, the circles of latitude that mark the southernmost and northernmost solstices of the sun. Beneath the idea of a rhetorical turn as trope is the figure of a protogeometric turn or rotation that describes a periodic course of action. I am still making the point that the forms within every sentence are governed by the idea of the physical transformations, which I have classified as the polytropes of translation, rotation, and twist. For example, when Hayden White uses the phrase "Tropics of Discourse," he recalls the Latin root of *discourse* as "a running to and fro."[22]

The Latin word contains a faded metaphor. If you run to and from, you run alternately back and forth with limbs of propulsion in the old sequence of wavelike gaits. The figurative rhetoric of discourse as mediating and reflexive and meandering—as discursive—is part of a European hermeneutics whose method is indirect and whose rhythm is irregular. Set against this discursiveness is the structural anthropology of Claude Lévi-Strauss, who always sought a kind of protogeometric structure underlying the meanders of archaic societies. He avidly studied "split representation" in bilaterally symmetrical masks and tattoos, where hardedged angles and triangles are set against curvilinear arabesques.[23] These distorting representations, when painted or tattooed upon human bodies, served him as troubled reproductions of the "split personalities" that characterize the human split in life between culture and nature. I call attention to morphological distortions here, too, but now I am more concerned with showing relations among morphological distortions of bodies that come about as a result of a distorted or twisted syntax: how do the symmetric forms of hockers and other creatures of bilateral symmetry, with their rhythmic gaits "when talking walking," begin to reveal strange symmetrical creatures who are made by tropes, cavorting and twisting in the structures of sentences?

What is meant here by "made"? Consider this example of a tropical distortion in a sentence. In a discussion of metaphor, Pablo Picasso once riddled in a sentence, "Le cygne sur le lac fait le scorpion à sa manière [in his own fashion, the swan on the lake makes a scorpion]."[24] How can a swan "make" a scorpion? What inference about pattern is meant by the sentence? The metamorphosis from one kind of animal to another depends upon a transformation called *rotation* or *reflection symmetry*. One must imagine the inversion of the swan's outline upon a reflecting lake. If you invert the profile of a swan with its S-shaped curve (the characterizing form of Art Nouveau), the upside-down image combines with the right-side-up image into a new animal with bilateral symmetry. But instead of side-view profile, one sees a top-down full body. Another way to perceive the emergent pattern of new thought is to draw the profile of the swan with ink on a sheet of paper. If you fold the sheet of paper at the bottom of the swan while the ink is still wet, then the upside down blot of ink renders a new Rorschach test image, which looked like a scorpion to Picasso. This physical fold backward and then forward makes the visual rhyme. A water image transforms into a fire image at the fold. The fold is a literal fold-over of paper, through three-dimensional space, which syntactically rotates the image a half turn upon its own axis to achieve a new position. In terms of my model, diagrammed in Figure 3a, you *rotate* the image to get a new *position*. This inversion is not a tricky verbal pun, for the syntactic form can be reflected or rotated into a new structure in the animal style of a hocker. Note in Picasso's sentence that the lexical meaning of the words is adventitious— *swan* or *cygne, makes* or *fait.* The adventitious relation between swan and object exemplifies the famous theory of Saussure's that the relation between the Signifier and Signified is arbitrary. This is true about grammars and lexicons; "swan" has no closer relation to the image of the bird than "cygne." But the anatomy of the new beast is generated from the commonly shared patterns of bilateral symmetry. To decipher the riddle, one uses a physical syntax and real physical transformations common to the shared anatomies of all beasts on earth.

Rotation is a trope of motion that reveals forms of nature that are archetypes of enduring figures, not just newly minted metaphors. Recall that so many of the bodies drawn by Picasso were images that depend for their troubling grotesquerie upon the push-pull ambivalence between profile and the bilateralism of full face. Think of his Minotaur face imposed upon the body of the artist, like Bottom with the head of an ass. This ambivalence characterizes much primal art, as in the two profiles of the t'ao t'ieh joining in the sutures at the snouts. Or the Greek gorgon that combines a full face in bilateral symmetry with a profiled torso with running legs.

To interpret Picasso's sentence, I must turn "in my mind" one visual form into another pattern, where arguably all such coded instructions are in their old sequence of alternating rhythmic forms of motion. If I try to understand the sentence only by linguistics or by formal logic, I will fail to make sense of it. I need to match up the words for the kinds of animals with symmetrical icons of their body structures that I have stored in my memory, and then I need to use the symmetry transformations of rotation to find my clue. This is iconic reflection. Recall Pinker's celebration of visual thinking and the experiment of rotating many versions of the letter F. To decipher Picasso's riddle, one must rotate the sideways profile of the swan in order to visualize the top-down outline of the scorpion. So there is also a subtle shift in the dimensional point of view by which one sees the creatures: the profile of the swan can be seen on a two-dimensional plane, but you need to rotate the profile through a third dimension to get the top-down picture of the scorpion. It is this rotation, plus the rotation in point of view, that makes the trans*form*ation. You have to combine two rotations to get it. And if you combine two different kinds of rotations, you have tacitly twisted points of view into three space without attending to the fact that the new beast is also a new twist.

Picasso's riddle tells me nothing about symmetrical transformations of units that make up changes in sentences. Its appeal is to Cubist tricks of points of view, which fuse profile and full face. But by way of a sentence he does resurrect the idea of a visual syntax of articulated animal anatomies, the subject of Chapter 3. To put the issue more pointedly now, can one imagine a physical relation between syntactical transformations, considered as tropes, and the shared physiology of bodies, which might be described in sentences? Once, when he was defending his kind of structuralism, Lévi-Strauss quoted D'Arcy Thompson, the naturalist who applied mathematical and geometric permutations to biological form: "In a very large part of morphology, our essential task lies in the comparison of related forms rather than in the precise definition of each; and the deformation of a complicated figure may be a phenomenon easy of comprehension, though the figure itself has to be left unanalyzed and undefined."[25]

In order to test an inherent relation between syntax and symmetry in sentences, let me try an indirect method of finding the deformation of structure. Although the patterns of a sentence—its symbolic logic, its calculus, its parts of speech—have been exhaustively studied since linguistics became a discipline in the eighteenth century, a regular and precise patterning for the generation of Universal Grammar has eluded these rigorous methods. I want to study certain deformations in the common syntax of sentences that reveal, I believe, their relations to symmetry through irregular shapes and deformed bodily structures. Wittgenstein introduces an everyday form of language that helps show these deformations:

"The problems arising through a misinterpretation of our forms of language have the character of *depth*. They are deep disquietudes; their roots are as deep in us as the forms of our language and their significance is as great as the importance of our language. Let us ask ourselves: why do we feel a grammatical joke to be *deep*? (And that is what the depth of philosophy is) [his italics]."[26]

There is a certain kind of riddling verbal pun that turns upon the close relation between syntactical transformations and the resulting shapes of grotesque bodies. Perhaps by way of a joke I can catch a deep sound of syntax in the act of generating a skewed semantics. Here, for example, is a trite pun about deformed morphologies:

> A Mandarin named Chan collected teak figurines. One morning he discovered to his chagrin that some of his figurines had been stolen. Looking around for clues, he noticed the tracks of a boy's footprints, leading inside and then outside. He decided that he would hide that night to find out whether the boy would return for more of his teaks. When the door quietly opened that night, he saw to his surprise a bear walk up to the curio cabinet and grab one of his teak figurines. So he said, "Stop, oh boyfoot bear with teak of Chan!"

This badly extended pun works only if one remembers John Greenleaf Whittier's sweet poem, which generations of school children were once forced to memorize:

> Blessings on thee, little man,
> Barefoot boy, with cheeks of tan!
>
> —"The Barefoot Boy" (1856)

The extended pun renders a new pattern of thought from an old rhyme. The pun shows how the fair body of a boy is variably transformed or deformed by the syntactic inversions of phonemes. The deformation of a boyfoot bear is also an example of split representation. How does the split figure result in terms of the transformations of a common syntax? If the syntax of any code can be marked by static states of order, position, and shape—as I mentioned in the Introduction—then the sequential order of the phrase "barefoot boy with cheeks of tan" has been translated and retained in the pun. Now it is an algorithmic set of internal rhymes: "boyfoot bear with teak of Chan." But the position of the consonants has been translated and rotated, while their *shape*–b, t, and ch—has been preserved. The riddle's transformation of syntactic forms of spoken letters morphologically deforms the deep structures of symmetrical physiology. The pun's morphological distortion works because of the rhymed inversion of the phrases. Perhaps Whittier's original rhymes and nostalgic theme, similar to that of "Abide with Me," prompted the parody.

Whoever invented this pun must have rotated the alliterating consonants and discovered with surprise that an odd antithetical creature was made, like Picasso's scorpion, when the phonemes were rotated into new morphemes—a boyfoot bear and a Mandarin named Chan. Whoever recognized the inversion then contrived a mini-plot and narrative to expose the surprising oddity in a punch line. The extended pun transforms an old rhyme into a new pattern of thought. Or perhaps this kind of syntactic transformation is a very old understanding: "Feet in the shape of bears are a common feature of Han bronze vessels from the first century B.C. onwards"(*CE*, 103). To append a set of bear feet to a vessel is to animate the container into a not-just-bronze order of articulated connections. For to embody the hypothetical space and time of an artwork is usually to animate it fictively with some kind of syntactic order from nature.

This joking deformation of a poem reveals Wittgenstein's *deep* relation between physical structures of syntax and anatomical structure. There seems to be a deep but tacitly unattended archetypal pattern of relations between the bilateral symmetries of bodies and the troubling transformations of verbal syntax. There seems to be a relationship between pattern detection in sentences and pattern detection of symmetrical hockers. Sentences about the world work at their basic levels only if the structure of a sentence and the structure of the bodies in the world have the same regularities of form, which can be transformed by innate syntactic instructions that translate, rotate, and twist units of discourse into body tropes.

In this kind of joke the disquieting setup is always a body that is morphologically distorted so as to make the syntactical displacements be revealed as tropical turns—what I call the symmetry operations of translation, rotation, and twist. I cite one more example of a stereotypical pun that is set up by the deep relation between the patterns of bodily form and syntactical rearrangement of letters. Virtually all jokes about bodies are unseemly, as Freud noted in his famous work on jokes and the unconscious, so I hope I may be forgiven for repeating a joke about Others. When a clairvoyant Lilliputian escaped from jail, what was the newspaper headline?

"Small medium at large." Again, as in the boyfoot bear joke, the units of clothing size have been formally translated, remaining almost unchanged under translation. The substitution of "at" for "and" allows the grammatical idiom "criminal at large" to be substituted for the clothing sizes, giving rise to the dual meaning of "large." Yet the syntactic line of reading the series, as a translated series, has remained phonemically invariant. But the clothing sizes in series have been surreptitiously changed into a tacit grammatical sentence, in which the "is" has been omitted in the style of newspaper headlines.

I notice that the joke works as a pun because the clashing, arbitrary, double meanings of "medium" have been fused into a sentence with "small" and "large" antithetically juxtaposed in the grammar of an American newspaper's sentence. By way of this example, one sees that it is useful to separate the deep natural syntax from the grammatical syntax of a sentence. The natural syntax enacts the physical transformations of forms of sounds. Under translation, the shapes, orders, and positions of the sounds have been retained as invariants. And in this joke, the tacit visual message of reading newspaper headlines is also assumed. The physical media of sound and sight are carriers of the syntax. But the grammatical peculiarities of the American newspaper headline make it work, too. The series of phonemic syntax and grammatical order work at variance with one another to make the joke. But in understanding the joke, we usually attend to the pun upon "small medium" without attending to the syntax and grammar that structure the joke. The important point is that syntax and grammar are usually so twisted together that one does not ordinarily attend to their different principles of construction. So the pun can double back upon clairvoyance. The small clairvoyant who is invented to pose the joke is a fascinating brief description of character. In his novels, Charles Dickens invented a number of insightful small people whose clairvoyance arose from their off-angled point of view on the assumptions of size and scale of normality assumed by big folks. Although the pun itself is egregious, it still reveals a new pattern of thought, in this case, a new awareness of the deep structure of the natural syntax.

This kind of split representation works because of the phenomenon of split reference in metaphor. "The mind has mountains," is not interpreted literally.[27] It asks us to recall iconic images of mountain ranges with valleys and peaks, and to combine that recollection with one's mind to be seen as a remote landscape. It reminds that the mind is not a spatial kind of thing; it is processionally more like tune. There are not literally "patterns in the

mind," the title of Jackendoff's interesting book. There must be neural patterns in the brain. While a brain is a physical entity, and while there are
pattern-generating codes of syntax "in" the brain that propel us in the "old
sequence," mind is not spatially located in truth—only in metaphorical
sentences. Split language is capable both of self- and other reference
because of mutual inferences about the structure of the world and the
structure of sentences. This unexpressed commonality between syntax in
sentences and bodies in the world is the reason why one can speak of split
reference in the first place. By way of the metaphorical sentence, mind is
imaginatively widened to be as deep and as high as all outdoors. One's
mind is no longer a small medium, but patterned clairvoyance at large, if it
is unleashed from the regular conventions of grammatical syntax and seen
as mental landscape "in" a hypothetical space.

Literature and art are the vast arenas in which these unconscious
tropes of natural syntax are played and exposed by warping conventional
shapes into embodied surprises that block normative expectation.
Although the sounds of words imply a promise between the writer and
reader, the promise, as Susan Stewart suggests, will be "broken" in different
ways. Often a plot will turn upon the sudden realization that a heroine or
hero has been playing by one set of rules, only to discover, very late in the
plot's development, a covert code that is really prescribing another set of
patterned actions, as Oedipus learns too late. At the end of the play, for example, Oedipus, whose name means "swollen foot," learns that he belongs
to a category that includes a boyfoot bear—he is a swollen-footed monster
who is deformed because he thought that chance ruled; whereas, for
Sophocles, Fate (*moira*) was physically shaping his actions.

Phonemes as Tropics

In my discussion of puns, I feature the morphological distortions of
symmetrical bodies but the essence of the joke is that the rhythmic sequence of sounds remains essentially the same: barefoot boy with cheeks of
tan—boyfoot bear with teak of Chan. The phonological point is that only
certain sequences of sounds are permissible in a language, not because of
grammatical rules, but because of invariants of sounds composed together
in rhythmic contrasts. Grammatical orders are superimposed upon
the sound-right sequences of sound-shapes. The "deep" significance of the
puns involves the eventual realization that the rhyming sound sequences
generate a necessary but nutty meaning. Puns sound right, but they mean

wrong. Languages seem built upon the inherent but limited patterning of physical sound shapes in pleasing rhythmic orders.

To test some of these sound clusters as symmetrical groups of wave forms, I move from puns to poems. Ordinarily, the syntactic transformations flow conventionally in most poems, for beneath each symmetrical transformation is the Heraclitean formula for the flow of the wave. It has already been suggested here that sound waves and electromagnetic waves are the vehicles for carrying these messages, with light their constant measure, and body symmetries modified an old sequence to help propel limbs through other kinds of waves.

Below, for example, is a poem with fairly conventional diction, but lovely in its unassuming way of syntactically *propelling* two lovers together. Martin Gardner quotes Robert Browning's lyric as an example of how "reflection" symmetry creates sound effects.[28] For Gardner, the rhyme scheme of *abccba—abccba* repeats the feeling of sea wash back and forth. That is, the end rhymes, composed in two symmetry groups of three units, repeat forward, but they repeat forward in inverse order because the second group is a rotation of the first group's order:

> "Meeting at Night"
>
> The gray sea and the long black land:
> And the yellow half-moon large and low;
> And the startled little waves that leap
> In fiery ringlets from their sleep,
> As I gain the cove with pushing prow,
> And quench its speed i' the slushy sand.
>
> Then a mile of warm sea-scented beach:
> Three fields to cross till a farm appears;
> A tap at the pane, the quick sharp scratch
> And blue spurt of a lighted match,
> And a voice less loud, through its joys and fears,
> Than the two hearts beating each to each!

The meter and rhyme scheme reenact the motion of sea waves tacitly impelling the force of the lyric sentences. The formal motion of the waves is reenacted by the syntax, meter, alliteration, and rhyme of the short, monosyllabic, Anglo-Saxon words in rhythmic sequence. Browning has chosen many conventional one-syllable words that have been cobbled down like sea stones through use. The effect of their passage through consciousness in a melody is achieved by the remembrance of physical feeling—of sea wash pushing and pulling us back and forth.

In order to be more exact about this formal motion of back and forth while being impelled forward—Coleridge's pattern of the wise serpent—I want to describe the lyric in auditory terms. Structural linguistics could be of preliminary help in understanding some of the subassemblies of Browning's artistry in making the sound impel the sense of love's pairing. Forgive the baldness of these opposing yet "meeting" dyads: gray-black, sea-land, large moon-little waves, lower-leap, cove-prow, wet sand-dry field, tap-scratch, match-dark, joy-fear, and paired heartbeats. All the dyads move physically to join one another, to meet, which is the subject of the title. I might begin to construct a semantic space out of a matrix of opposing dyads that seem universal in cultural anthropology, according to Jakobson: with sunlight versus darkness, female and male, light and heavy sounds, or vowels and consonants that seem to accompany them (*SSL*, Vol. 8, 195).

In Browning's poem all the senses are at play. Even taste is synaesthetically present in "quenched." Think for a moment about the senses as if they were receptors that help the brain transform different kinds of sensuous touching. One of the five senses synaesthetically includes all the others. All touch, for they are all impulsive. Even the taste of "quenched" is a touch of the tongue. The sense of touch is a direct impulse, which is mediated by symbols only when one reads about touch with words such as *quenched* or *heartbeats*. All the senses immediately receive their signals from waves that touch sensory membranes. One feels and senses the forms of these waves without comprehending them as such. A touching sign is what C.S. Peirce called an *index*, one that depends upon physical contiguity for sending and receiving. If all signs are carried by waves that touch us tacitly, then all communicating is bodily feeling.

Because Browning's poem is a love lyric, it ends touchingly with paired heartbeats that sound louder than the speaking voice. Love also vibrates in a harmony of two movements. Perhaps it would not be too reductive to remind, with Weyl, that the human heart is not a bilaterally symmetric valentine. It is an asymmetrical twist (*S*, 687). But "two hearts beating each to each" means that love and physical frequency are here in phase, beating back and forth, *abccba—abccba*.

When the lovers join anonymously but tenderly, they meet as part of a range of natural pairings. But the pairings move together with other physical impulses in a melody of pulsations. The lovers' hearts beat momentarily with other anonymous natural forces. In this poem, love is the touch of joined energies, just as the sea touches the land, as the prow touches the sand, like the tapping on the pane, like the scrape and spurt of the match, like the voice less loud than the heartbeats.

Can you tell by which naturally periodic force the lovers have timed their meeting? "The yellow half moon large and low" indicates that the moon is lowering and will touch and fall below the horizon. Not clock time but the moon's *rotation* is the natural measure for the meeting. Their tryst is timed so that one lover can catch the last of the moonlight in order to see his way ashore by way of the fiery ringlets of sea glint. They expect to meet furtively after moonset in a darkness that will require the light of a match. As Whitehead features the hymn of time and tide, in "Abide with Me," Browning builds on one's prescience about the great rhythmic forces of process and stasis, those things that go and those things that temporarily abide, those anonymous forces that sustain us, and that retreat from us, even though we have no ordinary sense of their sway. The great physical frequencies go on in their formal periods, like the rhymes of "light" and "night," visual images with identical sound frequencies, "ight" sounds, separated in meaning only by a twist of the tongue that differentiates the consonants of *l* and *n*.

"Meeting at Night" also helps clarify the tropical forms of syntax. In order to review, I set forth the tropes as transformations for turning words into new groups:

1. Translate units along a sequential line to achieve *order*.
2. Rotate units to achieve antithetical *positions*.
3. Twist or fold units together into new *shapes* of phonemes and morphemes.

Starting with Aristotle, it has long been known in the history of linguistics that words can seem to refer "outward" to concepts and also they can refer "inward" to the forms of sentences.[29] As Morton Bloomfield says, the first role of words in language is "to convey meaning: they name; the second convey grammatical function: they connect or relate" (*SP*, 309). He explains how these kinds of words can be loosely aligned with Jakobson's distinction between metaphor and metonymy, one on a vertical plane of combined word meanings, the other on a horizontal plane of contiguous connectors and integuments in a sentence string. Bloomfield's main effort in this essay is to rethink most poetic tropes as syntactic devices that relate inwardly to the conventions that connect words within sentences. For him, rhyme, meter, and alliteration refer to the metalinguistic connectors within poems. By way of these tropes, poems seem to be tacitly referring to their own principles of linguistic construction. In sum, poems seem to be almost always self-referring because they are about their own language.

The off-rhymes in Creeley's poem may be self-referring, but I do not think that the alliteration in Browning's poem is so. Instead, the undulatory rocking of the rhymes and the alliterative patterns in the vowels and consonants seem to underscore the natural scene of sea wash, weather, and love. The rhymes and alliterations are tacit, for they do not call attention to the sentences as artful forms. And the grammar is oddly repetitious, but neither does it seem to be about the poem or the poet. What is remarkable to me is the poem's lack of what Bloomfield calls explicit grammatical directions *inward* toward the construction of the sentences. It is a spatial and temporal description of seacoast and farm, but it has very little grammatical sophistication. What is the subject, and where is the verb of the first sentence? There are few coupling verb forms that would give grammatical precision to the two sentences. A few articles suffice: *the, a,* and *And* are repeated over and over. Instead of the grammatical order of subject-verb-object, a few noun clauses introduce the two sentences that comprise the two stanzas, and a couple of odd misuses of verbs—*gain* and *quench* and *appears,* all of which are written as prepositional clauses. "As I gain," "till a farm appears." Why are the verbs so oddly couched? Is not desire the drive of verb tense in love lyrics? *Desire* is a ritualized conceptual word for a feeling of lovely anticipation that is expressed forward by the future tense of "whither." In a Darwinian sense, musical language is the very vehicle of desire and design; it is the vehicle of courtship and promise. But so far the poem does not seem very ritualistic.

One "Then" suffices to indicate that the successions of clauses and "Ands" are also a temporal succession of scenes being traversed. They are a series of impressions loosely and perhaps hastily glimpsed as the lover moves on his furtive quest. (It is furtive because of the scary loudness of the scratched match.) One does not even know until the last line that the poem relates a quest. Only then does one understand that it is a love lyric. For the first nine lines, one reads cumulatively forward through the phrases and the oddly located subjects and verbs, inferring forward, looking for the object, so to speak, of the sentences' impelling. What kind of meeting is it? One seems to be carried by the loosely paratactic sentence construction on a descriptive journey whose teleonomy or destination is not realized until one reads, at the very end, "two hearts beating each to each." The two hearts beating are just as anonymous as all the other forces that are rhythmically beating and meeting and in nature. Undoing Browning's restraint, I call your attention to the distant rhyme between "beating" and the "Meeting" of the title.

Alliteration as Symmetry Groups or Patterns

Although the two sentences are loosely paratactic in grammar, without carefully constructed Ciceronian antitheses that might balance the coupled dyads meeting, still the transformations of sounds as consonants and vowels are carefully repeated. Look closely at the alliteration in "Meeting at Night." There is the syntax of symmetry groups. First pick out and sound out the groups of consonants. Which sounds are repeated forward? Which sounds are translated along the one-dimensional line of the poem's impelling? There are three primary groups of alliterating consonants: *l* sounds suffuse the first stanza, *ch* and *sh* sounds are repeated through the second stanza, and *th* and *t* sounds run throughout. But the consonants alliterate in groups: long, black land, yellow half moon large and low, startled little waves that leap, ringlets from their sleep. (Always, too, there is the symmetry-sharing of nature's circles ringing in poems; a visual ring is transposed to a verbal ring because they are both carried by waves.) Look closely at the patterns of alliterations along the lines. Within some of the lines there are groups that are patterned in three sounds, two sounds, or paired sounds, in patterns similar to the groupings of Anglo-Saxon alliterations:

> Three fields to cross till a farm appears.—*tf tf*
> And quench its speed i' the slushy sand.—*chss sshs*
> A tap at the pane, the quick sharp scratch—*pp ss*
> And a voice less loud through its joys and fears,—*ll ss*
> Than the two hearts beating each to each!–*tt ch ch*

The major groups of consonants—*l*, *s*, *ch*, and *t*—harmonically converge in the last two lines, where the couple meet.

What are symmetry groups? In mathematics they are permutation groups. In physics they are mathematical descriptions of quantum transformations. In poems they are pairs and triplets and quatrains of sound patterns that move and transform the gaits. The measure of tiny feet. They control the speed and pace of the meaning. As Alexander Pope said, "the sound must be an echo of the sense." Oftener quoted the less understood, the sound measures the sense in symmetry groups. Yet as I inferred from studying the physical rhymes in Creeley's poem, I am moving toward the reverse principle: if the syntax of music actually precedes speech, in terms of Darwin's evolutionary context, then meaning follows or echoes the limited patterns of sound shapes, not the other way round, as Pope has it. The natural syntax of sounds is an anonymous, first, physical language, which carries the message tacitly and rhythmically in a limited cluster or group of possible patterns.

Syntactic Translation

Let me rephrase Bloomfield's argument about the syntactic relations of certain words. Although most would grant that there are certain words that are primarily syntactic words of relation—such as *this, the, an, in, of*—still the conventions of grammatical construction are not *in* the words themselves. If the operations of sentence construction were those listed, then meter, rhyme, alliteration, or any trope of *repetition*, as Bloomfield calls these poetic devices, would be grouped under *translation* along the sequence of a sentence. Even the parallel constructions of certain subordinate or coordinate clauses are translations of the words along the sequential (or horizontal) direction of the sentence's eventual meaning. For example, in "Meeting at Night," there are two verses, each composed of one loosely constructed sentence. Both sentences are related grammatically by the cumulating repetitions of the word *And* at the beginning of some lines. The idea of accumulation is not a metaphor in this use. To the extent that sounds have physical density, in terms of their resonance, they are truly quantities. The repetition of the three phonemes that compose the "and" sounds are quantitative accumulations of physical units, and not just the repeating of coordinate conjunctions. In poems and songs, and even in prose, the repetition of sounds is a physical translation of quantities.

Syntactic Rotation

I have already listed words that Browning sets in dyadic opposition to one another, so as to reinforce meeting as physical touching. The meaning of the great primal words, such as Browning's *joy* and *fear*, requires their opposing contexts to yield intensity. And the great poets knew that one emotion, too intensely experienced, always goes over into its opposite, as when John Keats in "Ode to Melancholy" bursts "Joy's grape." But is this not more of a semantic contrary, and not a syntactic rotation? If to rotate is a strict syntactic inversion, then it must turn forms so as to distinguish a new position and order, as Aristotle reordered NA to AN. We have seen that rotation is not just a semantic rotation; the mouth and tongue roll the sounds in the mouth phonetically, too. So to find a strict syntactic rotation or a reflection, one must recognize the reversal of two or more forms, like the two forms of letters. Here is a neat example whose source I have been unable to locate:

> Do and you shall be—Sartre
> Be and you shall do—Camus
> Do-be-do-be-do—Sinatra

The syntactic forms rotate and alternate in a new rhymed order, and so a new pattern of thought emerges. The third line reduces the French existential question of being versus doing to American jive talk. But at the same time, the pop refrain reorders the nonsense phonemes into a third wave of combined patterning that alternates doing and being, sense and sound, into a thread of melody. Two snappy but reductive sentences are transformed into a new mnemonic unit at the end of the line. The sense echoes the sound, for language is carried by the warbling waves. In the third line, musical syntax triumphs over grammatical meaning. It enacts the thesis of this chapter, for physical sound patterns exist and the meaning is reduced to nothing but the pattern of the sequence.

This three-part repetition of *grammatical* constructions is technically known as a *tricolon*, as in the tripartite construction of Thomas Hobbes's state of life in nature: "nasty, brutish and short." Or in Abraham Lincoln's "of the people, by the people, for the people." Or in the old saying:

> Once is happenstance,
> Twice is coincidence,
> Thrice is the Devil's hand.

The repetition forward of the three parallel grammatical constructions also is a cumulative gathering of rhythmic sounds, reinforced by the repetitions of *ce* sounds at the beginning and end of the lines. The third time reveals the shape of the twisting trope, the Devil's hand.

In "Meeting at Night" the deep structure of reflection or rotation is the rhyme scheme *abccba—abccba*. For Gardner, this set of rhymes evokes the formal motion of the wave form. This is my prime thesis, starting with the model of the old sequence: the form of the wave, its troughs and crests, backward and forward, is the archetype for syntactic forms of moving proportion.

Syntactic Twist

When Ezra Pound said "Rhythm is a shape," he was not being metaphorical.[30] To review, let us say that rhythmic speech is a *solid* flow of alternating stressed and unstressed sound waves, punctuated by silent intervals, whose *shapes* are transmitted in a rhythmic line, and enacted physically in an elapsed time by somebody and received by somebody else. While speech acts follow one another in a temporal order, they are not exclusively one-dimensional; speech acts are three-dimensional wave fronts that crest and dip as helical wave fronts. So speech acts are rhythmic acts of four-dimensional shapes that crest and fall and follow one another in a rhythmic time. Sounds give a rhythmic shape to a morpheme, as in Koch's example of *FATH*-er. Unlike the twisted lines in a drawing, you cannot *see* the twists of sound; you only hear them. The shapes of speech acts are literal shapes because speech acts are a series of modulations of wavy puffs and billows, as the Dogon taught. Twist is an over-and-under transformation of some carrier wave in three space and in time, so twist in symmetry theory is a necessary function of the transforming act that renders "shape" in three-dimensional entities. With respect to speech, a twist is a cognitive transformation that rounds the sound into shape, as in the pleasantly twisted sound shapes of the phonemes that make up the rhythm of the word pur-*LOIN*. Every speech act is necessarily a *twist* or fold of phonemic sounds into morphemic groups.

Bloomfield shows that "deviation" from the conventions of grammatical order is common in poems. To violate or stretch the common definitions of words by placing them in ungrammatical order blocks the reader's or hearer's expectations and thereby extends the range of conceptual meaning. I explored this concept by discussing poems of Emily Dickinson and Robert Creeley. Hence this deviation in poetic grammatology is of a kind with D'Arcy Thompson's idea of morphological distortion. Bloomfield quotes lines from Dylan Thomas and e. e. cummings as examples. He quotes the last line of *Finnegan's Wake*: "the." This malformation of conventional usage cues one to the trope of twist as a grammatical deviation. To misplace a word, or several words, from customary position or order is to misshape the expected format of the spoken sentence into a new format, which makes one turn back and reexamine the ordinary tacit construction of sentences.

Because the word *twist* connotes in itself the operation of deviation and deformation, some of its various uses need to be mentioned. A tongue twister is a phonetic twist, as in "How many sheets could a sheet slitter slit, if a sheet slitter could slit sheets?" Here the blade and tip of the tongue literally twirl and twist around the mouth cavity to avoid an embarrassing slip of the tongue. A phonetic twist causes a semantic slip. A syntactic twist is very hard to catch; it, too, is best seen as a deformation of convention. In a picture on a two-dimensional plane, one can see the illusion of a twisted line over and under and through the imaginary third dimension, as in the scudding clouds in Blake's "Act of Creation." Although one cannot see the twist in the grammar of a spoken sentence, one can better see its deviation of expected grammar, when it is written: "This sentence no verb." This is clearly a self-referential sentence, one that calls attention to its own elements of composition. It also belies the grammatical rule; for a construction without a verb is not a sentence. Phonetic, syntactic, and semantic structures are interrelated through the translations, rotations, and twists of words. If information is high when expectation is low, then syntactic twists are the basic transformations that block conventional expectation and move us toward a new pattern of thought. Syntactic deviations make us check back on the prejudgmental assumptions of our conventions in the sentence. It seems to be the operation of twist in the syntax of a sentence that lets one ask formidable questions of self reference—the meaning of meaning.

A Twist of the Imagination

Writing about the poetry of his friend William Carlos Williams, who had recently died, Kenneth Burke quotes a sentence from one of Williams's early works: "There is nothing that with a twist of the imagination cannot be something else."[31] By way of that sentence, Burke introduces Willliams's mobility of thought, his "constant transformations" of ideas into physical metaphors, a poetic mind that, for instance, might speak of pathology as a "flower garden." When Williams speaks of twist here, he renders the physical fold or warp that characterizes shape under torsion as if it were the transforming ability of the imagination itself turning or troping. Williams renders the physicality of twisting into the very trope of transformation. A metaphorical twist embodies concepts. I have suggested that thinking, inference, or cognition itself is a patterned act of transformation of some transmitting carrier waves into new patterns of images, by which one sees or hears, not the waves, but instead the associated ideas, such as the image of a tree, or farm, or lettuce, or asparagus. In this context, poetic metaphor is a sort of *second-order* transformation that lets me construe a cognitive poetics and not a cognitive science.[32] I have maintained that the physical order of syntax carries the message even as it is transformed by a grammar into signification. So any kind of figurative language, any trope, that uses the physical acts of translation, rotation, and twist as if it could imaginatively transform concepts, such as pathology, into putative bodies, can be seen as a second-order transformation.

Hence Williams's *metaphorical* twist of the concept "imagination" transfigures imagination itself into a faintly embodied shape. Twist seems to be the primal trope of embodiment in sentences. The person who imagined a "boy-foot bear with teak of Chan" gave bodily shape to an impossible concept by imaginatively transforming the syntactic positions of phonemes. Similarly, by way of three repeated occurrences, coincidence is revealed as the faintly embodied *shape* of the Devil's Hand. Although Robert Frost said that all thinking is metaphorical, it may be useful to draw a distinction between the fundamental but unaware transformations of cognition, and the self-aware transformations of metaphor, jokes, hypericons, and other tropes: I do not want to lose sight of the fact that the physical order of transmitting waves tends to get occluded when the only physiology that one can find in poetic tropes is the physicality of concepts such as "pathology," seen as a flower garden. In other words, because many of us know so much about poetic metaphor, we may forget what Williams as physician never lost sight of—that bodies and embodiments are not the same kind of creatures. Bodies touch us, and by way of the five senses we touch other bodies in sending gestures and strokes and caresses, for nature is that part of the message which strokes the physiological signal or sign. Embodiments are tropes that give second-order shape to abstract concepts: "Time has wings."[33] The healer and caresser of bodies, Williams as poet and physician never wanted readers to lose contact with the nature of bodies. As he said in *Spring and All*, "the work of imagination is not 'like' anything but is transfused with the same forces that transfuse the earth."[34] I take it that this is claimed in the same spirit as the shih painters and poets, as well as Fenollosa and the sculptor Henry Moore, who adapted their lines, strokes, and shapes from the earth's transfusions. I take some time to spell this out because the word *twist* is such a familiar trope or archetype that it is easy to lose sight of its first-order role as cognitive transformer of rhythmically shaped sounds.

I conclude that syntax in language has a natural substructure that is transfused with the earth's rhythms, and that poets transform those strokes, turns, and shapes to their own hypothetical embodiments. (For example, in Old English the word for poet was "scop," related to modern English "shaper.") As Wittgenstein said:

> If the formation of concepts can be explained by facts of nature, should we not be interested, not in grammar, but rather in that in nature, which is the basis of grammar? Our interest certainly includes the correspondence between concepts and very general facts of nature. (Such facts as mostly do not strike us because of their generality.) But our interest does not fall back upon these possible causes of the formation of concepts; we are not doing natural science; nor yet natural history—since we can also invent fictitious natural history for our purposes.[35]

CONCLUSION

Symmetry, Conservation, and an Aesthetic Point of View

Synopsis

I conclude by pointedly asking, Why does symmetry theory help describe the changes of syntax in language? I believe that I have demonstrated *how* symmetry theory serves to reveal a physical syntax in the acts of transformation, but *why* does it seem to be inherently related to the composition of rhythmically patterned sentences? How should a natural syntax, as defined by symmetry and construction, be considered from an aesthetic point of view?

I have claimed that syntax in natural language belongs to a larger group of orderly transformations in the physical world, and I have correlated physical orders of syntax to the archetype of wave transformations. That is, I have shown that the three kinds of modulations in wave forms—transverse, horizontal, and torsional wave motion—coincide with the transformations of symmetry theory: translate, rotate, and twist. These few basic strokes of composition, which derive from the archetype of wave forms, serve as signs of transformation in several arts. I have also conjectured that a generative syntax in language must be part of a larger system of wave patterning in the world. Some, however—perhaps even some literary theorists!—may ask, how do symmetry transformations actually relate to the common orders of changes in the physical world? What is it about symmetry that makes its transformations helpful to highlight compositional principles of a natural syntax? How does symmetry help to order things in nature? Because natural syntax, as I broadly define it, comprises the elements of orderly transformation in the physical world, symmetry may be seen as the regularly patterned set of permutations within a syntactic group that occur in enacting any physical composition or assemblage.

For physicists, symmetry is a function of the conservation of energy.[1] Wherever one finds a symmetry one finds a conservation principle, and vice versa. The forms of symmetry contribute to the conservation of energy. Because conservation of an energy is a *balanced* state in which the particles are held in tight equilibrium, symmetries of rotations and spins are said to hold the components in place, much like a child's spinning top gyroscopically maintains its verticality by rotating on the tiptoe of its own axis. So when I speak of 'composition' at large, I mean the elements of composition that stabilize physical quantities of the world together into periodic orders of harmonious spacing.

Finally, I describe here the phenomena of symmetry breaking and symmetry sharing. Whenever a symmetrical form is broken, its equilibrium is disturbed by some asymmetrical interruption, a dissymetry. Because things tend elastically to return to equilibrium, they return in symmetrically patterned shapes, like the shapely fall of rain drops, tear drops, or the splash of spilt milk. Wherever a dissymetrical object plops into still water, ringlets break across the surface, carrying a form of energy to the margins of the pond. Wherever the brain of a fish drives ripples through its body, the wave forms disturb the surrounding water and propel the fish through a complementary series of alternations and contractions of the water molecules. Although waves can be read as signs of a disruption, they can also be interpreted as preliminary signs that the system will return elastically to stability or equilibrium.

Thus within a dynamic medium, the basic sign of symmetry breaking, transference, and sharing, seems to be a wave form. Wherever symmetry breaks from one dynamic medium to another, the form that carries the energy is some aspect of a wave. As we have seen throughout this book, a wave form is that helical rhythmic pattern that transfers energy from the metal of xylophones to air, from air to water (as in sonar), or from a lyre to an ear. Because wave forms traverse different physical media, they seem to be the basic carriers through any field. Similarly, any disturbance of an air wave by a speech act will be shaped by symmetrical transformations into the formal patterns of a common syntax. The great laws of form must be transmitted in formal waves whose formulas derive from structural stabilities of bodies in different phases of transition. In this Conclusion the interdependence of symmetry and conservation is described within several art forms by the few basic strokes of composition that seem to suffice as an anatomy of natural syntax.

Push and Pull as Alternating Points of View

A sentence, composed of rhythmic speech acts, is a composition of phonemes physically assembled into periodic order and read as words, phrases, parts of speech. I have suggested that when you study the composition of a sentence, you can look at its grammar or you can look at its syntax, but you can't focus on both at once. Even in reading poems aloud, when you are led by the voice phrasing the syntax of phonemes, you feature one aspect or the other, but not both simultaneously. This twofold composition has been a premise of linguistics and semiotics since Saussure compared a linguistic sign to two sides of a sheet of paper, one part being the Signifier and the other the Signified. Now the physical relation between conservation and symmetry has the same kind of alternating switch in perspective as one of its features, at least when the composition is being described in plain language.

For example, crystallographers were early students of symmetry theory. To many amateurs, the crystalline elegance of snowflakes is one of the early intuitions that the physical world snows everywhere in hexagonal economies of lovely order. This intuition about symmetry sharing is felt by almost everyone who knows snow. A crystallographer like Shubnikov can say that symmetry "has two opposing aspects: transformation (change) and conservation (invariance); the set of transformations which keeps something invariant is its symmetry group" (*SSA*, ix). In this definition, the symmetry transformations actually hold the figure in invariance, so that symmetry is doing the conserving. This definition folds the bicameral point of view under symmetry.

Physicists, however, tend to include symmetry under conservation laws in their definitions; for instance, they speak of the invariance of electrical charge in terms of spin or angular momentum. In Chapter 3, I quoted Feynman as saying that Poincaré was the first thinker to study the equations of physical laws by way of symmetries in space-time. Here, for example, Edelman explains how the conservation of quantities is "*formally*" correlated with principles of symmetry: "the conservation of momentum corresponds to the symmetry of space under translation. The conservation of angular momentum corresponds to the symmetry of space under rotation. The conservation of energy corresponds to the symmetry of time under reversal of direction. (Time reversal cannot actually be carried out, but the physical laws can be checked for their invariance under such operations)" (*BA*, 202). Note in Edelman's sentence that the last conservation law—conservation of energy—seems to require a more explicit symmetry transformation, unnamed by Edelman, if it is to coincide with translation and rotation in the previous sentences. I have shown that the symmetry operation in space-time transformations is a *twist*, and that its sign is a helical wave format, which only apparently reverses direction. So to complete Edelman's parallel construction, one could theorize that the conservation of energy corresponds to the symmetry of space-time under a twist.

Furthermore, in one of his later essays, "$E = mc^2$," Einstein shows in one of his own drawings how conservation laws, including his famous formula for energy, are essentially principles of balance or equilibrium.[2] His own rough line drawing of the swinging semicircular arc of a pendulum illustrates how balance or energy conservation is described in a symmetrical arc of partial rotation. When the pendulum stops, it hangs straight down, with its vertical axis implicitly delineating a bilateral symmetry according to the law of gravity. He observes that friction stops the pendulum's velocity, and that friction converts into heat (E, 51). Over time, following Carnot, it gradually became clear that the production of heat should be equated with the expenditure of energy. As Einstein considers the implications of his own diagram, he remarks that "from time immemorial" the relation between heat and friction has been preserved in the Indians' use of fire-making drills (E, 51). This is the same tool as a bow lathe mentioned in Chapter 3, in which I described the relation between translation, rotation, and twist.

Einstein thinks that the energy transformations in a pendulum use the same principles of symmetry sharing as the rotary motions of a fire drill. He drew the pendulum as if it were swinging off its vertical axis. When a pendulum hangs straight down, like an architect or surveyor's plumb line, its up-down line is similar to the vertical axis of a bow lathe. The energy transformation by transference from one physical medium to another can be thought of as a symmetry sharing because the rotation of the vertical stick transforms friction and heat into wave patterns of fire and light.

The archetype of a bow or *arcus* is also the elastic archetype of push-pull. One pulls and pushes the bow back and forth, working the shoulder and elbow joints in a horizontal motion, in order to make the stick rotate around its own vertical axis. The bow string is wound around the vertical axis in a helical design that transforms and displaces but also conserves the energy in another locale. Symmetry is being transferred and shared, but it is the symmetry of energy in rhythmically shaped packets or groups of wave patterns. By means of one's ball-and-socket joints, one exchanges and displaces packets of force and form. So a meditation on a bow lathe is a sensorimotor meditation about the symmetry sharing between work and heat and fire and light.

Wherever energy is transferred to another region, no energy is totally lost. Where it is lost entropically in one region, it is conserved elsewhere, but it must be transformed into other forms of symmetries, as with the transferences of power that occur when bow and the lyre are plucked. Here too, as with the workings of Dennett's crane, one sees that the transfer of energy from one form to another is accomplished by the transformations of *translation, rotation,* and *twist.* The energy of mass, velocity, friction, and heat are interrelated through their transformations from one kind to another by virtue of this conservation law, while the symmetry is illustrated by the arcs and rotations and twists that accomplish the displacements. The back and forth action-reaction of the bow lathe is constructed upon the same compensatory principle of elasticity as the bow and the lyre. In fact the back and forth motion is but a slow version of the reverberations seen when a taut string is plucked upon a bent back limb. The common principle of elasticity, mentioned in Chapter 1, may be applied to symmetry theory by extending Newton's third law of motion: for every action that chaotically breaks up a symmetrical design there is an equal and opposite reaction that conserves the energy in another locale. This principle of transfer and transformation has been called a "symmetry compensation law." As Shubnikov and Koptsik say, "If symmetry is reduced at one structural level, it arises and is preserved at another" (*SSA*, 348). I hope by now that a word like "law" may be, in my text at least, translated as a distribution principle about rhythm. This compensatory principle aims toward a balancing or stabilizing state which is at the heart of a phrase like "the conservation of energy." Since the principle of compensation is also one of transference, it may be included in the statement of Trefil and Hazen cited earlier, that all of the different forms of energy are "interchangeable" (*TS*, 73). We have seen that the forms of energy are interchangeable by means of a transferring wave motion, whose formal elements are seen as symmetries. All these axioms about wave energy may be found in the actions of the bow lathe.

The interrelation between conservation and symmetry implies that these two terms are complementary perspectives taken upon the inherent atomic structures of a matter under scrutiny. (Recall from Chapter 2 the surmise of the metallurgist C.S. Smith on complementarity as a function of "levels" of observation.) When I say that symmetry transformations of *translation, rotation,* and *twist* may be derived from the helical form of a wave, I mean that conservation and symmetry are functions of the periodically oscillating atomic wave packet that is under scrutiny. I do not mean that symmetry is a transcendental cookie cutter, but that conservation and symmetry are alternately perceived as inherent properties wherever there is matter. All of the different elements of the periodic table of elements allow for the chunking of molecules into compounds of latticed symmetries. Organic chemistry is the study of these living lattices, and Martin Gardner's chapter on "Carbon" is an excellent review of this building block of life (*AU*, 109–121).

Again, when seen from another point of view, matter is a form of wave. As Stephen Hawking has said, "quantum mechanics tells us that particles are in fact waves, and that the higher the energy of a particle the smaller the wavelength of the corresponding wave."[3] Since each element of the periodic table, including carbon, oscillates according to the character of its atomic order, each element has its own characteristic wave shape. The quantum shift in point of view means that an oscillating wave frequency is the measure of its periodicity for each atomic particle, while the symmetry of its atomic structure is a freeze-frame of its rhythmic motion. From one point of view an atom of carbon looks like a static tetrad; from another point of view it looks like a whirligig. In between it may look like an aesthetic weave of potential energy, a pregnant moment, which could be interpreted as a spatial thing or as an incipient event, according to the point of view of one's heuristic intent. These inverse points of view, as we have seen, are not just reciprocal, because one must twist and weave through a third dimension to solve either for one or the other.

My aim is not to review the intricacies of symmetry and conservation in these sciences. The mathematics is certainly far beyond my competence. But by suggesting that symmetry transformations are essential in composing homeostatic states in the physical and organic world, I can also suggest that the symmetry transformations of phonemes in sentences can lend themselves to conserving the physical balance of any rhythmic sentence as it is being spoken and as it is being heard. When I perceive and convert waves, I receive certain forms of information that are coming to me in frequent packets. The very impression of wave forms upon my senses is the tacit awareness that I am receiving physical information composed of certain kinds of elements whose character is definable by conservation and symmetry. To describe the rhythmic order of a sentence, I need to describe its elements of harmonic measure, its musical syntax, perhaps its frequency and amplitudes, and not just the grammatical elements of subject-verb-object. I think that I have shown in the last chapter that the rhymed natural syntax of the poetic line actually moves the thought and completes Plato's idea, quoted in the Introduction, that a sentence, in its weaving together of subject and verb, is an assertion that gets us somewhere. In other words, I suggest that the use of symmetry transformations is no fluke, but

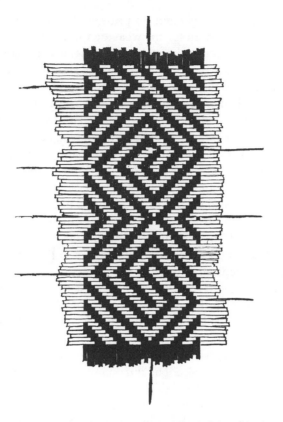

Figure 19. Weave of a Greek Fret. Drawing from *The Origin of Art* by Gene Weltfish, Plate 20.

that symmetry theory regularizes in protogeometric orders what we intuitively know about composing in the arts. So I end with a few examples that feature alternative points of view about symmetry and conservation in their elements of composition, which also show that they share a common natural syntax.

Consider first a homely example of weaving. Figure 19 illustrates what has been called a "Greek Fret," but it is also called a "meander," or running spiral pattern, that is common to most weaving cultures.[4] Its design has been found carved on mammoth ivory dating from the late Paleolithic era. Gene Weltfish illustrates many such weaves to suggest that the numerous patterns of weaving were probably the origins of similar patterns in other art forms, such as designs painted on pottery. Weaving, for Weltfish, was the origin of art. The pattern for the weaving motion in Figure 19 consists of the weft or woof (white horizontal strands) being placed under two and over three strands of the warp (black horizontal strands). The pattern then expands outward from the center point by one warp strand until the outer edge is reached; it then reverses itself by one warp strand back to the center. This kind of instruction is sometimes called a "process" description. It is a recipe of successive injunctions to actions in regular steps: "Do this now and that afterward." But look at the overall pattern of over-and-under white and black strands. A weaver could check a "blueprint" of the completed design, drawn in the sand or on a sheet of graph paper. Unlike the recipe, this kind of visual diagram (Figure 19) is not a process description of successive directions. It seems a static panorama of the whole symmetrical product. A blueprint lets one see how the light and dark strands correspond in spatial arrays. (Adjudicating sunlight and shadows is one way we see the third dimension of depth, by chiaroscuro.)

During composition, the weaver can check the emergent design for mistakes in the plait. In this way, the sequential recipe of enactment rehearses the narrative, while the overall visual design, the illusion of a running meander, retrospectively reveals the plot (plait). The interwoven form seems to reciprocate between injunctions in sequential time and between correspondent images in space. The illusion of a running fret twisting between succession and correspondence provides the most common archetype for space-time, its interwoven curvature.

In Chapter 2, I introduced creation myths with a Dogon account of spirals and weaves as the essence of language. Just how completely the technique of weaving permeated archaic cultures may be understood by quoting a realization of Marcel Griaule about the evolution of the Dogon Word. Remember that among the Dogon people, the first word was like a twisted, pendulous strand of fiber. Griaule was told that there were three stages in the evolution of the Word, all of which were associated with stages of weaving. The first stage was the primitive helicoid breath. The second stage in the composition of the Word was associated with a more sophisticated set of combinations, which was likened to warp and weft of horizontal and vertical acts of weaving. But the third and final stage was associated with the very sophisticated drum language of the Dogon, which I have not described at all.

The reason why drums are associated with weaving is that the tympanum of the drum is lashed to the barrel in sophisticated lacing and knots.[5] You must visualize with your inward eye the angular lashing around a circular drum shape to get an idea that the symmetrical angles of lashing, distributed around the body of the drum, actually conserve the tension of the drum skin so that it can reverberate. Here is Griaule's reflection:

> The first imperfect Word was associated with a technical process, simple in character and no doubt the most archaic of all processes, which had produced the most primitive form of clothing made of fibre. The fibre, which was neither knotted nor woven, flowed in a wavy line, and it might be said therefore to be of one dimension. The second Word, less restricted than the first, arose from weaving, done on a wide warp crossed by vertical threads forming a surface, that is to say, having two dimensions. The third Word, clear and perfect in character, took shape in a cylinder with a strip of copper winding through it, that is to say, in a three-dimensional figure. (*CO*, 67)

The over-and-under rhythm of weaving seems to unify the runnel patterns of elements from early to late, from Plato's weave of subject and verb to quantum physics. This is the twist of natural syntax. What is most instructive in this meditation on the development of the Word in Dogon language is that the acts of composition are physical acts of weaving that exist through the several dimensions of the physical world. The elements in composing a set of implements in one, two, and three dimensions are associated with the development of their Word. As Griaule also realized, "These three technical processes . . . all proceeded by following a line, either undulating or zigzag . . ." (*CO*, 67). The zigzag, or off-angled S curve, is invariantly the sign of an overall stepped-fret design. All these processes begin with an actual twist of a line, over and under, which is the first step in the topology of knotting. *In many elementary acts of composition, you translate an element along a one-dimensional line, you rotate it on a plane so as to augment the surface, and you twist it through three dimensions to shape it in the round. These are the primal transformations of natural syntax.*

This interwoven dimensionality of physical things is the simplest beginning of physical composition. It comprises a sufficient anatomy of physical syntax. The sufficient elements of composition in this anatomy in turn can describe the transformations in weaves, speech acts, and drums. Symmetry, conservation, and dimensionality are concepts whose principles are interrelated in elementary compositions. Now I can refer to the earlier diagram of compositional instructions, which includes the several dimensions of symmetry transformations (Figure 3A). If weaving is a series of periodic twists, then any sentence is similarly composed of an off-angled (or fractional) point of view that weaves syntax and grammar into an assertion.

To reinforce this association of weaving with the composition of carrier elements in the physical world, consider the tools described in a Navajo account of the origin of weaving. This story was collected in the 1890s:

> Spider Woman instructed the Navajo women how to weave on a loom which Spider Man told her how to make. The crosspoles were made of sky and earth cords, the warp sticks of sun rays, the heddles of rock crystal and sheet lightning. The batten was a sun halo; white shell made the comb. There were four spindles: one a stick of zigzag lightning with a whorl of cannel coal; one a stick of flash lightning with a whorl of turquoise; a third had a stick of sheet lightning with a whorl of abalone; a rain streamer formed the stick of the fourth, and its whorl was white shell.[6]

These zigzags (like Issa's lightning flash and Griaule's zigzag line discussed in Chapter 1) signify the design in whorls of colored shapes. The four traditional colors are here, as well as several kinds of lightning weather, and they probably are to be associated with the four directions of the loom, which has its own bilateral symmetry of horizontals and verticals. The four basic minerals are here too. Abalone shell, a prized trade item found even in Anasazi and Mimbres ruins, flashes yellow iridescent, while black cannel coal burns bright in spurts when lit. One can see how thermodynamic spurting whorls are structured in the very tools that send their energy through the weaves. The symmetrical elements from the Four Directions of the earth and air comprise the natural syntax of weaving.

The Eurocentric term "Greek Fret" may be described as a pair of interlocking zigzagged spirals that coil out from a point or node with arms or legs radiating indefinitely outward into the Four Directions. For example, after mentioning serpent cults in Bali, India, and Mexico, Covarrubias says, "American two-headed serpents often assume the form of a double spiral, curvilinear or angular, a motif identical with the *lei-wen*, the lightning pattern basic to the design on all Chinese bronzes" (*EJS*, 45; drawings, 54). You can see how the form of the zigzag reenacts the physical force of lightning. Everywhere the ancients saw this zigzag as a symmetry-breaking and a symmetry-sharing, in which the archetypal sign of violent interruption is also an index of the return to thermal equilibrium, much like a sneeze may be a sign of an incipient cold but also immediately serves to restore a thermal equilibrium to the body after a shiver of chill. Furthermore, Baldwin cites still another book by Covarrubias, in which is described the amazing stepped-fret stone designs on walls and columns in the pyramid at Mitla: "there are over twenty varieties of patterns in these [wall] panels, all based upon a *single motif*, the stepped spiral called *xicalcoliuhqui*, perhaps the most characteristic of

Middle-American aboriginal art motifs, derived from the stylized head of the 'Sky Serpent,' and thus a symbol of Quetzalcoatl" (his italics).[7] This remarkable convergence of patterns occurs across the world as a common syntax of topological composition—of weaving and stepping strands of fiber over and under, with variations of translation and rotation to enact the recipe. The first step in the natural syntax is a *Twist*. The terms *Greek Fret* or *swastika* or *lei-wen* or *xicalcoliuhqui* are socially constructed parts of indigenous lexicons and grammars, but the composition of the syntactical elements is a commonplace group of physical pattern variations. In these primal arts a variant of the stepped fret was distributed through the world as a common design under different names. In itself it represents a distribution principle of interrelated arms, legs, branches. Its commonly shared design of the distribution of fretted patterns is an example in the primal arts of a symmetry compensation law.

In pursuit of conservation and symmetry as compositions of force and form, it is worthwhile to diverge from weaving momentarily in order to learn a bit more about the connotations of lei-wen. In Chapter 2, I mentioned in passing that the motif was associated with the t'ao t'ieh. Lin Yutang describes the terms separately, together with their ideographs, under the introductory description of "Esthetic notions borrowed from physical objects in general":

> *wen*: originally "grains of pebbles, ripples on water, waving lines of objects (e.g., brocade)," now meaning "literature." The fundamental idea is the natural lines of movement or beauty of lines and form, and when applied to writing, it refers to the movement of one's thoughts and language. We speak also of the "whirls" or "eddies" of one's literary composition (*wenchang p'olan*), describing the curious over-lapping and back-and-forth twists and turns of the author's thoughts. (*IL*, 437)

For Yutang, the aesthetics of Chinese art and literature derive from the dynamic patterns of nature, but autumnal trees, for instance, which are pinched by cold and "busily occupied in slowing down their breath and conserving energy" are not dressed in stark patterns for the appreciation of artists (*IL*, 394). Their wen, or aesthetic lines, are a function of conserving physical energy. In this ancient theory also, conservation and symmetry are interrelated through design. That a literary composition reenacts natural twists and turns is now, I hope, understood as an archetype, not a metaphor, of back-and-forth, elastic push-and-pull patterns of force and form. Note that the waving lines of objects are exemplified in the striations of brocade, which is what one sees as the pleasing off-angled point of view upon the intricacies of the woven silk. Weave and wave are always formally implicated. *Wen* is a serviceable word for the appearance of natural patterns in the grains of things. So this excursus from the woven Greek Fret is only apparently a diversion. One can see that force and form are drawn from the dynamics of physical objects in order to describe literary composing.

To return explicitly to symmetry and conservation, I note that a principle of conservation is not necessarily connected with a woven Greek Fret. The intended use of the weave—whether for a basket to hold liquid or grain, or for cloth to hold in warm energy—is to condense energy against dissipation over time by interlacing a container. The first conservation law, that energy is not lost wherever it might be distributed, may be accomplished without the Greek Fret. But some kind of zigzag or curvilinear fret will necessarily emerge off-angled from any regularly sequenced code for plaits over and under, as in the waving lines of Chinese brocade. In this kind of traditional use, conservation and symmetry are correlated through the habitually skilled movements of eyes and hands in an economy of means. Their commonsense meanings in everyday use are still rooted in the language of contemporary physics. Conservation is counted by economical recipes for saving over time, sometimes despite time, and symmetry is measured as a set of visual, geometrical, and mathematical transformations that favor fittingness in space.

Again, the tacit assumption, even in the most archaic accounts, is that the act of physical composition conserves energy through symmetrically stepped fretwork. And the most complete form of composition is one that includes the three dimensions of space, plus the ringer of a temporal dimension. It is a ringer because, as Feynman suggested, following Einstein's scientific revolution, there is said to be no absolute distinction between space and time. But space-time is a push-pull phenomenon of off-angled points of view: to solve for space, one must use time; to solve for time, you must use a spatial entity, but you cannot solve for both at once. (For instance, Time = Distance / Speed.) In what follows, I test this idea about elementary composition with a few more examples from different dimensions of crafting.

To check these perspectives on composition, consider another kind of folk assemblage, "coppo" roof tile (Figure 20). In ancient Greece, it was already a popular style, for it can be seen as the roof structure of houses that were painted on early Greek pots. Widely used around the Mediterranean, the tile remains popular for adobe roof reconstructions in California and the Southwest. Each roof tile is a half cylinder open at both ends. A two-dimensional drawing of a half cylinder would begin with an arc of rotation. Its act of composition in the round is a process of stacking the clay tiles

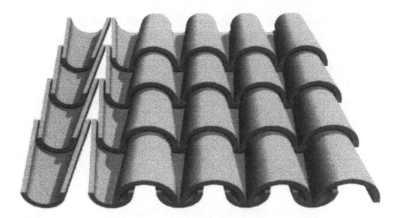

Figure 20. Coppo Roof Tile. Drawing by Anthony Rozak.

over and under one another in layers. The process is a kind of weaving with clay shapes in three dimensions. Its recipe is to start at the lower edge of the roof and cement a few half cylinders horizontally in a row so that they curve concavely up to the sky. Then one interlocks a few half cylinders by fitting them upside down over the concave tiles so that convexity is up and concavity is down. Two layers of tiles are interlocked by alternating the upside down elements with right side up elements. In terms of symmetry theory, one recalls that every displacement is either a translation, rotation, or twist. So the symmetrical recipe might sound like this: for each layer of half cylinders, rotate the next layer 180 degrees and translate each one of the tiles one half space horizontally so that it fits into the space made by the opposite curve beneath it. The tiles fret horizontally and vertically and in depth in a loose but effective mesh. And their weight orients them up and down in a gravitational interlock. In Figure 20, the tiles have been tapered at one end to ensure a snug fit.

But if you look obliquely at the total patterned array on a coppo rooftop, orthogonally, as from a third-dimensional point of view, you can see that the design is like a twisting wave. The pleasing rhythm of the design (wen) is literally a derived side-effect of the intention to channel water down a slope. The symmetry and the conservation are interlocked. Depending upon the point of view, whether seen as a process or a product, as recipe or blueprint, or as a third form, an aesthetic design (wen), the composition will "mean" something different. Its conservational end is to construct a minimum stack of invariant units so that they will channel rain in least angles off the roof slope. The water follows a symmetry law of translation down the gravitational slope of the channeled ducts. The recipe is a syntactical code for generating or stacking physical signifiers economically. But the off-angled side-effect can also be seen as an aesthetic symmetry, with illusory running ripples (*wen*). That phenomenon exemplifies what I mean when I say that conservation and symmetry are off-angled points of view upon an event in space and time. Recipe, blueprint, but also curvilinear stroboscopic effect—these are the points of view I pursue.

For a poetic example, Gary Snyder's "Riprap" lays out words like stones in cosmic arrays.[8] A head note to the book says that riprap means "a cobble of stone laid on steep slick rock to make a trail for horses in the mountains":

> Lay down these words
>
> Before your mind like rocks.
> placed solid, by hands
>
> In choice of place, set
>
> Before the body of the mind
> in space and time:
>
> Solidity of bark, leaf, or wall
> riprap of things:
>
> Cobble of milky way,
> straying planets,
>
> These poems, people,
> lost ponies with
>
> Dragging saddles—
> and rocky sure-foot trails.
>
> The worlds like an endless
> four-dimensional
>
> Game of *Go*.
> ants and pebbles
>
> In the thin loam, each rock a word
> a creek-washed stone
>
> Granite: ingrained
> with torment of fire and weight
>
> Crystal and sediment linked hot
> all change, in thoughts,
>
> As well as things.

Like coppo roof tile, riprap is laid in stabilizing steps against the grain of an inertial slope. Its compositional end is a utilitarian desire to redistribute gravitational force so that one can walk up and down a slippery hill. Here too, a metonymic pattern is overlaid contiguously and successively down the hill, horizontally, against the slope of gravity, so as to interweave the order of things and the order of words into an aesthetic point of view of fire and weight that can poetically see worlds as a four-dimensional game of "Go." If you imagine the grid of a "Go" board, like a checkerboard, upon which a hundred black and white stones have been arranged in tactical patterns, then you can see that Snyder has remanded the game to its origins, as a divining board, but in this case the divination is accomplished by a dilation from a small pattern of pebbles to a cosmic model. In the case of "Go", the totality of arranged pebbles may be seen as having elegant patterns of symmetries (*wen*). But the intent was to attack and defend (as in Yutang's definition of *shih* as "battle formation"), to conserve power, to lay each stone in a reticulated network of orders of things. Here, the syntactic act of assembling words into spatiotemporal, meaningful orders is superimposed upon the physical act of laying riprap. Laying stepped riprap and laying roof tile is like laying the stones in "Go." All are designed for conservational purposes, with lines of communication and resistance that are products of their syntactical ordering codes, but the resulting total display may be aesthetically pleasing. There is an equivalence and an equilibrium as a result of the superimposition. Conservation seems to be a physical state that depends upon the physical act of assembling pieces of elemental stuff, into sequential, horizontal, and four-dimensional arrangements, whose stepped-fret result is a stability formulated by symmetries. Where one works to conserve energy, one uses repetitive patterns that will appear symmetrical. With "each rock a word," the equivalence allows the syntax of the poetic sequence to become the forms of the four dimensions of the world. One stacks up words like rocks against the slippery slope of gravity.

Even though noone knows what gravity is exactly, everyone always *feels* its force and compensates. As Coleridge said:

> In every voluntary movement we first counteract gravitation, in order to avail ourselves of it. . . . Most of my readers will have observed a small water-insect on the surface of rivulets, which throws a cinque-spotted shadow fringed with prismatic colors on the sunny bottom of the brook; and will have noticed, how the little animal *wins* its way up against the stream, by alternate pulses of active and passive motion, now resisting the current, and now yielding to it in order to gather strength and a momentary *fulcrum* for

a further propulsion. This is no unapt emblem for the mind's self experience in the act of thinking. There are evidently two powers at work, which relatively to each other are active and passive; and this is not possible without an intermediate faculty, which is at once both active and passive. (In philosophical language we must denominate this intermediate faculty in all its degrees and determinations, the IMAGINATION.)[9]

This wonderful archetype for the alternating stream of thought exemplifies in itself the new standard of German transcendental thought, an aesthetic third way of thinking, here called "Imagination." The aesthetic of judgment is that third kind of thinking that, for Kant, mediates contingent probability and rational logic.[10] Here, however, I am less interested in the history of the question than in the Heraclitean archetype of a physical current that propels action and thought. How does thought "get us somewhere" against the necessity of gravitation? In addition there is a refractive point of view that prismatically lights the water bug's five-pointed shadow along the bottom, and then a third point of view that effortfully twists the image to an imaginative trope for voluntary action impelled by countervailing thought.

Whenever one worries over the relative differences between the inevitability of the physics of motion as averse to voluntary human action, Coleridge's poetic point of view can be recalled. His "momentary fulcrum" is a kind of elastic jujitsu, of riding with the natural force in order to twist and avail it, like the old Taoist, quoted in Chapter 1, who lived in and by the flowing water. This kind of momentary physical fulcrum, this elastically moving proportion in space and time, sustains an uplifting twist of aesthetic liberty won momentarily from gravitational necessity—at least that seems to be its "twist of imagination," as Williams would call it. Coleridge's trope also helps make my point about conserving energy through symmetry. Coleridge does not argue that there is a necessary connection between symmetry and conservation principles, but he does note that a shadow of a pattern is projected on the bottom in a fivefold shape. "Cinque" is a descriptive label that is usually botanical nomenclature for the symmetrical distribution of leaf petals, as in the flower cinquefoil. While the slant of sunlight is bent as it travels through water, its light is also broken by the body of the insect. But the light waves carry the broken patterns that one sees as symmetrically distributed shadows projected onto the bottom. In sum, Coleridge's extended trope for Imagination suggests that thought itself, at least the original kind of thinking that falls under the term "imagination," is not a stand-alone lexical concept, but truly a rhythmic pattern of thought because patterned symmetry is to be found everywhere.

If all motion through three space is an asymmetrical twist, as we saw in Chapter 3 about the sun's symmetries, then the organic act of locomotion is a rhythmic counterforce to the pressure of gravity. And so imagination is an intermediate faculty twisted from physical necessity by a riprap of natural syntax. Imagination is an elastic form of thought that is everywhere characterized by "leaps." One must move obliquely against the grain of a slope to work toward one's ends. These very solid acts of weaving strands in several dimensions, of laying tile, of placing riprap, of setting words in their syntactic orders, can be seen off-angled from an aesthetic point of view as voluntary acts that are for Coleridge and Snyder parts of cosmic patterning.

Symmetry Sharing

Everywhere one looks there are to be found reflections of symmetrical pattern. Coleridge makes the projected design of a water bug part of a shared assumption among his perceptive readers. "Most of my readers will have observed" such motes and mites of pattern. That tacit assumption is the basis for Weyl's list of the most distantly related objects that share symmetrical organization—stars, snowflakes, flowers, veils, hearts, quantum energy, light, and gravitation. How is it that every body is correlated by actions of symmetry? Early in the twentieth century, the physicist Pierre Curie speculated that any physical body will take shape in the simplest symmetry available to it by means of the energy available to it in its local region.[11] A shape is shaped by the simplest actions of symmetry and conservation principles working together in that locale. This simplicity principle is called by Matila Ghyka a principle of "Least Action." From atoms to asteroids, all compounds will take regular periodic shape in their simplest forms, according to their physical principles of composition—that is, the natural syntax of conservation and symmetry. These optimal shapes, like silver firs and fiddlehead ferns and cinquefoil, are thus expressions of physical stability. The atomic shapes are twisted and woven and spun together into stable simplicities of relative hardness and smoothness and elasticity. Each organic symmetry is a temporizing or "momentary fulcrum." Any physical act of composition, even the phonemes of sentences, is governed by these optimal measures. Everywhere one looks in the far reaches of the universe these periodic oscillating measures can be found correlating energy by symmetry into simplest local shapes, like the bow and the lyre. Think of a saline sea that dries into salt crystals, or think of the Anasazi ladle squeezed into Y-shaped branches; all these simple shapes assume an optimal form that is mathematically related to geometry by conservation principles. Pierre Curie also wrote that an interfering break or disruption of symmetry in one place necessarily prompts a reciprocal adjustment toward a redistribution of stable symmetries in optimal forms at another level or in another place.

This kind of "symmetry breaking" at one level, only to appear as "symmetry sharing" at another level or locale, is the leading hypothesis of Stewart and Golubitsky.[12] If you dent a plastic ping pong ball, the dent should appear as a spherical concavity. In other words, when a simple state is disrupted, another symmetry will appear nearby in another transformation. The sunlight, which is broken on the surface of Coleridge's brook as light moves by the form of a wave from air to water, projects the shadow of a five-spot symmetry on the sunny bottom. Broken at one level, a symmetrical image, much like a flower, is transformed at another level. Although "symmetry breaking" is a new explanatory theory, the observed phenomenon that patterns adjust into other patterns is everywhere, as old as art and poetry: "Abide with me / Fast falls the eventide." So, for example, symmetry sharing provides another way of thinking about the patterns of the most famous haiku of all: the surface of the still pond is broken, but one doesn't look at the symmetrical shared rings of rivulets; instead, one attends to the phonemic interrupting sound of the frog's "plop," with the correlative pattern of thought that the sound doesn't call attention to itself, or to the frog, but rather to the surrounding silence.

Like Coleridge's water bug, a leaping frog avails the push of gravity in order to surmount it. The enormous but tacit weight of gravitation is governed on earth by our sun, and its sunlight lets us see the bug, the frog, and the trees. But what, more exactly, is sunlight, and how is light a function of principles of conservation and symmetry? Light is the portion of the electromagnetic spectrum that can be seen with the naked eye. The electromagnetic spectrum is usually depicted as a diagram that displays by temporal measures of frequencies all the known physical waves that surround and pass through us (see the diagram by Trefil and Hazen, *TS*, 158). They include radio waves, X rays, microwaves, infrared, ultraviolet rays, and gamma rays. None of them exceeds the speed of sunlight. All of these waves, with their different frequencies on the spectrum, are transmitters in the telecommunications revolution. Because every object in the universe oscillates wave frequencies according to its physical principles of composition, the radiant energy of every object can be seen as a sender of messages.

The object I see as a sycamore tree is interfering its radiant energy upon the carrier wave of light. This total environment of frequencies makes up all the transmitters of messages that we use to send and receive.

According to Trefil, the sun itself is a gigantic transformer: each second, it converts about 600 million tons of hydrogen into helium.[13] But the sun is gigantic, so scientists reckon that billions of years will pass before it burns out entropically. Sunlight is a symmetrical by-product of energy breaking and sharing elsewhere. How so? Sunlight comes from helium; its Greek root is *helios*. More precisely, sunlight is the energy that gets converted when four protons of hydrogen come together and fuse into a new element, the nucleus of helium. The first conservation law says that energy remains constant under any transformation. Although energy may be redistributed (in symmetries), nothing is lost in the exchange. Since the mass of a helium nucleus is less than the weight of the four protons of hydrogen, the difference is *equalized* in the energy by-product, which is sunlight and heat. When the sun shines on us, this is the energy by-product that we see and feel. It touches us in radiant waves. The conversion of mass into energy radiates to us as symmetrical forms, as angular rays of sunlight. Just as the sun is itself a transformer of energy, so is the cerebral network a transformer of oscillating wave frequencies into images of objects.

That slanting phenomenon of symmetry sharing is the shape that Emily Dickinson feels in her famous poem:

> There's a certain Slant of light,
> Winter Afternoons—
> That oppresses, like the Heft
> Of Cathedral Tunes—
>
> Heavenly Hurt, it gives us—
> We can find no scar,
> But internal difference,
> Where the Meanings, are—
>
> None may teach it—Any—
> 'Tis the Seal Despair—
> An imperial affliction
> Sent us of the Air—
>
> When it comes, the Landscape listens—
> Shadows—hold their breath—
> When it goes, 'tis like the Distance
> On the look of Death—[14]

Note in the first stanza the synaesthesia of light, sonority, and heft, which lend an impression of human embodiment to the feeling of despair that is sent by the slant of light. As one reads the poem, one sees how the light lends shape, in symmetry sharing, to the feeling that transforms the speaker and "us."

This symmetry sharing of sunlight, this redistribution of symmetrical shapes at every interruption, is necessary to understand my thesis about the commonality of natural syntactic elements. Dickinson assumes "we" share it. Throw a stone into still water; it follows the arc of its impetus aligned with the effect of gravity. The plop breaks the plane symmetry of the water surface, but the circular ringlets redistribute the energy in symmetrical forms across the surface to the rim.

> And the startled little waves that leap
> In fiery ringlets from their sleep
> (Browning, "Meeting at Night")

Symmetry sharing is at play here, too. The plane surface of a pond or a sea is not really a flat plane, but ever so slightly convex because the formal arc of the convex curve follows the gravitational bulge of the earth's curvature at that place, and at every place, like the curve of Emerson's transcendental eyeball. At each break in a dynamic pattern, the moving form of a helical wave is reenacted and re-formed. Since any interference of one pattern of shapes causes a new symmetry to appear elsewhere in another optimal transformation, there are to be found breaks that shape into new patterns at another level, like the crackling glaze on a Chinese jar. In a dynamic medium such as sound waves, the spherical shapes must always follow the redistributed form of the wave—translating, rotating, and twisting. In Vitruvius's ampitheatre, the spherical sound waves are amplified by the gradual rise of the semicircular steps. The propelling momentum of the carrier signal is shaped into significant shapes by symmetry sharing. The geometrical forms—arcs, circles, spirals—are derived from the propelling forms of waves.

If symmetries are replenished everywhere when broken, how then does this equilibrating phenomenon apply to speech? If spoken syntactical shapes always interfere with the form of the original carrier wave, then they redistribute the symmetrical shapes of sound around the mouth cavity by twists of tongue and slips of the lips into new shared shapes. I quoted Olson in the Introduction as saying "speech is the solid of verse." Now one knows that the feeling of "sculpted" solidity is not just a metaphor, as Pinker aptly expressed it; the shapes of speech acts are more or less dense solids, which follow the principles of symmetrical organization in wave formats.

Seen as a developmental process, a spoken language is the sum of all the ad hoc decisions that a people make when they learn how to speak it first. Every child learns one's mother tongue tacitly without the rules being explained. Everybody has a slightly different style of speech because of those learning tricks. It's like a game in which each player assumes that the rules are known but in which the assumptions are slightly different for each player, even while each player is attuned to the slight shifts in usage that suggest the dialect is shifting and its rules are modifying. This thesis is proven only by those linguists able to study the gradual changes in the sounds of a language like the diachronic vowel shifts in Anglo-Saxon.

Every child remakes the rules of the game, and they are only changed at large by all of the participants over a long time. In a discussion of rhythm in narrative Julia Kristeva asserts my premise in this way, "Children learning a language first learn the intonations indicating a syntax structure—that is, melody or music—before they assimilate the rules of syntactic formation. Intonation and rhythm are the first markers of the finite in the infinity of the semiotic process. . . . "[15]

The first sounds that a baby ordinarily attends to are the singsong inflections of a parent's baby talk. Listen for the lilt of a small child's first efforts to string words together. One hears the inflected syntax of a limited number of sound clusters moving the grammar. Any child who learns to speak will be able to shape the limited possibilities of sound sequences by this kind of symmetry sharing. By means of these commonly shared physical transformations, a child needs no other innate rules of composition to form thoughts; any first inflections of syllables will be innately patterned as groups of rhythmic sound-shapes. No parent teaches a child the *rules* of grammar; instead, one sound at a time, a rhythmic word such as *OR*-ange is spoken, prosodically learned by frequent repetition, and then algorithmically combined in a dialogue of symmetrically shaped sounds, which are usually associated with the sensorimotor shape, smell, color, and tactility of the fruit:

> Oranges and lemons,
> Say the bells of Saint Clement's
> When will you pay me?
> Say the bells of Old Bailey.
> When I grow rich,
> Say the bells of Shoreditch

In this English nursery rhyme, the sound waves of the bells have peculiarly significant shapes that carry over as shaped sounds to the synaesthetic embodiments performed in the rhyming sentences. Here the sounds of the bells and the words are composed and explicitly understood as shared symmetries. That synaesthetic symmetry sharing, which jumps from one medium to another, is the art of this nursery rhyme. The sounds govern the sense too, for the bells send a message appropriate to their neighborhoods. Old Bailey, for instance, housed the debtors' prison; hence, "pay me." Nursery rhymes are the very primers of a singsong set of symmetries that generate the limited repertoire of sound sequences as embodied words. One learns how to generate sentences wave after wave.

Fearful Symmetry

William Blake's *Songs of Innocence and Experience* are visionary nursery rhymes. Many of them are composed as a dialogue of questions and answers, such as those a child might ask: "Little Lamb, who made thee?" The same questioning governs "The Tyger." When Blake's awful creator wrenched elements of symmetry from all over the universe to twist his tyger into shape, that composing of a beast was based upon a tacit assumption of appropriating elements of symmetry from the physical world and recomposing them into a fearful energy. Blake was grasping and clasping the dreadful forces of the philosophy of natural law. A terrific cosmic patterning of thought, a fearful symmetry of conserved energy is the subject of Blake's "Tyger."[16] This tyger's fearful symmetry is a terrifying consequence of the idea that we are pattern-generating animals, for the devouring tyger shares in the symmetries of the cosmos and eats the bodies of other commonly organized energies. In the natural world of fang and claw, predators feed off the symmetries of other bodies. Even a vegetarian lamb nibbles up cinquefoil.

I have been making a case for the idea of consonance as a rhyme between the physical strokes of a medium and the presumed principles of nature that the artful strokes reenact. I have also described certain interrelated transformations of an articulated syntax—the symmetry transformations of translation, rotation, and twist—as if they were reenactments of the forms of motion in waves. These commutations are nowhere better reenacted than in the following famous poem about composition itself. The speaker in the poem imagines how the anatomy of a cosmic tiger is composed from the elements of symmetry wrenched from around the universe, and the speaker imagines that composition by way of a few elementary physical operations. As you read, recall from Chapter 3 the primal symmetry of a hocker's body as it orients itself with the Four Directions. Gradually, as you read, you conceive in your mind's eye the shape of the beast being twisted into existence. But ask yourself whether you see it or not. A twist of the imagination, it enacts an anatomy of syntax:

> Tyger! Tyger! Burning bright
> In the forests of the night
> What immortal hand or eye
> Could frame thy fearful symmetry?

First notice that the poet's questions about the composition of the beast are always framed so that fear of the beast is turned back upon the eye and hand of the creator, the composer. If the beast is so terrible, what about its Maker? It seems to be a poem about the consequences of seizing physical force and transforming symmetry into a terrible embodiment of energy. Who framed those alternating black and gold stripes that glow and flicker and stalk fearfully in one's night visions? How did it come about that the sun's energy burns at night? These are the alternating stripes that ripple upon the bilateral symmetry of a gaited hocker. Like Lévi-Strauss's tattoos of geometrical meanders upon human bodies, the vertical stripes ripple and alternate in a rhythm with the alternating motions of the haunches and joints of the beast as it paces through the night in the "old sequence" of propulsion. But the pace is alternately lit by the golden stripes burning alternately against the black stripes. Who made the "darkness visible," as Milton described hell in *Paradise Lost* (1, 63)?

> In what distant deeps or skies
> Burnt the fire of thine eyes?
> On what wings dare he aspire
> What the hand dare seize the fire?

The brief allusions to Icarus's wings and Prometheus's theft remind that the composer's fire and heat, this energy, come from the sun, or maybe from the deep. The "what" and "when" questions now start to repeat in a singsong tempo of quantitatively accumulating sounds.

> And what shoulder, & what art,
> Could twist the sinews of thy heart
> And when thy heart began to beat,
> What dread hand, & what dread feet?

In three space, the combined symmetrical transformations of translation plus rotation converge in a twist, a fold, or a warp. All the framer's rotator joints—shoulder, hand, even feet—twist the sinews of the heart. I noted that in much archaic art, the power of haunches was delineated with S-curved meanders that reenact the momentary fulcrum of power. But note that the "art" is the combined rhythmic beat of the framer's dancing hands and feet, which begin the heartbeat. This "twist" is the act that conserves energy in periodic beats. Heart and hand and feet and art all beat together through the magic of parallel grammatical constructions that repeat each other but defer the predicate. The rhythmic beat of the heart and the beat of the pacing feet of the beasts is witches' meter: "Double bubble, boil and trouble."

> What the hammer? What the chain?
> In what furnace was thy brain?
> What the anvil? What dread grasp?
> Dare its deathly terrors clasp?

An iron-age smithy, like Hephaestus's, was the place that made tools that handle and displace the fire and work and heat, tools that make tools, metatools, such as anvil and furnace; but the "what" questions finally ask about grasping and clenching, not with anvils and with displaced tongs as prostheses, presumably, but with bare hands. One feels seized by the hands, but that clasping occasions the nightmarish fear of being seized and devoured by a tyger's clasping claws and teeth. Note that "thy" is a sacred possessive that refers to the tyger, and that "its" is applied to the composer. "Thou" always attributes a sacred fear for the tyger as addressee. Deathly terrors seem first to be properties of the grasping framer, an "it" or a "he."

> When the stars threw down their spears
> And water'd heaven with their tears,
> Did he smile his work to see?
> Did he who made the Lamb make thee?

Perhaps one imagines that starlight is synaesthetically transformed into slanting spears of tearful rain in surrender to the creation of this fearful symmetry. Did the Hephaestus-like smith smile to see that his spear work, which he presumably made on his anvil, has left in its wake a night sky spotted and mutilated as if it were the landscape of a battle's aftermath? Did the victor smile as remorseless Achilles did when he chopped down the submissive and tearful Hector? Could he also have been the Maker of the Lamb? Here the symbolic contraries of force are again summoned, the

Prolific and the Devouring, like the tiger and fawn of the Scythian brooch and Penelope's gift. The question presupposes but opposes Milton's great paean in *Paradise Lost*: "These are thy glorious works, Parent of Good, Almighty, thine this universal Frame" (4, 154).

The poem may be read as a palinode that shifts the universal frame from Milton's Almighty to Newtonian gravity. Tears and spears, light and rain, both drop down in shaped symmetries according to the tacit but universal frame of gravitation. The poem reframes the evidence of natural theology; it questions the appropriations of natural law. How is energy built up? What are its steps of assemblage? What is the recipe for composing a tyger? To what end is it conserved? Why such symmetry for an innocent or indifferent brute? Is there a geometry of reciprocal atonement in this fearful symmetry?

Notice how the rhythm of the poem is built up quantitatively. Blake wrote his verse in a linear dimension of repeated and variant rhymes. He drew a picture of a tabby tiger on a plane sheet of two dimensions, and as an engraver of his own pictures and poems, he twisted his own shoulder, his wrist, his fingers at different times so as to motor the lines on metal in three space. As a framer or creator or composer, Blake himself had to appropriate the principles of natural law to create his "art." But the poem's moving pattern of thought (*shih*), though built of the laws of energy, symmetry, and gravity, moves into a visionary dimension of rhythmic and haptic periods. The periods of the poem enter our inner lives. One's own childlike imagination is made to feel the dread heart, hand, and feet of the maker's rhythms. It is all haptic, all touch; not mind, but "brain." Although the maker in the poem is described as reaching through the universe to frame this fearful symmetry, it does not seem to be a transcendental mind or ultra-consciousness. Instead, in keeping with the idea of a palinode against the order of natural law, the maker is only known through the natural forces it consolidates in the beast. Its forces pervade the universe pantheistically, but they do not seem to go beyond. Despite the graphic description of the elements that go into the making of the beast, it may be that this Tyger is not meant to be seen, or at least not seen clearly in this visible darkness. Instead one feels haptically the metonymic buildup of the animal from the making of the eye, heart, brain in a rhythmic sequence, like riprap. One envisions the beast but does not see it because the poem is more haptic and dance-like than optical. Feeling at its joints and its parts, one only glimpses the fearful symmetry of the whole. Maybe one cannot see the fearful symmetry of the beast or of the composer, for to see a night-stalking tyger is like trying to see the darkness of the trees when a light has been turned on them. For the fearful symmetry is a sort of camouflage that allows the tyger to be embedded in the cosmos, barely visible in the forests of the night, like those children's pictures that hide the figures of rabbits, chickens, ducks, and lambs in the leading lines of the landscape. The symmetry sharing is camouflaged so that it is not so much seen as it is heard as a natural syntax.

Blake summoned a fearful symmetry that one can't quite see. Its composition amounts to a revelation or apocalypse, as Northrop Frye taught in his book, *Fearful Symmetry*.[17] Yet if Blake summons a "momentary deity" here, it is the cosmic tyger that is twisted into a shape that one can't quite see.[18] And the self-referential questions that turn back upon the maker are such that one infers the composer only in the principles of the natural philosophy of symmetry and energy. The composer is not quite personified, even as the tyger is addressed. As Jonathan Culler notes, apostrophe is the rhetorical figure of addressing some body or thing that doesn't really listen (*LT*, 77).

So Blake formulates aesthetic and ethical questions about the conservation of energy by symmetry. No matter that the first law of conservation of energy had not yet been scientifically phrased. Blake studied and scorned Voltaire and Rousseau. "Energy is Eternal delight," he wrote in "The Marriage of Heaven and Hell." Why is so much fell energy composed with such fearful symmetry in such a terrifying beast?

How can a rhythmic body be designed to kill and devour? In Kristeva's discussion of poetic rhythm she makes the poet's use of rhythm itself to be the disruptive element which works against the stabilizing desire of the State; that is why fascists of all stripes wanted to exclude poets from their societies:

> Murder, death, and unchanging society represent precisely the inability to hear and understand the signifier as such—as ciphering, as rhythm, as a presence that precedes the signification of object or emotion. The poet is put to death because he wants to turn rhythm into a dominant element; because he wants to make language perceive what it doesn't want to say, provides it with its matter independently of the sign, and free it from denotation. For it is this *eminently parodic* gesture that changes the system (her italics, *DIL*, 31).

But in Blake's palinode even rhythm can be appropriated.

To end the poem, Blake repeats the first stanza, with the fit substitution of "dare" for "could." If one thinks in terms of inferential patterns of repeated thoughts, discussed in Chapter 5, then the last stanza *rhymes* with the first, almost exactly. It is translated forward exactly, except for one word, which gets attention because of the supplement of "dare" for "could." Here, the difference of one word reorients the whole algorithm. Discussing this poem, Culler seems almost like Koch and Hass when he says, "The foregrounding and making strange of language through metrical organization and repetition of sounds is the basis of poetry" (*LT*, 80). When reading the rhythmic periods of the last stanza, one recollects backward, even as one repeats forward in translation. Instead of thinking of Blake's brilliant grammatical composition, think about the repeated stanza as a carrying forward of rhymed accumulations of quantifiable sounds. Here is Culler again: "It is the scandal of poetry that 'contingent' features of sound and rhythm systematically infect and affect thought" (*LT*, 80). At this stage of the argument, one can agree that sound works by contact and contiguity; but if "contingency" means by chance, then poets usually work by means of the accidents of sounds in search of the promise of an impending rhyme. We are being carried somewhere we have never been.

Usually when one speaks of parallel constructions, or of antithesis, or of a balanced prose style, one thinks only grammatically or rhetorically. But always underlying the grammar is the physical measure of the sound of natural syntax, which gives quantifiable shape, density, and heft to the words by means of the relative density and lightness of the air molecules that are themselves differentially shaped by speech acts. So "balance" is always at least in part the physical ballast of sound shapes set against each other in symmetry groups. For an example, take another look at Coleridge's use of "balance" in the passage about the wise serpent, quoted in the Introduction. There the idea is of quantitative balance of words in their locations in the sentence, which carries over to the balance between freedom and propelling necessity. In "The Tyger" then, Blake's fearful symmetry is not just the visual shape of the beast, which is difficult to see, but it is primarily the "second language" of the physically beating words of the poetic line. And what is the beat? It is the pattern of double trochees, which is presumably the heartbeat of the tyger and the framer, and the propelling pattern of alternating contraction waves, that is to say, the old sequence itself, repeating itself, wave after wave.

Natural law is now more accurately seen as the rhythmic repetition everywhere of what amounts to cosmic habit. The order of things is not so much the order of static things, but in Blake's composition, Nature is a moving proportion of fearful symmetries repeated everywhere. Just as one does not teach a child the laws of grammar, but repeats the words in

inflected patterns, so nature repeats itself everywhere in endless symmetries, as the drops from an icicle extend its length wherever there is a brief
thaw. Words are learned by children as accordion instruments that repeat
everywhere their common symmetries, like the sunburst patterns repeated in the pods of sunflower seeds

The process of reading Blake's "Tyger" seems to be a rhythmic process
of repetition and emergence, during which, as Plato said, the sentences get
us somewhere. When we read these little songs, however, we are asked once
again to be as children: "I a child—thou a lamb." But if it is a threatening
nursery rhyme, it takes us as children somewhere we had certainly never
been before. Children may search for one set of things, only to find so many
other beastly things visited upon them unannounced: the arrival of unpre-
dicted sexuality, the hints of a ferocious social order, the complicity of par-
enting—the troubled themes of *Songs of Experience.* Here the thing is a
tyger we can't quite see, but we feel it taking shape in the world. The repetitions of natural syntactic patterns carry us into an emergent idea, perhaps
a thing from a supernatural order, which we did not know we would want
to seek out. In this case, the arrival of a supernatural beast comes by the
magic of reading, or by the magic of hearing words read to children in new
orders. The *twists* of sound and sense are also the twists of sequential compositions that gradually give heft and feeling to the fearful symmetry of the
beast. Twisting is its natural syntax, its fearful symmetry.

If the elements of the cosmos are commonly patterned by repeated
symmetry sharing, then we are forced to share the assumption that Blake's
inner life has been formulated by a rhythmic beast composed from the energies of the external world, and that it is now transferred as an experienced
part of our own inner lives. Is not the "inner life" always staged as an external metaphor? If that question is answered affirmatively, then it seems
that there is no other language than that of *deixis.* Blake's beast—his not-
quite-seen vision now ours—is one that suggests that the inner life is existentially patterned by the syntactic anatomy of a fearful cosmos and that we
commonly must share those accordion metaphors of inner life transported
from the external world. As children, are we always searching for something
that we suspect or hope may not be there? But where is that beast that is
being summoned by the hypnotizing act of reading or hearing a nursery
rhyme? As children we were glad, I suppose, that there was nothing hidden
under the bed and about to seize us. But as adults, reading and thinking
and acting as grown-up children, we are perhaps content that we are
searching for something that is never there.

In the 1930s, the semiotician Charles Morris said that it is common for us to search for "things," only to find classes and concepts, which may have few or no members within the class. When we refer to things, he said, we are more accurately referring to classes: "This distinction makes explicable the fact that one may reach into the icebox for an apple that is not there and make preparations for living on an island that may never have existed or has long since disappeared beneath the sea."[19] Similarly, he thought that syntax could best be understood as a formal set of logical relations as taught by (the early) Rudolf Carnap. In this 1930s view, syntax would not be encumbered by a metaphysical search for meaningful things. Unlike my own definition of an existential syntax inherently following the form of a wave, wave after wave, a purely logical syntax is like Whitehead's algebra, a function of hypothetical sentences untroubled by metaphysics. Logical positivism has been a rich school of thought in its applications of a pure syntax of symbols. But Blake's daring was of another order entirely. He was not fearful of classes or archetypes, or genres, or logical sentences. And he was not looking for a thing in the refrigerator or in the cosmos. Blake's haptically rhythmic patterning is based upon the substructural belief of the Enlightenment that there is a systematic Order of Nature out of which the tyger must be composed. That all-pervasive but unseen energy is what is fearful for Blake. Not that there is no order of nature out there, but that the order itself was there as a terrifying symmetry, and that it could be systematically appropriated to be rhythmically visited upon us, as children and adults. Perhaps we are also of that order, tygers as well as lambs; our bodies and inner lives framed by a fearfully rhythmic symmetry. And the Order invades our childlike inner life of vision, dream, nightmare by order and method of some more terrifying creator about whom and about which we had no say.

Some might say that these conjectures about natural syntax are ill founded because there is nothing there to be found except symbolic classes, that every sign system is just socially constructed. I believe, on the contrary, that perhaps we may have been looking for the wrong kinds of things, that even the inner life is patterned by the natural language of the cosmos: "The mind has mountains." I have attempted to describe a natural syntax of a few transformations whose archetypal motions follow the form of a wave. For those who think that nature has no order to be found, or for those who think there can be no commonly shared order that can be found, or for those who say that syntax is to be sought in an abstract logic of formal relations, I cannot say more. But I believe I have demonstrated that there can be found an inherent old sequence of alternating wave motions, which allows for an eventual fitness within a Darwinian hypothesis that language evolved from the sexual selection of musical tones among animal calls. If one looks for a natural syntax camouflaged as a grammar, one may find a set of transformations that everywhere reenacts principles of symmetry and conservation. Finally, this thesis of a common natural syntax may be ended with another poem, that kind of rhythmic thinking that promises that readers will find not so much what they were looking for, but perhaps more than they expect. Let Walt Whitman end it:

> A song of the rolling earth, and of words according,
> Were you thinking that those were the words, those upright
> lines, those curves, angles, dots?

> No, those are not the words, the substantial words are in the
> ground and sea,
> They are in the air, they are in you.

> Were you thinking that those were the words, those delicious
> sounds out of your friends' mouths?
> No, the real words are more delicious than they.

> Human bodies are words, myriads of words,
> (In the best poems re-appears the body, man's or woman's body,
> well-shaped, natural, gay,
> Every part able, active, receptive, without shame or the need of shame.)[20]

NOTES

Preface

1. New York: Simon & Schuster, 1995. Hereafter cited as *DDI*; see pp. 381–400. For an overview of the linguistic wars in the context of evolution, see Gary Cziko, "The Evolution, Acquisition, and Use of Language," in *Without Miracles: Universal Selection Theory and the Second Darwinian Revolution* (Cambridge: MIT Press, 1995), 179–212.

Introduction

1. In order to feature the "creative aspect of language use," Noam Chomsky discusses Goethe's *Urform* and *Urpflanze* as "innate" organic forms within the context of German Romanticism in *Cartesian Linguistics: A Chapter in the History of Rationalist Thought* (New York: Harper & Row, 1966), 19–24, 89n. In order to justify his own theory of deep-structural form, Chomsky found it necessary to write a brief history of the question of the innateness of linguistic form, beginning with Descartes, and he featured the idea of innate organic forms as a theory that was shared by biologists and creative writers. For Goethe, the helical form was an organic manifestation of polar forces in rhythmic pulsation. For an analysis of Goethe's views about physical and organic polarity, see Karl Vietor, *Goethe the Thinker* (Cambridge: Harvard University Press, 1950), 98–103. For a brief history of uses of vibrations between poles in terms of Romantic philosophy and science, see James H. Bunn, "The Attractive and Repulsive Poles of Coleridge's Science and the Question of a Dialectic in Nature," in *Afterimages: A Festschrift in Honor of Irving Massey*, eds. William Kumbier and Ann Colley (Buffalo: Shuffaloff, 1996), 74–92.

2. Johann Wolfgang von Goethe, *Italian Journey, 1786–1788*, trans. W. H. Auden and Elizabeth Mayer (New York: Schocken, 1968), 401–2.

3. Vitruvius, *The Ten Books on Architecture*, trans. Morris Hicky Morgan (New York: Dover), 281—319; hereafter cited as *TBA*.

4. See James H. Bunn, *The Dimensionality of Signs, Tools and Models: An Introduction* (Bloomington: Indiana University Press, 1981). There I used the words *torque* and *torsion* to describe an oblique transformation from one dimension to another.

5. Ernst Cassirer, *The Problem of Knowledge: Philosophy, Science and History Since Hegel*, trans. William H. Woglum and Charles W. Hendel (New Haven: Yale University Press, 1950), 139–40. Quoted by Charles W. Hendel in his introduction to Cassirer's *The Philosophy of Symbolic Forms*, vol. 1, trans. Ralph Manheim (New Haven: Yale University Press, 1953), 31; hereafter cited as *PSF*.

6. Hermann von Helmholtz, *Popular Scientific Lectures*, trans. H. W. Eve (New York: Dover Publications, 1962), 5.

7. Lancelot Law Whyte, "Towards a Science of Form," *Hudson Review* 23 (winter 1970–71): 613–32.

8. Herbert Read, "The Creative Process," in *The Forms of Things Unknown: Essays Towards an Aesthetic Philosophy* (New York: Horizon Press, 1960), 59.

9. Stephen Jay Gould, in *Ontogeny and Phylogeny* (Cambridge: Harvard University Press, 1977), says, "But I share D'Arcy Thompson's conviction that complex organic patterns usually can be reduced to fewer and simpler generating factors" (395). See also his "D'Arcy Thompson and the Science of Form," *New Literary History* 2 (winter 1971): 229–58. See also Eric Lenneberg, *Biological Foundations of Language* (New York: John Wiley, 1967), with an appendix by Chomsky entitled "The Formal Nature of Language," 397–442.

10. Jerry Fodor, *The Language of Thought* (Cambridge: Harvard University Press, 1979), 103.

11. See most recently George Lakoff and Mark Johnson, *Philosophy in the Flesh: The Embodied Mind and Its Challenge to Western Thought* (New York: Basic Books, 1999). In their ambitious overview, the thesis of our identity as beings with minds neurally embodied with motor controls is posed as a challenge to most Western philosophical principles. See especially chapter 22, "Chomsky's Philosophy and Cognitive Linguistics," for a description of Universal Grammar as an innate syntax uninfluenced by evolution. See also George Lakoff, *Women, Fire, and Dangerous Things: What Categories Reveal About the Mind* (Chicago: University of Chicago Press, 1987), and also Mark Johnson, *The Body in the Mind: The Bodily Basis of Meaning* (Chicago: University of Chicago Press, 1987). For a favorable discussion of Chomsky, which seeks to integrate Lakoff and Johnson's theories into his idea of brain function, see Gerald M. Edelman, *Bright Air, Brilliant Fire: On the Matter of the Mind* (New York: Basic Books, 1992), 241–52: hereafter cited as *BA*.

12. Robert Frost, "West-Running Brook," in *The Poetry of Robert Frost: The Collected Poems, Complete and Unabridged*, ed. Edward Connery Latham (New York: Holt, Rinehart & Winston, 1969), 257–60.

13. Paul Davies and John Cribben, *The Matter Myth* (New York: Touchstone: Simon & Schuster, 1992). Chapter 1 is called "The Death of Materialism." The relationship between waves, particles, and the principle of least action is discussed in Paul Davies, *Superforce: The Search for a Grand Unified Theory of Nature* (New York: Simon & Schuster, 1984), 233–35. Hereafter cited as *SFS*.

14. A. R. Ammons, "Swells," in *A Coast of Trees: Poems* (New York: W. W. Norton, 1981), 2–3. For an influential overview of Ammons's work, see *A. R. Ammons: Modern Critical Views*, ed. Harold Bloom (New York: Chelsea House, 1986), hereafter cited as *AA*. A brief study of Ammons's relation to science appears in Jerome Mazzaro's chapter "Reconstruction in Art." For a longer study about Ammon's efforts to acquaint science with poetry, see Steven B. Schneider, "The Poet and the Scientist," in *A. R. Ammons and the Poetics of Widening Space* (Rutherford, N.J.: Fairleigh Dickinson University Press, 1994), 19–29.

15. Ralph Waldo Emerson, "Self-Reliance," in *Selected Essays*, ed. Larzer Ziff (New York: Penguin, 1982), 201. This volume is hereafter cited as *SE*.

16. Karl Pearson, *The Grammar of Science* (1892; reprint, London: Everyman, 1937), 218; hereafter cited as *GS*.

17. Noam Chomsky, *Language and the Problem of Knowledge: The Managua Lectures* (Cambridge: MIT Press, 1988), 4; hereafter cited as *LM*.

18. Steven Pinker, *The Language Instinct: How the Mind Creates Language* (New York: Harper Perennial, 1994), 238; hereafter cited as *LI*.

19. Carl Gans, "Locomotion Without Limbs," *Natural History* (February and March 1966): 11–16, and 36–41; hereafter cited as *LWL*. For an extensive study of animals living and moving in water, see Steven Vogel, *Life in Moving Fluids: The Physical Flow of Biology*, 2d ed. (Princeton: Princeton University Press, 1994), 68.

20. Max Black, *Models and Metaphors: Studies in Language and Philosophy* (Ithaca: Cornell University Press, 1962), 239–40.

21. Stephen C. Pepper, *World Hypotheses: A Study in Evidence* (Berkeley: University of California Press, 1942), 91–92. For the idea that an archetype is distinct from a metaphor—that is, an archetype such as the sun or the wheel is a trope of connection with physical motions of periodicity—see Philip Wheelwright, *The Burning Fountain: A Study in the Language of Symbolism* (Bloomington: Indiana University Press, 1954), 123–46.

22. I take this passage from David Thistlewood, *Herbert Read: Formlessness and Form, An Introduction to His Aesthetics* (London: Routledge & Kegan Paul, 1984), 100. Herbert Read quotes Moore in *The Philosophy of Modern Art* (New York: Horizon Press, 1953).

23. Quoted in Thistlewood, *Herbert Read*, 128; from D'Arcy Thompson, *On Growth and Form* (Cambridge: Cambridge University Press, 1942), 357.

24. Alfred North Whitehead, *Process and Reality: An Essay in Cosmology* (New York: Free Press, 1969), 240; hereafter cited as *PR*. Because nature was left abandoned in tatters by many of the most influential French philosophers of the post-WWII generation, excluding Gaston Bachelard, George Poulet, and M. Merleau-Ponty, I have returned often to English-speaking philosophers of the 1930s such as Whitehead, Russell, and Dewey for guidance.

25. Heraclitus's sayings are collected in *The Presocratic Philosophers: A Critical History with a Selection of Texts*, 2d. ed., eds. G. S. Kirk, J. E. Raven, and M. Schofield (London: Cambridge University Press, 1983); hereafter cited as *PP*. The first assertion is translated, "Men should try to comprehend the underlying coherence of things: it is expressed in the *Logos*, the formula or element of arrangement common to all things" (186). Hereafter, passages from this text are cited as *PP*.

26. From *The Collected Dialogues of Plato*, eds. Edith Hamilton and Huntington Cairns (New York: Princeton University Press, 1963), *Cratylus*, 436b. Hereafter, citations from this edition are noted at the end of the passage. Socrates' critique of the Pythagorean concept of the soul as a tempered attunement of bodily proportions, much like the "attunement of a musical instrument," is to be found in *Phaedo* 86a-88. See R. Burger, *Phaedo: A Platonic Labyrinth* (New Haven: Yale University Press, 1984), for a discussion.

27. Paul Ricoeur, *The Rule of Metaphor: Multidisciplinary Studies of the Creation of Meaning in Language*, trans. Robert Czerny (Toronto: University of Toronto Press, 1977). For the idea that the sentence, not the semiotic sign, is the basic unit of linguistic predication, see Ricoeur's discussion of Benveniste and Strawson (69–73). He also briefly describes an "archetype" as a network of relations, and he mentions Black and Pepper in this regard (244).

28. Chapter 3 of Chomsky's *Language and Mind* (New York: Harcourt Brace, 1968) is called "Linguistic Contributions to the Study of Mind (Future)." For Chomsky's brief discussion of "abduction," as Peirce's notion of a guessing instinct based upon adaptation, see 78. For a discussion of asymmetry in the basic combination of subject and predicate, see P. F. Strawson, *Subject and Predicate in Logic and Grammar* (London: Methuen, 1974), 18–20. Chomsky also occasionally refers to this asymmetry.

29. In *Art and Visual Perception: A Psychology of the Creative Eye* (Berkeley: University of California Press, 1954), 167.

30. For an overview of the physiological beginnings of language in association with the development of hand tools, see J. Z. Young, *An Introduction to the Study of Man* (New York: Oxford University Press, 1971), 494–518; hereafter cited as *SM*. He explores Chomsky's thesis of formal universals—the syntax of nouns, verbs, and tenses found in all languages—as inherited traits (494–98). Roman Jakobson also endorsed this association, according to Jacob Bronowski, "Human and Animal

Languages," in *A Sense of the Future: Essays in Natural Philosophy*, ed. Pietro E. Ariotti (Cambridge: MIT Press, 1977), 104–31; hereafter cited as *SF*. See also Bunn, "An Encomium for Hands," in *The Dimensionality of Signs*, 48–62. The idea that language derives from hand gestures is an old one, but it has been persuasively renewed by David Armstrong, William Stokoe, and Sherman Wilscox in *Gesture and the Nature of Language* (Cambridge: Cambridge University Press, 1995); the authors show that syntax in language may well have derived from the motions of gesture in three space, much like the syntax for American Sign Language discovered by Stokoe. Were my study of wave forms a strict study of linguistics, I would follow their line of thought more extensively.

31. This illustration is the frontpiece to Blake's *Europe*. See Martin Butlin, *The Paintings and Drawings of William Blake* Vol. 1 (New Haven: Yale University Press, 1981), 147–48. Butlin discusses the relation of the design and symbolism of the compasses to Blake's earlier figure, Urizen, and to his design of "Newton," which apparently derived from the frontispiece of Motte's famous translation of Newton's *Principia* (1729).

32. This passage is quoted in Erwin Panofsky, *Perspective as Symbolic Form*, trans. Christopher S. Wood (New York: Zone Books, 1991), 146n.

33. Claus Westerman, *Genesis 1–11: A Commentary*, trans. John J. Scullion (Minneapolis: Augsburg Publishing House, 1984), 34.

34. René Thom, *Structural Stability and Morphogenesis: An Outline of a General Theory of Models*, trans. D. H. Fowler (Reading, Mass.: Benjamin/Cummings, 1975), Illustration 7.

35. Charles D. Drewes and Charles R. Fourtner, "Helical Swimming in a Freshwater Oligochaete," *Biological Bulletin* 185 (August 1993): 1–9. Hereafter cited as *HS*.

36. *Biographia Literaria* in *The Collected Works of Samuel Taylor Coleridge*, eds. James Engell and W. Jackson Bate (Princeton: Princeton University Press, 1983), II, 14.

37. Aristotle, *Metaphysica*, 2nd ed., ed. W. D. Ross (Oxford: Clarendon Press, 1963), 985b. Hereafter cited as *M*. For an excellent discussion of this passage, considered in its historical context, see Norma E. Emerton, *The Scientific Reinterpretation of Form* (Ithaca: Cornell University Press, 1984), 99. For analysis of this and other passages of extant writings by Democritus, see Kirk and Raven, "The Atomists: Leucippus of Miletus and Democritus of Abdera" (*PP*, 402–33). For a comparable experiment in this kind of "mental rotation" of abstract images through symmetry-axes of three dimensions, see Steven Pinker, *How the Mind Works* (New York: Norton, 1997), 280–84.

38. James S. Trefil and Robert M. Hazen, *The Sciences: An Integrated Approach* (New York: John Wiley, 1995), 73–75. This is an admirable textbook, designed for a basic college course, which integrates several sciences.

39. See, for instance, Edwin Hartman, *Substance, Body, and Soul: Aristotelean Investigations* (Princeton: Princeton University Press, 1977): "According to both the

Categories and the *Posterior Analytics,* the world is reflected in the language in which we talk about it; so the syntactical relation between 'log' and 'white' in the sentence 'The log is white' faithfully represents the predication of whiteness of a log in reality— 'The white is a log' is a sign of bad ontology as well as bad grammar," 10.

40. Quoted in part by Susanne K. Langer, *Philosophy in a New Key* (New York: Mentor Books, 1948), 71; from Bertrand Russell, *Philosophy* (New York: W. W. Norton & Company, 1927), 265.

Chapter 1

1. *The Odyssey of Homer,* trans. Richard Lattimore (New York: Perennial Library, 1965), XXI, 11, 59. Hereafter passages quoted by page number are taken from this edition and are cited as *O*. See also *Odyssey, A New Translation,* trans. Albert Cook (New York: W. W. Norton, 1967).

2. Simone Weil, *The Iliad, or the Poem of Force,* trans. Mary McCarthy and Dwight Macdonald (Wallingford, Pa.: Pendle Hill, 1956), 15.

3. "A Gossip on Romance," ed. William Lyon Phelps, *Essays by Robert Louis Stevenson* (New York: Charles Scribner's Sons, 1918), 224–25.

4. In "Hermeneutic Ellipses," Werner Hammacher discusses Friedrich Schleiermacher's use of oscillation as a trope for vibrating between the general and the particular. Collected in *Transforming the Hermeneutic Context: From Nietzsche to Nancy,* eds. Gayle L. Ormiston and Alan D. Schrift (Albany: SUNY Press, 1990), 190–91. The trope of vibrating between two poles was commonly used by British and German Romantic writers. In the mid-eighteenth century, the sinuous or meandering "line of beauty" was the compositional principle featured by William Hogarth, who derived it from nature and who applied it to the visual arts. *The Analysis of Beauty,* ed. Ronald Paulson (New Haven: Yale University Press, 1997).

5. Richard J. Bernstein, *The New Constellation: The Ethical-Political Horizons of Modernity/ Postmodernity* (Cambridge: MIT Press, 1992), 9, 41–42; hereafter cited as *NC.* Bernstein in turn cites Martin Jay's discussion of Theodor Adorno's *Kraftfeld* in *Adorno* (Cambridge: Harvard University Press, 1984), 14–15. See especially Martin Jay's *Force Fields: Between Intellectual History and Cultural Critique* (New York: Routledge, 1993). For an analysis of Derrida's indeterminacy with respect to some contemporary paradigms, see Alexander Argyros, *A Blessed Rage for Order: Deconstruction, Evolution and Chaos* (Ann Arbor: University of Michigan Press, 1991). For an extended comparison of Neils Bohr's thinking with Jacques Derrida's, see Arcady Plotnitsky, *Complementarity: Anti-epistemology after Bohr and Derrida* (Durham: Duke University Press, 1994).

6. Gerald L. Bruns, *Hermeneutics Ancient and Modern* (New Haven: Yale University Press, 1992), 248–49. For a study of the perturbations of waves in contemporary accounts of order within chaos theory, see Floyd Merrell, *Simplicity and Complexity: Pondering Literature, Science, and Painting* (Ann Arbor: University of Michigan Press, 1998), 112–122.

7. Willard Bascom, "Ideal Waves," in *Waves and Beaches: The Dynamics of the Ocean Surface* (New York: Anchor, 1960, 1980), 25. René Thom's illustration of a breaking wave form, mentioned in the last chapter, is taken from this book.

8. Trefil and Hazen, *The Sciences: An Integrated Approach*, 143. Hereafter cited as *TS*.

9. Ben Bova, *The Beauty of Light* (New York: John Wiley & Sons, 1988), 125.

10. James H. Prout and Gordon R. Bienvenue, *Acoustics For You* (Malabar, Fla: Robert E. Krieger Co., 1990), 57, Figure 4.2. Hereafter cited as *AY*.

11. Michael I. Sobel, *Light* (Chicago: University of Chicago Press, 1987), 195.

12. Seamus Heaney, *Selected Poems 1966–1987* (New York: Noonday Press, 1990), 177.

13. Zitkala-Sa, "The School Days of an Indian Girl," *American Indian Stories* (Lincoln: University of Nebraska Press, 1921). Excerpts are reprinted in *The Essay Connection: Readings for Writers*, ed. Lynn Z. Bloom (Boston: Houghton Mifflin Co., 1998), 255–62.

14. Denny Carter, *Henry Farny* (New York: Watson-Guptill Publications, 1978), 186. Hereafter cited as *HF*.

15. *Charles Burchfield: The Charles Rand Penney Collection*, ed. Julie Hartenstein (Baltimore: Smithsonian Traveling Exhibition, 1978), 68. Figure 75.

16. "Heidegger's Silence" is Bernstein's chapter title in *The New Constellation* describing Heidegger's strange defense of his Nazi role. Heidegger's discussion of projection as interpretation appears in *Being and Time*, sections 31 and 32. Collected in *The Hermeneutic Tradition: From Ast to Ricoeur*, eds. Gayle L. Ormiston and Alan D. Schift (Albany: SUNY Press, 1990): 115–44. Hereafter cited as *HT*. See also Martin Heidegger, *The Question Concerning Technology and Other Essays*, trans. William Lovitt (New York: Harper Colophon Books, 1977), 4–6. Hereafter cited as *QT*. For an important study of the story itself as a "projection," see Mark Turner, *The Literary Mind* (New York: Oxford University Press, 1996), 5–6 and throughout.

17. Ernst Cassirer, *The Myth of the State* (New York: Doubleday Anchor, 1946), 65, 356. Hereafter cited as *MS*.

18. Heraclitus, fragment 51, *PP*, 192. For a discussion of Heraclitus's adage as a *palintropos* that implies a riddle about the "law of contradiction," see Albert Cook, *Myth and Language* (Bloomington: Indiana University Press, 1980), 78–83. For an analysis of different meanings of *polytropos*, see Pietro Pucci, "The Proem of the *Odyssey*," *Arethusa* 15 (1982): 39–62. Octavio Paz's *El Arco y la Lyra* has been translated as *The Bow and the Lyre: The Poem, The Poetic Revelation, Poetry and History*, trans. Ruth L. C. Simms (Austin: University of Texas Press, 1973). For his discussion of Heraclitus's fragment as a polemical struggle within the universe, with the human being as the convergence between earthly and divine forces, see 183–84.

19. For a defense of Plato's base in the physical world, see John Wild, *Plato's Modern Enemies and the Theory of Natural Law* (Chicago: University of Chicago Press, 1953). In chapter four, Wild writes a history of natural law, with Plato positioned as the

founder of Stoicism, the philosophy of those who follow the relations between natural law and moral philosophy.

20. Charles Olson, "Projective Verse," in *Selected Writings*, ed. Robert Creeley (New York: New Directions, 1966), 16. Hereafter cited as *PV*.

21. Michel Serres, "The Natural Contract," in *Critical Inquiry* 19 (Autumn 1992): 2. See also *The Natural Contract*, trans. Elizabeth MacArthur and William Paulson (Ann Arbor: University of Michigan Press, 1995). Serres' experience of returning to the land and the sea is reminiscent of the earthiness of Gaston Bachelard, especially *The Poetics of Space*, trans. Maria Jolas (Boston: Beacon Press, 1969). In his Forward (ix), Étienne Gilson defines Bachelard's "material imagination [which] is attracted by the elements of permanency present in things."

22. Hugh Kenner, *The Poetry of Ezra Pound* (London: Faber & Faber, 1951), 87. Quoted by Donald Davie, *Articulate Energy: An Enquiry Into the Syntax of English Poetry* (London: Routledge & Kegan Paul, 1955), 41. Hereafter cited as *AE*. Davie's title aptly expresses the subject of my inquiry.

23. Hannah Arendt, *Between Past and Future: Six Exercises in Political Thought* (New York: Viking, 1968), 221. Quoted by Jay, *Force Fields*, 82. For the most exhaustive study of projection into others' work shoes, see Jacques Derrida, *The Truth in Painting*, trans. Geoff Bennington and Ian McLeod (Chicago: University of Chicago Press, 1987), 255–382.

24. For the well-known account of Odysseus's deceit, see Max Horkheimer and Theodor W. Adorno, "Odysseus or Myth of Enlightenment," in *Dialectic of Enlightenment*, trans. John Cumming (New York: Continuum Books, 1972), 43–80.

25. Jacques Derrida, "White Mythology," trans. F. C. T. Moore, *New Literary History* 6 (1974): 5–74. "White Mythology" is also included in *Margins of Philosophy*, trans. Alan Bass (Chicago: University of Chicago Press, 1982), 207–72. Hereafter cited as *WM*. For a critique of Derrida's "hermeneutics of suspicion," especially via this essay, see Paul Ricoeur, *The Rule of Metaphor*, 284–89. For a concise introduction to the larger hermeneutic issue, see Jonathan Culler, *Literary Theory: A Very Short Introduction* (New York: Oxford University Press, 1997), 68–69; hereafter cited as *LT*. A thorough study of Derrida's grammatology is Gregory Ulmer's *Applied Grammatology: Post(e)-Pedagogy from Jacques Derrida to Joseph Beuys* (Baltimore: Johns Hopkins University Press, 1985).

26. Kostas Axelos, *Alienation, Praxis, and Techne in the Thought of Karl Marx*, trans. Ronald Bruzina (Austin: University of Texas Press, 1976), 74–75. Marx quotes Heraclitus in *Capital: A Critique of Poltical Economy*, trans. Ernest Untermann (New York: Modern Library, 1936), 118. Hereafter cited as *C*.

27. For a discussion of Midas, Marx, exchange, and money, see Marc Shell, *Money, Language, and Thought: Literary and Philosophic Economies from the Medieval to the Modern Era* (Berkeley: University of California Press, 1982), 58 and throughout.

28. One of the most contested terms in social philosophy, *Aufhebung* is discussed in Paul Ricoeur's critique of ideology and science, *Lectures on Ideology and Utopia*,

ed. George H. Taylor (New York: Columbia University Press, 1986), "Althusser," 112–15. For the strategies of social sublimation, see Anthony Wilden, *Man and Woman, War and Peace: The Strategist's Companion* (London: Routledge & Kegan Paul, 1987); hereafter cited as *MN*. For his use of *Aufhebung*, see 246.

29. Alfred North Whitehead, *Science and the Modern World* (New York: Free Press, 1925, 1967), 31.

30. For a summary of Hans-Georg Gadamer's use of tradition as a laudable prejudice, dependent upon Heidegger's hermeneutic circle of the fore-structures of interpretation, as well as Jürgen Habermas's critique of this kind of prejudice, see Paul Ricoeur, "Hemeneutics and the Critique of Ideology," in *HT*, 298–334. Also see the editors' introduction.

31. See Habermas's review of *Truth and Method*, *HT*, 232. A defense of natural law as natural right is Ernst Bloch's critical history, *Natural Law and Human Dignity*, trans. Dennis J. Schmidt (Cambridge: MIT Press, 1987).

32. To feature aesthetic judgment is to favor Kant's *Critique of Judgement* and his dialectics of nature as ends and means. John Dewey's *Art and Experience* (New York: G. P. Putnam's Sons, 1934, 1958) remains the magisterial treatment of the everyday reciprocity of "esthetic recurrence" with natural energies. Hereafter cited as *AE*. In Dewey's tradition, see Richard Schusterman, *Pragmatist Aesthetics: Living Beauty, Rethinking Art* (Oxford: Blackwell, 1992). For an essay to re-found philosophy upon nature, see Robert Cummings Neville, *Recovery of the Measure, Interpretation and Nature* (Albany: SUNY Press, 1989). More recently, see Robert S. Corrington, *Ecstatic Naturalism: Signs of the World* (Bloomington: Indiana University Press, 1994) for four definitions of nature by way of the American pragmatist tradition, especially Peirce's prospect of thirdness.

33. See William James's famous chapter, "Habit," in *Principles of Psychology*, vol. 1 (New York: Henry Holt Co., 1890), 104.

34. E.H. Gombrich, *The Sense of Order: A Study in the Psychology of Decorative Art* (Ithaca: Cornell University Press, 1979), 121–26; hereafter cited as *SO*.

35. According to Jean Baudrillard, "All the repressive and reductive strategies of power systems are already present in the internal logic of the sign, as well as those of exchange value and political economy." In *For a Critique of the Political Economy of the Sign*, trans. Charles Levin (St. Louis: Telos Press, 1981), 163.

36. For Freud, conscience suppresses aggression: "Civilized man has exchanged a portion of his possibilities for happiness for a portion of security." In *Civilization and Its Discontents*, trans. James Strachey (New York: W.W. Norton, 1962), 62.

37. "Custom, then, is the great guide of human life. It is that principle alone, which renders our experience useful to us, and makes us expect, for the future, a similar train of events, with those which have appeared in the past." David Hume, *An Enquiry Concerning Human Understanding*, ed. Eric Steinberg (Indianapolis: Hackett Publishing Co., 1977), 27. William Godwin, "The Voluntary Actions of Men Originate in Their Opinions," *Political Justice*, ed. Isaac Kramnick (New York: Penguin, 1973): Book I, V. Godwin divides those opinions between self-love

and benevolence in order to pose the question of voluntary action: Book IV, X. For Peirce, habit turns into belief, "that upon which a man is prepared to act," a premise of pragmatism. See "Pragmatism in Retrospect: A Last Formulation," in *Philosophical Writings of Peirce*, ed. Justus Buchler (New York: Dover, 1955), 270. See especially James's chapter, "Habit": "Habit is thus the enormous fly-wheel of society, its most precious conservative agent. It alone is what keeps us all within the bounds of ordinance, and saves the children of fortune from the envious uprisings of the poor" (vol. 1, 121). For Gadamer's rehabilitation of prejudice and the Enlightenment's "prejudice against prejudice," see Habermas's review, *HT*, 236–38.

38. Arnold Toynbee, *A Study of History: A New Edition* (New York: Weathervane Books, 1972), 34.

39. John Locke, "Of Power," *Essay Concerning Human Understanding* (Oxford: Oxford University Press, 1975), 239. Though admitting that *power* is a heuristic term by which we blank its essence, Locke did not confuse the term with the fact of its force. For a contemporary study of the heart's rhythms, as well as the wave dynamics that create arrhythmia, see Arthur T. Winfree, *When Time Breaks Down: The Three-Dimensional Dynamics of Electrochemical Waves and Cardiac Arrhythmias* (Princeton: Princeton University Press, 1987). Winfree describes and draws computer-generated models of three-dimensional waves that ripple through the heart's muscular tissue under the electrical charges that incite the heart's rhythmic pumping of blood: "Electrical impulses in nerve membrane or heart muscle propagate without attenuation, unlike sound waves or ripples from a raindrop striking water. . . . Each little fiber of heart muscle contributes to the wave as it goes by, restoring it to full strength as in a Pony Express relay" (102–03). Hereafter cited as *WT*.

40. "Things participate in signs as signifiers, not as referents." A thesis ascribed to Augustine by Tzvetan Todorov, *Theories of the Symbol*, trans. Catherine Porter (Ithaca: Cornell University Press, 1982), 40. One of the clearest expositions of Saussure's basic concepts of the Signifier and Signified, with respect to the history of earlier French theories of language, is Hans Aarsleff, "Taine and Saussure," in *From Locke to Saussure: Essays on the Study of Language and Intellectual History* (Minneapolis: University of Minnesota Press, 1982), 356–71.

41. For an extensive discussion of Marx's use of Base and Superstructure as the syntax of the historical process, see Hayden White, *Metahistory: The Historical Imagination in Nineteenth-Century Europe* (Baltimore: Johns Hopkins University Press, 1973), 303–09. See V. N. Volosinov, "Concerning the Relationship of the Basis and Superstructures," *Marxism and the Philosophy of Language*, trans. Ladislav Matejka and I. R. Titunik (Cambridge: Harvard University Press, 1986), 17–24. See also Terry Eagleton, "Base and Superstructure in Raymond Williams," in *Raymond Williams: Critical Perspectives*, ed. Terry Eagleton (London: Polity Press, 1997), 165–83.

42. For a clear discussion of Lacan's formulation of the semiotic bar in relation to Derrida, see Ulmer, *Applied Grammatology*, 208. For Kaja Silverman, the subject

of semiotics is the play of the "orchestrated" differences between the dual codes of the signifier and the signified; *The Subject of Semiotics* (New York: Oxford University Press, 1983). Silverman discusses the dualist mediations of Saussure, Jacobson, Lévi-Strauss, Barthes, Lacan.

43. Henry Sussman, personal communication. See especially, Sussman, *The Hegelian Aftermath: Readings in Hegel, Kierkegaard, Freud, Proust, and James* (Baltimore: Johns Hopkins University Press, 1982), passim.

44. For a history of the logic of the sign, beginning with Aristotle, see Todorov, 20–30. For a discussion of metaphor as Ryle's category mistake, see Ricoeur, *Rule of Metaphor*, 252; for Ricoeur's critique of Heiddeger's turn that "the trans-gression of meta-phor and that of meta-physics are but one and the same transfer," see 280. For the most extensive use of Russell's logical types, see Gregory Bateson, "A Theory of Play and Fantasy," in *Semiotics: An Introductory Anthology*, ed. Robert E. Innis (Bloomington: Indiana University Press, 1985), 131–47.

45. Ferdinand de Saussure, *Course in General Linguistics*, trans. Wade Baskin (New York: McGraw-Hill, 1966), 112.

46. Stephen K. Land, *From Sign to Proposition: The Concept of Form in Eighteenth-Century Semantic Theory* (London: Longman, 1974), 14.

47. Heinrich Zimmer, *Myths and Symbols in Indian Art and Civilization* (Princeton: Princeton University Press, 1946), 146.

48. In his *Principles of Chinese Painting* (Princeton: Princeton University Press, 1959), 52, George Rowley discusses "consonance" as the repetition of rhythmic likeness among apparently unrelated natural forces. I shall stress this concept in the next chapter. For a wider discussion of the illustration of lohans within the Buddhist tradition of Chinese painting, see Wen C. Fong, *Beyond Representation: Chinese Painting and Calligraphy 8th–14th Century* (New York: The Metropolitan Museum of Art, 1992), 267–69, 348–63, where several of the lohans are depicted as being seated while reading sutras. Hereafter cited as *BR*. Orignally called "Lohan on Lotus" by the Walters Art Gallery, it now entitles this image more neutrally as "Traveling on a Raft." Hiram Woodward, the curator, describes it: "Chang Ch'ien, a minister of the 2nd century B.C., is said to have sailed up the Yellow River, past the paradise of the Queen Mother of the West and ultimately to the Milky Way. The eighth-century poet Li Po was later thought to have made a similar journey, and it is he who may be shown here, with a book in his lap" (Personal communication, May, 2000).

49. Ricoeur, *Rule of Metaphor*, 70–73.

50. Toynbee, 244.

51. Arnheim, *Visual Thinking* (Berkeley: University of California Press, 1969), 62. Denise Levertov, "To the Reader," from *The Jacob's Ladder* (1961), collected in *Poems 1960–1967* (New York: New Directions, 1983), 1.

52. So pervasive was wave theory in early Chinese thought, according to Joseph Needham, that its norms inhibited development of modern science there. See his *Science and Civilization in China: Physics and Technology* (Cambridge: Cambridge University Press, 1962), 7.

53. In *Behind Appearance: A Study of the Relations between Painting and the Natural Sciences in this Century* (Cambridge: M.I.T. Press, 1970), 133, Waddington cites Hans Hofmann's theory of push and pull for arguing the mutual relations between arts and sciences in this century: see also Hofmann, *Search for the Real and Other Essays* (Cambridge: M.I.T. Press, 1948), 40–45.

54. Jerome Bruner, *Actual Minds, Possible Worlds* (Cambridge: M.I.T. Press, 1986), 51. Also Paul Feyerabend groups these topics together: incommensurability, Bohr, Gestalt illusions, and Gombrich, in *Against Method, Outline of an Anarchistic Theory of Knowledge* (London: Verso, 1975), 225–26. For an introduction to quantum field physics in relation to modern literature, see N. Katherine Hayles, *The Cosmic Web: Scientific Field Models and Literary Strategies in the Twentieth Century* (Ithaca: Cornell University Press, 1984), chapter one; for the gestalt switching effect of focusing only upon one image at a time, see 175–76. For the most recent and the most complete analysis of gestalt switching effects, which enable dialectical or contrary readings of one image, see W. J. T. Mitchell's "Metapictures," in *Picture Theory: Essays on Verbal Theory and Visual Representation* (Chicago: University of Chicago Press, 1994), 35–81.

55. Walter Benjamin, "N [Re the Theory of Knowledge, Theory of Practice]," in *Benjamin: Philosophy, History, Aesthetics,* ed. Gary Smith (Chicago: University of Chicago Press, 1989), 60. This passage is the first sentence of Jay's *Force Fields,* 1.

56. Sigmund Freud, "Beyond the Pleasure Principle," in *A General Selection of the Works of Sigmund Freud,* ed. John Rickman (New York: Doubleday Anchor, 1957), 162. This essay is replete with metaphors from Fechner's psycho-physics and from contemporary thermodynamic principles of least action and energy conservation.

57. As Kant said, the imagination spreads "itself over a number of kindred representations that arouse more thought than can be expressed in a concept determined by words. They furnish an *aesthetical idea,* which for that rational idea takes the place of logical representation; and thus, as their proper office, they enliven the mind by opening out to it the prospect into an illimitable field of kindred representations." *Critique of Judgement,* trans. J. H. Bernard (New York: Haffner Library, 1951), 158.

Chapter 2

1. Robert Graves, *The Greek Myths,* vol. 1 (Baltimore: Pelican Books, 1955), 27. Several of these creation myths are included in Barbara C. Sproul's useful volume, *Primal Myths: Creating the World* (San Francisco: Harper and Row, 1979), hereafter cited as *PM.*

2. For an introduction to these basic Derridean terms, especially *trace* as *differance,* see Gayatri Chakravorty Spivak's Preface to Jacques Derrida, *Of Grammatology* (Baltimore: Johns Hopkins University Press, 1976), xiv–xviii. For an appreciation of Derrida's use of such tropes as interlace, Moiré effects, and other

flourishes, see Ulmer, "Op Writing: Derrida's Solicitation of Theoria," in *Applied Grammatology*. For an analysis of Derrida's weaving of "interlace" and dialectics, see Rodolphe Gasché, *The Tane of the Mirror: Derrida and the Philosophy of Reflection* (Cambridge: Harvard University Press, 1986) 98–105; for his discussion of Derrida's "archetrace," see 189–94. For a discussion of *archetrope*, a term I also like, see Franklin R. Rogers, *Painting and Poetry: Form, Metaphor, and the Language of Literature* (Lewisburg: Bucknell University Press, 1985), 62–80.

3. See llya Prigogine and Isabell Stengers, *Order Out of Chaos: Man's New Dialogue with Nature* (New York: Bantam, 1984), for a challenging reinterpretation of thermodynamics as an agency for building ordered complexity in reaction to entropic decline. For their use of "symmetry breaking" as the mediator between chaos and order, see 160–70. In *Simplicity and Complexity*, Merrell stresses Prigogine's use of symmetry breaking as the element that helps achieve order out of chaos, 108–12. For Merrell's discussion of the vexing periodicity of the continuous wave in current literature, see his discussion of Italo Calvino's *Mr. Palomar*, 70–78; also 51–58. For an earlier definition of energy as energy-change, see P. W. Bridgman, *The Nature of Thermodynamics* (Gloucester, Mass.: Peter Smith, 1969), 85–115.

4. Marcel Griaule, *Conversations with Ogotemmeli* (New York: Oxford University Press, 1965). Hereafter cited as *CO*.

5. Robert Creeley, *Life and Death* (New York: New Directions, 1998), 41.

6. Genevieve Calame-Griaule, *Words and the Dogon World*, trans. Dierdre La Pin (Philadelphia: Institute for the Study of Human Issues, 1968), 220, 229.

7. A. David Napier, *Masks. Transformation and Paradox* (Berkeley: University of California Press, 1986), xxiii.

8. For Victor Turner's ideas of liminality, see *The Ritual Process: Structure and AntiStructure* (New York: Aldine, 1965), 95. For literary applications of Turner's work, see *Between Literature and Anthropology: Victor Turner and Cultural Criticism* (Bloomington: Indiana University Press, 1990).

9. E. H. Gombrich, *The Sense of Order: A Study in the Psychology of Decorative Art* (Ithaca: Cornell University Press, 1979), 121–26. I am much indebted to Gombrich's analyses of the acanthus and the Greek fret and other S-curved variants. See especially chapter IX, "Designs as Signs."

10. See Michel Serres, "Lucretius: Science and Religion," in *Hermes: Literature, Science, Philosophy*, eds. Josue V. Harari and David Bell (Baltimore: Johns Hopkins University Press, 1982), 112–14.

11. Marija Gimbutas, *The Language of the Goddess: Unearthing the Hidden Symbols of Western Civilization* (San Francisco: Harper Collins, 1989), 59. She takes the drawing from one in L. Bernabo Brea's *Sicily Before the Greeks*, rev. ed. (New York: Frederick A. Praeger, 1966), 104.

12. For a good survey with excellent reproductions, see *From the Lands of the Scythians, Ancient Treasures of the Museums of the U.S.S.R. 3000 B.C.–300 B.C.* (New York: Metropolitan Museum of Art, 1975). See also Frank Jettmar, *Art of the*

Steppes (New York: Crown, 1964), 71. Also, a bronze ritual vessel made in China about the sixth or early fifth century B.C.E. is ornamented with serpents and tigers: "At about this time a tiger with spiral nose, crescentic feet, and marked with spirals on shoulder and haunch was becoming popular in the art of the nomads on the north-west borders of China and on the Central Asian steppes. It is debated whether this connexion of Chinese art with territories beyond her borders was a case of influence accepted or exerted; in either case the exchange was fruitful for both traditions." See *The Chinese Exhibition: The Exhibition of the Archaeological Finds of the People's Republic of China* (London: Times Newspapers Ltd., 1973), 84.

13. Theodore Andrea Cook, *The Curves of Life: Being an Account of Spiral Formations and Their Application to Growth in Nature, to Science, and to Art: With Special Reference to the Works of Leonardo da Vinci* (1914; reprint New York: Dover Publications, 1979), 166–69.

14. Herbert Read, *The Contrary Experience: Autobiographics* (New York: Horizon Press, 1963), 345.

15. Franz Boas, *Primitive Art* (1927; reprint New York: Dover Pub., 1955), 20–21, 36–38, 40, 51; hereafter cited as *PA*.

16. Diane Wolkstein and Samuel Noah Kramer, *Inanna: Queen of Heaven and Earth: Her Stories and Hymns from Sumer* (New York: Harper & Row, 1983), 7, 179–80. Hereafter cited as *I*. The art illustrations in the book were compiled and interpreted by Elizabeth Williams-Forte. I am indebted to Diane Christian for introducing me to *Inanna*.

17. John Ruskin, "The Bird and the Serpent," in *The Queen of the Air*, vol. 19 of *The Complete Works of John Ruskin*, eds. E. T. Cook and Alexander Wedderburn (London: George Allen, 1907), 68. Hereafter cited as *QA*.

18. John Ruskin, "Living Waves," in *Deucalion*, vol. 26 of *Works*, 1–56.

19. For a discussion of the reversals in these two poems, see Harold G. Henderson, *An Introduction to Haiku* (New York: Doubleday, 1958), 16.

20. Anthony Stevens, *Archetype: A Natural History of the Self* (London: Routledge & Kegan Paul, 1982), 162–64.

21. John Ruskin, *Works. Fors Clavigera*, vol. 27, 404–05. Hereafter cited as *FC*. Ruskin provides illustrations of several Greek Frets as labyrinths. For a discussion of Ruskin, but also for analysis of labyrinth as archetype, see Angus Fletcher, "The Image of Lost Direction," in *Center and Labyrinth: Essays in Honor of Northrop Frye*, eds. Elinore Cook, et al. (Toronto: University of Toronto Press, 1983), 329–46.

22. Langer, *Philosophy in a New Key*, 200.

23. *The Chinese Exhibition*, 53. Hereafter cited as *CE*. More recently, a color photograph of the jar appeared in Li Zehou, *The Path of Beauty—A Study of Chinese Aesthetics*, trans. Gong Lizeng (Beijing: Morning Glory Publishers, 1988), 21. There is also a useful discussion of the origins of the dragon as a snake clan totem, which eventually became China's national symbol, 16–21.

24. George Rowley, *Principles of Chinese Painting*, 2d ed. (Princeton: Princeton University Press, 1959). See "Sequence and Moving Forms," 61. Hereafter cited as *PCP*.

25. Richard Barnhardt, *Peach Blossom Spring* (New York: Metropolitan Museum of Art Catalog, 1983), 18.

26. Mai-Mai Sze, *The Tao of Painting. A Study of the Ritual Disposition of Chinese Painting*, vol. 1 (New York: Bollingen Series, Pantheon Books, 1956), 112. For mention of pines as young dragons coiled in gorges, see Mai-Mai Sze, *The Mustard Seed Garden Manual of Painting*, vol. 2, 101; hereafter cited as *MSM*.

27. Kenneth Clark, *Leonardo da Vinci: An Account of His Development as an Artist* (Harmondsworth, England: Penguin, 1958), 151. In *Beyond Representation*, Fong reproduces a different section of *Nine Dragons*, and he discusses dragons as rainmakers that Taoist masters might summon by way of paintings, 363–66.

28. Lin Yutang, *My Country and My People*, (New York: Reynal & Hitchcock, 1935), 319.

29. Lin Yutang, *The Importance of Living* (New York: John Day Co., 1937), 438–39. Hereafter cited as *IL*.

30. Lao Tzu, *Tao Te Ching*, trans. D. C. Lau (Baltimore: Penguin Books, 1963). Hereafter cited as *TTC*.

31. Cited by Alan Watts, *Tao: The Watercourse Way* (London: Jonathan Cape, 1976), 48; from *Chuang-tzu: Mystic Moralist and Social Reformer*, trans. Herbert A. Giles (Shanghai: Kelly & Walsh, 1926; New York: AMS Press, 1972). Watts argues that water waves were the main symbol of Tao.

32. Joseph Needham, *Physics and Physical Technology*, vol. 2 of *Science and Civilization in China* (Cambridge: Cambridge University Press, 1962), 7. Hereafter cited as *SCC*.

33. William Watson, *Style in the Arts of China* (London: Penguin, 1974), 29. Hereafter cited as *SAC*. For a discussion of the totemic ferocity of the beast, fearful to alien clans and protective to its own clan, see Le Zehou, 45–52. He also discusses its degeneration in form and in meaning to a mere decorative device.

34. Gombrich, *The Sense of Order*, 263–70.

35. Johan Huizinga, *Homo Ludens: A Study of the Play Element in Culture* (Boston: Beacon Press, 1955), 16. Hereafter cited as *HL*. The enacting of a natural syntax is not so much a mimetic theory; it is not like Richard Rorty's *Philosophy and the Mirror of Nature* (Princeton: Princeton University Press, 1979); instead, if *mimesis* means representation, then *methexis* is more like Ian Hacking's idea of intervention. These theoretical oppositions are discussed in his *Representing and Intervening : Introductory Topics in the Philosophy of Natural Science* (New York: Cambridge University Press, 1983).

36. The diffusion theory of similarities between Far Eastern and Western American designs is best summarized by Covarrubias as the so-called "Old Pacific Style," dating from as early as the third millennium B.C. Miguel Covarrubias, *The*

Eagle, the Jaguar, the Serpent: Indian Art of the Americas (New York: Alfred A. Knopf, 1954), 24–72. Hereafter cited as *EJS*.

37. Bradley Smith, *Mexico: A History in Art* (New York: Harper & Row, 1968), 51. For a careful scholarly analysis of the composite name Quetzal-coatl, as standing for various sky-earth-water attributes, see Dennis Tedlock, *Popol Vuh: The Definitive Edition of the Mayan Book of the Dawn of Life and the Glories of Gods and Kings* (New York: Simon and Schuster, 1985), 361–62. For a recent study of Quetzalcoatl as fertility god, see Enrique Florescano, *The Myth of Quetzalcoatl*, trans. Raúl Velásquez (Baltimore: Johns Hopkins University Press, 1999); for his comparison of Quetzalcoatl with Inanna, see 203–12.

38. *Four Masterworks of American Indian Literature*, ed. and trans. John Bierhorst (Tucson: University of Arizona Press, 1984), 3. Hereafter cited as *FM*.

39. Neil Baldwin, *Legends of the Plumed Serpent: Biography of a Mexican God* (New York: Public Affairs, 1998). For a description of one of the oldest images of the Plumed Serpent, prosaically called "La Venta Monument 19," see Baldwin, 17. Other scholars date this relief carving from the middle of the Olmec culture, about 900–600 B.C.E. For an analysis of this relief as a feathered-serpent deity, "ancestor" of later plumed serpents in Mesoamerican history, see Peter David Joralemon, "In Search of the Olmec Cosmos: Reconstructing the World View of Mexico's First Civilization," in *Olmec Art of Ancient Mexico*, eds. Elizabeth P. Benson and Beatriz de la Fuente (Washington: National Gallery of Art, 1996), 58.

40. Bernard De Voto, *The Course of Empire* (Boston: Houghton Mifflin Co., 1952), 40.

41. Laurette Séjourné, *Burning Water: Thought and Religion in Ancient Mexico* (New York: Grove Press, 1960), 94–99. Hereafter cited as *BW*.

42. Burr Cartwright Brundage, *The Phoenix of the Western World: Quetzalcoatl and the Sky Religion* (Norman: University of Oklahoma Press, 1982), 21.

43. Sadi Carnot, *Reflections on the Motive Power of Heat, and on Machines Fitted to Develop that Power*, trans. R. H. Thurston (New York: American Society of Mechanical Engineers, 1943), 43. Hereafter cited as *RM*.

44. See Miguel León-Portilla, *México-Tenochtitlán: Su Espacio y Tiempo Sagrados* (Instituto Nacional de Antropología y Historia, n.d.), 19.

45. Séjourné, *El Universo de Quetzalcoatl* (Mexico City: Fundo de Cultura Económica, 1962), 50–62. Hereafter cited as *UQ*.

46. W. J. T. Mitchell, *Picture Theory*. Chapter two is called "Metapictures." For circular arguments, circular figures, and logical quirks of self-reference, Mitchell cites Jon Barwise and John Etchemendy's *The Liar: An Essay on Truth and Circularity* (New York: Oxford University Press, 1989). See also Tyler Volk, *Metapatterns Across Space, Time and Mind* (New York: Columbia University Press, 1995), especially chapter 10, "Cycles."

47. Waley, *The Way and Its Power*, 195.

48. D. H. Lawrence, *Etruscan Places*, 2d ed. (New York: Viking Press, 1933), 99–100. Hereafter cited as *EP*.

49. D. H. Lawrence, *The Plumed Serpent* (New York: Knopf, 1926), 45. Hereafter cited as *PS*.

50. Balaji Mundkur, *Cult of the Serpent: An Interdisciplinary Survey of Its Manifestations and Origins* (Albany: SUNY Press, 1983), 275. For his discussion of the association of serpents with the beginnings of civilizations, see 274 and bibliography.

51. See Peter S. Stevens, *Patterns in Nature* (Boston: Atlantic Monthly Press, 1974), 190–206.

52. Gombrich, *The Sense of Order*, 137–38 and Index.

53. See J. J. Brodie, *Mimbres Painted Pottery* (Albuquerque: University of New Mexico Press, 1977). Also see *Mimbres Pottery of the American Southwest*, ed. Paul Anbinder (New York: Hudson Hills Press, 1983).

54. See Armin W. Geertz, *Hopi Indian Altar Iconography* (Leiden: E. J. Brill, 1987), 1–19, 24. For Benjamin Lee Whorf's analyses of prevalent verb forms that express "meanders" or "oscillations," see "The Punctual and Segmentative Aspects of Verbs in Hopi," in *Language, Thought, Reality: Selected Writings of Benjamin Lee Whorf*, ed. John B. Carroll, 51–54.

55. For a discussion of "Water Management," see William M. Ferguson and Arthur H. Rohn, *Anasazi Ruins in Color* (Albuquerque: University of New Mexico Press, 1987), 44–47. Perhaps the best introduction to the question of natural resources is in a book about a related people, the Mimbres, who lived to the south and west of the Anasazi: Brodie's *Mimbres Painted Pottery*, "The Physical Environment," 2–33. Brodie also provides the clearest overview of the relations among these people, including possible "foreign influences," notably the Casa Grande people, who lived in northern Mexico but who traded northwards regularly (57–103).

56. Ray A. Williamson, *Living the Sky: The Cosmos of the American Indian*, (Boston: Houghton Mifflin, 1984), 96.

57. Polly Schaafsma, *Indian Rock Art of the Southwest* (Albuquerque: University of New Mexico Press, 1980), 81–83, 198, 238. Hereafter cited as *IRA*.

58. *The Mustard Seed Garden Manual* in Sze's *The Tao of Painting*, vol. 2, 53. For a fuller description of the ways one Chinese-American painter taught basic calligraphic strokes, see James H. Bunn, "The Artist in the Classroom" in *James Kuo: A Retrospective* (Buffalo: Burchfield-Penney Art Center Catalog, 1997), unpaged.

59. Robert Hass, "Images" in *Twentieth Century Pleasures* (New York: Ecco Press, 1984), 293. See also *The Essential Haiku: Versions of Basho, Buson, and Issa*, ed. Robert Hass (New York: Ecco Press, 1994), 158.

60. Ernest Fenollosa, *The Chinese Written Character as Medium for Poetry*, ed. Ezra Pound (San Francisco: City Lights Books, 1936). Hereafter cited as *CWC*. Cited by Turner, 160.

61. Ernest Fenollosa, *Epochs of Chinese and Japanese Art*, vol. 2 of *An Outline History of East Asiatic Design*, rev. ed. (New York: Frederick A. Stokes Co., 1921), 6.

62. Stephen Owen, *Traditional Chinese Poetry and Poetics: Omen of the World* (Madison: University of Wisconsin Press, 1985), 84. Hereafter cited as *TCP*.

63. John Ruskin, *The Elements of Drawing in Three Letters to Beginners* (London: Smith, Elder, & Co., 1859), 121–22. Hereafter cited as *ED*. For a discussion of the helix and labyrinth as syntactic curves, as "logo-centric geometries," see Jay Fellows, *Ruskin's Maze: Mystery and Madness in His Art* (Princeton: Princeton University Press, 1981), 272. For a fascinating study of Goethe's use of "leading lines" in landscape, transposed to the developing lines of character, see Mikhail Bakhtin, "The *Bildungsroman* and Its Significance in the History of Realism (Toward a Historical Typology of the Novel)," *Speech Genres and Other Late Essays*, trans. Vern W. McGee, eds. Caryl Emerson and Michael Holquist (Austin: University of Texas Press, 1986), 10–59.

64. Cyril Stanley Smith *The Search for Structure: Selected Essays on Science, Art, and History* (Cambridge: MIT Press, 1981), 21; hereafter cited as *SS*.

Chapter 3

1. Samuel Pierpoint Langley, *The New Astronomy* (Boston: Houghton Mifflin Co., 1898), 75.

2. For Vitruvius's discussion of the proportions of the human body as they fit into the symmetries of a temple, see *TBA*, 72–73. See also *Vitruvius on Architecture*, 2 vols., trans. Frank Granger (New York: G. P. Putnam's Sons, 1931). Granger's edition includes the Latin text. Hereafter cited as *V*. Leonardo's image of "Vitruvian Man" was the unofficial emblem of the 1492 exhibition at the National Gallery of Art, Washington, D.C., in 1992. For a good reproduction of Leonardo's diagram, see *Circa 1492: Art in the Age of Exploration*, ed. Jay A. Levenson (New Haven: Yale University Press, 1991), 276. For Leonardo's own notes on Vitruvius, see 477. For a brief discussion of Leonardo's drawing of the "vortex" as a generative form of growth in nature, see 282–83.

3. George Herbert, "Man," lines 13–18, *The Temple* (1633) in *Works of George Herbert*, ed. E. F. Hutchinson (Oxford: Clarendon Press, 1941), 90–91.

4. Mark Verstokt, *The Genesis of Form: From Chaos to Geometry* (London: Muller, Blond & White, 1987), 72.

5. Michael Leyton, *Symmetry, Causality, Mind* (Cambridge: MIT Press, 1992), 3. Hereafter cited as *SCM*.

6. See the extensive archives of Carl Schuster's drawings, photographs, carvings, weavings, and other kinds of material culture, which have been published in a multivolume series under the general title *Materials for the Study of Social Symbolism in Ancient and Tribal Art: A Record of Tradition and Continuity, Based on the Researches of Carl Schuster*, 3 vols., 12 bk., ed. Edmund Carpenter (New York: The Rock Foundation, 1986–1988). Volume 3, *Rebirth*, consists of three books, the first of which is *Cosmic Games*. Chapters 34 and 35 in the first book are "Anthropomorphic Gaming Boards" and "Anthropomorphic Divination Boards," respectively.

7. James R. Newman, "Commentary on Symmetry" in *The World of Mathematics*, vol. 1 (New York: Simon & Schuster, 1956), 670. References to Weyl's *Symmetry* (hereafter referred to as *S*) are taken from this selection. For the origins and significance of geometry and its relation to physics, see especially in this volume William Kingdon Clifford, "The Postulates of the Science of Space," and Hermann von Helmholtz, "On the Origins and Significance of Geometrical Axioms." For an exercise in deciphering mathematical groups by way of an example of Greek Frets on an amphora, see Philip J. Davis and Reuben Hersh, "Group Theory and the Classification of Simple Finite Groups," in *The Mathematical Experience* (Boston: Birkhauser, 1981), 203–09.

8. Jean Piaget, *Structuralism*, trans. Chaninah Maschler (New York: Basic Books, 1970), 20–22. Also, Roman Jakobson and Linda A. Waugh quote this passage as an epigraph: "The important thing about a transformation is what it doesn't transform, i.e., what it leaves invariant." The passage is from S. S. Stevens, "Mathematics, Measurement, and Psychophysics," in *Handbook of Experimental Psychology* (New York, Wiley, 1951), 1–49. See the second chapter, "Quest for the Ultimate Constituents," in *The Sound Shape of Language*, in *Roman Jakobson: Selected Writings, Major Works, 1976–1980*, ed. Stephen Rudy, vol. 8 (Berlin: Mouton de Gruyter, 1988). Hereafter cited as *SSL*.

9. Emily Noether is usually credited for the first insights that connect the conservation of energy with symmetry. For a discussion of the importance of her breakthrough, see Robert P. Crease and Charles C. Mann, *The Second Creation: Makers of the Revolution in 20th-Century Physics* (New York: Macmillan, 1986), 186–89. For a critique of uses of the term *symmetry* in sociology of science, see Barbara Herrnstein Smith, "Belief and Resistance: A Symmetrical Account," *Critical Inquiry* 18 (autumn 1991), 113–39, especially 131–34.

10. For an early attempt to understand rhythm in this manner, see James H. Bunn, "Circle and Sequence in the Conjectural Lyric," *New Literary History* 3 (spring 1972), 511–26.

11. Martin Gardner, *The Ambidextrous Universe: Left, Right and the Fall of Parity* (New York: Mentor, 1969), 41. Hereafter cited as *AU*.

12. H. S. M. Coxeter, *Introduction to Geometry* (New York: John Wiley and Sons, 1969), 96. In *The Dimensionality of Signs, Tools, and Models: An Introduction* (Bloomington: Indiana University Press, 1981), a spatiotemporal twist was defined as "torsion" (16, 72).

13. Coxeter, 95–99. See especially Ian Stewart and Martin Golubitsky, *Fearful Symmetry: Is God a Geometer?* (New York: Penguin, 1992), 264–65. Hereafter cited as *FS*.

14. Miguel Covarrubias, *The Eagle, the Jaguar, and the Serpent: Indian Art of the Americas* (New York: Alfred A. Knopf, 1954), 39. Covarrubias also drew a colored representation of the bear, which he identified as a Tlingit house screen, and he said that the hole between the legs was a door; facing page 56. His drawings are superb. Famous as a caricaturist of 1920s and 1930s American personalities, Covarrubias

later became known as an excellent artist and avid cultural anthropologist. For an account of his life, see Adriana Williams, *Covarrubias*, ed. Doris Ober (Austin: University of Texas Press, 1994).

15. Some of Schuster's vast collection of hockers and other symmetrically limbed creatures are collected in volume 1, *Genealogical Patterns: Form and Meaning*. Covarrubias's drawing of the Northwest coast image is identified by Schuster as a Tlingit house screen, vol. 1, bk. 3, *Meaning*, 794, figure 1278.

16. Franz Boas, *Primitive Art* (New York: Dover, 1955), 252.

17. Hippocrates, the almost legendary founder of Greek medicine, pays close attention to joints, particularly with respect to the Scythians. For his analyses of their tattoos at the joints, see *On Airs, Waters, and Places*, trans. Francis Adams (The Internet Classics Archive, classics.mit.edu), part 20.

18. Anne Bradstreet, "The Author to Her Book," lines 14–17, in *The Works of Anne Bradstreet*, ed. Jeannine Hensley (Cambridge: Harvard University Press, 1967), 221.

19. Herbert Read, *Phases of English Poetry* (London: L & V. Woolf at the Hogarth Press, 1928). He quotes Francis Barton Gummere in his fascinating book, *The Beginnings of English Poetry* (New York: Macmillan, 1908).

20. William Carlos Williams, "Poem," in *Selected Poems*, ed. Randall Jarrell (New York: New Directions, 1963), 57.

21. Hugh Kenner, *The Pound Era* (Berkeley: University of California Press, 1971), 400.

22. Richard Feynman, *The Character of Physical Law* (Cambridge: MIT Press, 1965), 100. Hereafter cited as *CPL*.

23. Roger Caillois, "Dynamics of Dissymmetry," *Diogenes* 76 (winter 1971): 80. Hereafter cited as *DS*. For a brief but provocative survey of a possible "isomorphism" between the genetic code and the syntax of phonemes, see Roman Jakobson, *Main Trends in the Science of Language* (New York: Harper Torchbooks, 1970), 49–55. Jakobson cites both Francois Jacob and Emile Benveniste, who speak of the "syntax" of the genetic code as being like phonemes in speech.

24. Jacques Monod, *Chance and Necessity: An Essay on the Natural Philosophy of Modern Biology*, trans. Austryn Wainhouse (New York: Alfred Knopf, 1971). Hereafter cited as *CN*. For a critique of Monod's celebration of evolutionary chance, see Stuart Kauffman, *At Home in the Universe: The Search for the Laws of Self-Organization and Complexity* (New York: Oxford University Press, 1995), 94–98. Trefil and Hazen provide a precise and vivid description of the replication processes of DNA (*TS*, 541–51). Their chapter, "The Molecules of Life" (485–511), with molecules seen as modular building blocks, is a useful review for understanding the basic structures of molecular genetics. In their discussions, Trefil and Hazen speak of "geometry," not symmetry, but the emphases are similar. See especially Dennett, "Molecular Evolution," for his discussion of the nucleotide base as a syntax (*DDI*, 155–63).

25. Erasmus Darwin, *The Temple of Nature, or the Origin of Society, a Poem, with Philosophical Notes* (New York: T. & J. Swords, 1804).

26. Philip Wheelwright, *The Burning Fountain: A Study in the Language of Symbolism* (Bloomington: Indiana University Press, 1954), 124.

27. S.K. Heninger, Jr., *Touches of Sweet Harmony: Pythagorean Cosmology and Renaissance Poetics* (San Marino: The Huntington Library, 1974), 146–67. For a study of symmetries in rose windows, see Painton Cowen, *Rose Windows* (London: Thames & Hudson, 1979).

28. Stewart Culin, *Chess and Playing Cards* (Washington, 1898). For a discussion of Mimbres designs as rotations of the four directions, see J.J. Brody, *Mimbres Pottery: Ancient Art of the American Southwest* (New York: Hudson Hills Press, 1983), 110–14.

29. The literary theory of four seasonal myths is found in Northrop Frye's *Anatomy of Criticism* (Princeton: Princeton University Press, 1957), 158–242.

30. S.K. Heninger, Jr., *The Cosmographical Glass: Renaissance Diagrams of the Universe* (San Marino: The Huntington Library, 1977), 198n. Anthony Wilden in *The Rules Are No Game: The Strategy of Communication* (London: Routledge & Kegan Paul, 1987), 156–57, discusses some of these models.

31. Müller's contributions to myth studies are discussed in Ernst Cassirer's *Language and Myth*, trans. Susanne Langer (New York: Harper & Brothers, 1946). For a study of the symbolic meanings of the sun, sometimes in relation to triskelions and other S-curved variants of snakes, labyrinths, and Egyptian serpent boats, see H. Rudolf Engler, *Die Sonne als Symbol: Die Schlussel zu den Mysterien* (Kusnacht-Zurich: Helianthus-Verlag, 1962), 106–59. For discussion of the sun year in Greek religion, see Jane Ellen Harrison, *Themis: A Study of the Social Origins of Greek Religion* (Cambridge: Cambridge University Press, 1912), ch. 6.

32. A.S. Eddington, *The Nature of the Physical World* (Cambridge: Cambridge University Press, 1929), 130–31. In chapter 15, wave formation becomes the vehicle for his linking of science and mysticism.

33. For a series of drawings and photographs of bow lathes, see Schuster, vol. 3, bk. 1, *Cosmic Games*. Drilling holes and making fires are connected with the turning of the Axis Mundi. For a discussion of the "itinerant" as a worker who follows the "flow of matter," see Gilles Deleuze and Félix Guattari, *A Thousand Plateaus: Capitalism and Schizophrenia*, trans. Brian Massumi (Minneapolis: University of Minnesota Press, 1987), 409–10. This passage is in their chapter 12, "1227: Treatise on Nomadology—The War Machine." A line drawing of a nomad chariot introduces their mythology of the war machine. Their preoccupation with "flow" pervades their psychoanalysis of culture.

34. Schuster, vol. 3, bk. 2, *The Labyrinth & Other Paths to Other Worlds*, 264–70. Reproductions of mazes woven into Pima and Hopi baskets are also included therein, 316–17. For analysis of the labyrinth as archetype, see Angus Fletcher, "The Image of Lost Direction," in *Centre and Labyrinth: Essays in Honor of Northrop*

Frye, eds. Elinor Cook et al. (Toronto: University of Toronto Press, 1983), 329–46. For a discussion of Jorge Luis Borges's use of labyrinths, see Floyd Merrell, *Simplicity and Complexity: Pondering Literature, Science and Painting* (Ann Arbor: University of Michigan Press, 1998), ch. 3, 79–94.

35. Engler, *Die Sonne als Symbol*, 115.

36. See Tom Hoskinson, "Saguaro Wine, Ground Figures, and Power Mountains: Investigations at Sears Point, Arizona," in *Earth and Sky: Visions of the Cosmos in Native American Folklore*, eds. Ray Williamson and Claire E. Farrer (Albuquerque: University of New Mexico Press, 1992), 143. For a discussion of "dynamic dissymmetry" in Zuni folklore, see M. Jane Young, "Morning Star, Evening Star: Zuni Traditional Stories," in *Earth and Sky*, 75–96. For more discussion of Zuni dissymmetries, see Barbara Tedlock, "The Beautiful and the Dangerous: Zuni Ritual and Cosmology as an Aesthetic System," in *Conjunctions: Bi-annual Volumes of New Writing* 6 (1984): 246–65.

37. John Muir, *The Mountains of California* (New York: Barnes & Noble Books, 1993), 174.

Chapter 4

1. Quoted by S. Sambursky, *Physics of the Stoics* (London: Routledge & Kegan Paul, 1958), 23, who avers that the Stoics were first to theorize a disturbance in air shaped as spheres. Heraclitus's writings, particularly on the motions of the *logos*, were one of the foundations of the Stoic conception of the continuum.

2. *Kenneth L. Pike: Selected Writings*, ed. Ruth M. Brand (The Hague: Mouton, 1972), 129–43. Hereafter cited as *LP*.

3. I have already cited the differences between Dennet and Gould. Perhaps the most challenging celebrant of Universal Grammar is Ray Jackendoff, who proposes a Universal Grammar for visual thinking, for music, American Sign Language, and concepts themselves, *Patterns in the Mind: Language and Human Nature* (New York: Basic Books, 1994). I have learned much about patterning from this book. For further discussion, see the review article, James H. Bunn, "Universal Grammar or Common Syntax, A Critical Review of Ray Jackendoff's *Patterns in the Mind*," together with his reply, in *Minds and Machines* 10 (February 2000): 119–28, 137–47. For a general-purpose theory of linguistic pattern detection, which may not depend upon the use of innate linguistic modules, see *The Crosslinguistic Study of Sentence Processing*, eds. Brian Macwhinney and Elizabeth Bates (New York: Cambridge University Press, 1989). "We suspect that more general principles of pattern detection and distributional learning are sufficient for the task . . ." (26). My descriptions of pattern awareness and symmetry sharing lend credence to that suspicion.

4. John Lyons, *Language and Linguistics: An Introduction* (Cambridge: Cambridge University Press, 1981), 102. Hereafter cited as *LL*.

5. Lyons, *Introduction to Theoretical Linguistics* (Cambridge: Cambridge University Press, 1968), 209. Hereafter cited as *ITL*.

6. Jackendoff uses the term *Ur*-grammar, "Response to Bunn," in *Mind and Machines* 10 (February 2000): 129–35. I am following a different line of thought. For a recent discussion of the evolution of syntax as following a Darwinian transition from visual gestures to speech, see Armstrong, Stokoe, and Wilcox's "The Origin of Syntax: Gesture as Name and Relation" in *Gesture and the Nature of Language* (Cambridge: Cambridge University Press, 1995) ch. 7, 161–97.

7. Whitehead, "Mathematics and the Good," in *Essays in Science and Philosophy* (London: Rider, 1948), 82. Hereafter cited as *MG*. For his further discussion of algebra, symbolic logic, and the role of symbolic logic in aesthetics, see the essay "Analysis of Meanings," also collected in *MG*, 93–99.

8. It first appeared in *Gammer Gurton's Garland* (1813). See William S. Baring-Gould and Cecil Baring-Gould, *The Annotated Mother Goose* (New York: Clarkson N. Potter, 1962), 268.

9. Lewis Carroll, *Through the Looking Glass* (Baltimore: Puffin Books, 1948, 1963), 268–69. In Susan Stewart's *Nonsense: Aspects in Folklore and Literature* (Baltimore: Johns Hopkins University Press, 1978), discussions of Lewis Carroll's play can be found throughout. For a recent study of the sound-shape of language, see Hugh Bredin, "Onomatopoeia as a Figure and a Linguistic Principle," *New Literary History* 27 (summer 1996): 555–69.

10. Bertand Russell, *ABC of Relativity*, rev. ed. (New York: Signet, 1958), 127.

11. Russell is also linking his notion of numbers as groups of entities to relativity theory. For a quick discussion, see Whitehead, *MG*, 79.

12. Einstein, "Common Language of Science," in *Out of My Later Years* (Totowa, N.J.: Littlefield, Adams & Co., 1967), 107. Hereafter cited as *CLS*.

13. C. H. Waddington, "The Modular Principle and Biological Form," in *Module, Proportion, Rhythm, Symmetry*, ed. Gyorgy Kepes (New York: George Braziller Inc., 1966), 2–37. For Waddington, there is a "system of proportions" that seems to govern the lengths of related units, such as legs, fingers, and toes (35). This sense of proportions was also that of Vitruvius's arrangement of the human limbs, quoted in the previous chapter.

14. Allen Nussbaum, *Geometric Optics: An Introduction* (Reading, Mass.: Addison-Wesley, 1968), 2.

15. Norman Kemp-Smith, *The Philosophy of David Hume* (London: Macmillan, 1949), 116 ff. Cited in Anthony Flew, "Rationality and Unnecessitated Choice," *Naturalism and Rationality*, eds. Newton Garver and Peter H. Hare (Buffalo: Prometheus Books, 1986).

16. Lyons says that Goethe apparently invented the term *morphology* and that it was transposed by early linguists, *Introduction to Theoretical Linguistics*, 195. Hereafter cited as *ITL*.

17. Cassirer, *Problem of Knowledge*, 31.

18. P. F. Strawson, *Subject and Predicate in Logic and Grammar* (London: Methuen, 1974), ch. 1, "The Basic Combination." Hereafter cited as *SP*. The most complete study of *deixis* in terms of fictional space and time is *Deixis in Narrative: A Cognitive Science Perspective*, eds. Judith F. Duchan, Gail A. Bruder, and Lynne E. Hewitt (Hillsdale, N.J.: Lawrence Erlbaum Associates, 1995).

19. Mikhail Bahktin, *Speech Genres and Other Late Essays*, 71–72.

20. Rex Stout, *The Doorbell Rang: A Nero Wolfe Novel* (New York: Viking, 1965).

21. Melissa Bowerman, "Constructing Spatial Semantic Categories: A Crosslinguistic Perspective," lecture given at Center for Cognitive Science, SUNY Buffalo (April, 1998). See also "The Origins of Children's Spatial Semantic Categories: Cognitive vs. Linguistic Categories," in *Rethinking Linguistic Relativity*, eds. J. J. Gumperz and S. C. Levinson (Cambridge: Cambridge University Press, 1996). Bowerman's studies of language acquisition are discussed in Kathy Hirsh-Pasek and Roberta Michnick Golinkoff's *The Origins of Grammar: Evidence from Early Childhood Comprehension* (Cambridge: MIT Press, 1996), 26–27. Their chapter, "Theories of Language Acquisition," is an eminently clear analysis of the great divide between the camps, which they describe as "Outside-In Theories" versus "Inside-Out Theories," in which the former tag represents cognitive categories of pattern detection of cues from the environment in interaction with a learning brain, while the latter tag represents the Chomskyan base that language is not so much learned as discovered (11–51). One of their own conclusions, derived from their analysis of Melissa Bowerman and Soonjā Choi, is this: "Perhaps syntax (and not just semantic categories) is there from the beginning"(26). For a study in the ways in which syntax may suffice for a computer to understand natural language, see William J. Rapaport, "Syntactic Semantics: Foundations of Computational Natural-Language Understanding," in *Thinking Computers and Virtual Persons: Essays on the Intentionality of Machines*, ed. Eric Dietrich (San Diego: Academic Press, 1994), 225–73.

22. Lyons, *Language and Linguistics*, 78–79.

23. In *King Lear*, Edgar, disguised as Tom O'Bedlam, says, "Fie, foh, and fum,/ I smell the blood of a British man." *William Shakespeare: The Complete Works*, act 3, scene 4 (Baltimore: Penguin Books,1969), 174–75.

24. See Claude Lévi-Strauss, "Postscript to Chapters III and IV," in *Structural Anthropology*, 91. For another approach that triangularly associates the primary colors red, yellow, and blue with *a, i,* and *u,* see Wendy Steiner, *The Colors of Rhetoric: Problems in the Relation Between Literature and Culture* (Chicago: University of Chicago Press, 1982), 52–53.

25. *The Annotated Mother Goose*, 281. The authors say that this tongue-twister was included by Dr. John Wallis, a mathematician and grammarian, in his *Grammatica Linguae Anglicanae*. As a juvenile piece it first appeared in *Mother Goose's Quarto* (1813).

26. "Four Trees," *Complete Poems*, no. 742, 364.

27. See Kenneth Koch's discussion of "Anecdote of the Jar" in *Making Your Own Days: The Pleasures of Reading and Writing Poetry* (New York: Scribner, 1998), 117–18.

28. For a range of essays that study the new measures of sound in American poetry, see *Close Listening: Poetry and the Performed Word*, ed. Charles Bernstein (New York: Oxford University Press, 1998).

Chapter 5

1. Charles Darwin, *The Descent of Man and Selection in Relation to Sex* (New York: D. Appleton & Co., 1872). Hereafter cited as *DM*. For Pinker's discussion of the evolutionary possibilities of language development, including a discussion of Chomsky's theories, see "The Big Bang," chap. 11, *LI*, 332–69. For an attack on Konrad Lorenz's neo-Darwinian theory of territory and aggressive communication among animals, as well as a celebration of animals' rhythmic communication, especially bird song, as a passage to art, see Deleuze and Guattari, "1837—Of the Refrain," chap. 11, *Thousand Plateaus*, 310–50.

2. Donald Fleming, "Charles Darwin: The Anaesthetic Man," *Victorian Studies* 4 (1961): 219–36. Collected in *Darwin: A Norton Critical Edition*, ed. Philip Appleman (New York: W. W. Norton, 1970), 587–88. See also Herbert Spencer, "The Origin and Function of Music," in Essays, *Scientific, Political, and Speculative*. (1868), II, 449. London.

3. James Burnett, Lord Monboddo, *Of the Origin and Progress of Language*, 6 vols. (London: T. Cadell, 1792), vol. 1, 469.

4. Rousseau and Herder's discussions may be found in *Two Essays On the Origin of Language: Jean-Jacques Rousseau and Johann Gottfried Herder*, trans. John H. Moran and Alexander Gode (Chicago: University of Chicago Press, 1966). For a discussion of Shelley's use of harmonic and analytic language, see James H. Bunn, "Shelley's Method in *A Defence of Poetry*," *English Romanticism: Preludes and Postludes*, eds. Donald Schoonmaker and John A. Alford (East Lansing: Colleagues Press, 1993), 97–114. Ernst Cassirer also thought that the origin of language was not in the domain of logic but in feeling, in "poetry, the mother tongue of humanity" (*Language and Myth*), 33.

5. Rousseau, "Essay on the Origin of Languages which treats of Melody and Musical Imitation," *Two Essays*, 62–63.

6. Robin Dunbar, *Grooming, Gossip, and the Evolution of Language* (Cambridge: Harvard University Press, 1996), 132–51. Hereafter cited as *GG*. Although many feminists have resisted the recourse to nature, and especially Darwin, see alternatively Elizabeth Grosz, "Darwin and Feminism: Preliminary Investigations for a Possible Alliance," *Australian Feminist Studies*, vol. 14, no. 29 (1999), 31–45.

7. Müller's jocular tags for the different theories of origins of language are cited by Gode in his introduction to Rousseau and Herder's essays.

8. For a discussion of rhythm and arm and leg motion, see Curt Sachs, *The Wellsprings of Music*, ed. Jaap Kunst (The Hague: M. Nijhoff, 1962), 114.

9. Herbert Read, *English Prose Rhythm* (Boston: Beacon Press, 1952), 146–47. See also George Saintsbury, *A History of English Prose Rhythm* (London: Macmillan & Co., 1922). Hereafter cited as *EPR*. Saintsbury also wrote an authoritative study of traditional meter, *Historical Manual of English Prosody* (London: Macmillan & Co., 1912). Susan Stewart says that this book "remains invaluable for its historical perspective on issues of meter" in "Letter on Sound," *Close Listening*, 48n. Studying a wide range of modern lyrics, Stewart insightfully explores an implied promise between a poet and reader by way of the drive of sounds: "I propose that the sound of poetry is heard in the way that a promise is heard" (46).

10. Darwin, *Descent of Man*, chap. 21.

11. Sten Grillner, "Neural Networks for Vertebrate Locomotion," *Scientific American* (January 1996): 64–69. Hereafter cited as *NN*. Stewart and Golubitsky discuss central pattern generators in terms of the symmetry of gaits (*Fearful Symmetry*, 199–203).

12. D. Margoliash, E. S. Fortune, M. L. Sutter, A. C. Yu, B. D. Hardin, and W. A. Dave, "Distributed Representation in the Song System of Oscines [songbirds]: Evolutionary Implications and Functional Consequences," *Brain, Behavior, & Evolution* 44 (4–5)(1994): 247–64.

13. Kenneth Koch, *Making Your Own Days: The Pleasures of Reading and Writing Poetry* (New York: Scribner, 1998), 19. Hereafter cited as *MD*.

14. Robert Hass, "Listening and Making," *Twentieth-Century Pleasures: Prose on Poetry* (New York: Ecco Press, 1984), 112. Hereafter cited as *LM*.

15. I have not been able to locate this passage in Browning's work; it is quoted in Gummere's *Beginnings of Poetry*, 30. This book is a fascinating study of the communal beginnings of poetry in rhythmic action.

16. Stanislaw M. Ulam, *Adventures of a Mathematician* (New York: Scribner, 1976), 180–81. Quoted and discussed by William H. Calvin, *The Cerebral Symphony: Seashore Reflections of the Structure of Consciousness* (New York: Bantam Books, 1990), 278.

17. Calvin, 25. Shelley explains this idea more thoroughly in his essay, "On a Future State," *The Complete Works of Percy Bysshe Shelley*, vol. 6, eds. Roger Ingpen and Walter E. Peck (New York: Gordian Press, 1965), 205.

18. Robert Creeley, "Sonnets," *Echoes* (New York: New Directions, 1994), 40–41. This sonnet is briefly discussed by Albert Cook, "New Music in Poetry: The Senses of Sound," unpublished manuscript, 13–14. A related essay of Cook's is "Metrical Inventions: Zukofsky and Merwin," *College Literature* (October 1997), 70–83.

19. When George Lakoff and Mark Johnson speak of "orientational metaphors," they feature up-down, in-out, front-back, on-off, deep-shallow, central-peripheral pointers; for example, to be happy is to be up; to be sad is to be down; also consciousness is an up metaphor, while unconsciousness is down. See "Orientational

Metaphors," *Metaphors We Live By* (Chicago: University of Chicago Press, 1980), 14–21.

20. L. B Meyer, *Music, the Arts, Ideas: Patterns and Predictions in Twentieth-Century Culture* (Chicago: University of Chicago Press, 1967), 288. Hereafter cited *MAI.*

21. Speaking of memes as small patterned units of replicating memory, Dennett asks, "What is it about acronyms, or about rhymes or 'snappy' slogans that makes them fare so well in the competitions that rage through the human mind?" (*DDI,* 359). Other examples of memes are catchy tunes, such as the opening bars of Beethoven's Fifth Symphony (*DDI,* 344–46). The term *meme* was coined by Richard Dawkins; his study of memes and genes as basic units of replication is found in *The Extended Phenotype: The Gene as a Unit of Selection* (San Francisco: W. H. Freeman, 1982), 89–92. J. Z. Young earlier spoke of "mnemons" as "classification neurons," or grouped units of memory in *An Introduction to the Study of Man* (New York: Oxford University Press, 1971), 252. For Young, this new kind of mnemonic grouping function of the cortex "constituted in effect the invention of a wholly new biological phenomenon, the transmission of detailed forecasts by codes other than genetic" (497).

Among phonologists there is ongoing research into the ways in which prosodic inflection may prompt cues for organizing larger syntactic units such as phrases and clauses. Especially among those studying the ways in which infants and children learn a native language, this kind of theory is sometimes called *prosodic bootstrapping.* For a fascinating overview of recent research about this new use of prosody, see Peter W. Jusczyk, "How Attention to Sound Properties May Facilitate Learning Other Elements of Linguistic Organization," *The Discovery of Spoken Language* (Cambridge: MIT Press, 1997), 137–66. I am much indebted to my colleague Jim Swan for helping locate this kind of work. For contemporary research into the ways that the hearing of just a few notes can prompt recollection of the whole song, see Willie Wong and Horace Barlow, "Pattern Recognition: Tunes and Templates," *Nature,* 404, 6781 (April 27, 2000): 952–53. Apparently, an initial hearing of a melody forms a mental template, which is later matched against parts of a tune subsequently heard.

22. Hayden White, *Tropics of Discourse: Essays in Cultural Criticism* (Baltimore: Johns Hopkins University Press, 1978), 2.

23. Claude Lévi-Strauss, "Split Representation in the Art of Asia and America," *Structural Anthropology,* trans. Claire Jacobson and Brooke Grundfest Schoepf (New York: Anchor Books, 1967), 239–63. For a brilliant reading of the ethnographer's troubled rhetoric about split representation, and the apparent symmetries of order that characterize the Caduveo people, see Carol Jacobs, "Architectures of Oblivion—Lévi-Strauss," *Telling Time: Lévi-Strauss, Ford, Lessing, Benjamin, deMan, Wordsworth, Rilke* (Baltimore: Johns Hopkins University Press, 1993), 41–66.

24. Pablo Picasso, *Picasso on Art: A Selection of Views*, ed. Dore Ashton (New York: Viking, 1972), 128. The need to visualize Picasso's visual metaphor is stressed by Franklin R. Rogers, *Painting and Poetry: Form, Metaphor, and the Language of Literature* (Lewisburg: Bucknell University Press, 1985), 120.

25. Lévi-Strauss, "Rebuttal," *Structural Anthropology*, 324.

26. Ludwig Wittgenstein, *Philosophical Investigations*, trans. G. E. M. Anscombe (Oxford: Basil Blackwell, 1967), 48. The most important study of Wittgenstein's discussion of philosophical language and avant-garde poetry is Marjorie Perloff's *Wittgenstein's Ladder: Poetic Language and the Strangeness of the Ordinary* (Chicago: University of Chicago Press, 1996). For an extended study of Creeley's grammar, and his attendance to Wittgenstein's grammar, see 181–200. For Perloff's study of the sounds of free verse, as well as new experiments in syntax, sounds, and typographical optics, see Perloff's "After Free Verse, The New Nonlinear Poetics," *Close Listening*, 86–110.

27. For a discussion of split reference as category mistake, see Paul Ricoeur, *The Rule of Metaphor: Multidisciplinary Studies of the Creation of Meaning in Language*, trans. Robert Czerny (Toronto: University of Toronto Press, 1977) 9–43, and throughout. Also see my review of *Patterns in the Mind*, in *Minds and Machines*. In a section called "The Terrible Pun," Susan Stewart says, "Puns are 'terrible' and 'awful' because they split the flow of events in time." *Nonsense*, 161.

28. Martin Gardner, *The Ambidextrous Universe: Left, Right, and the Fall of Parity* (New York: Mentor, 1969), 44–45. See *The Complete Poetic and Dramatic Works of Robert Browning* (Cambridge: Riverside Press, 1895), 170. Symmetry groups of words and metrics have been studied in Russian poetry since the 1900s. See A. V. Shubnikov and V. A. Koptsik, *Symmetry in Science and Art*, trans. G. D. Archard (New York: Plenum, 1974), 350–50.

29. Morton Bloomfield, "The Syncategorematic in Poetry: From Semantics to Syntactics," *To Honor Roman Jakobson: Essays on the Occasion of His Seventieth Birthday*, vol. 1 (The Hague: Mouton, 1967), 309–17. Hereafter cited as *SP*.

30. Quoted by Harvey Gross, *Sound and Form in Modern Poetry: A Study of Prosody from Thomas Hardy to Robert Lowell* (Ann Arbor: University of Michigan Press, 1964), 134. Taken from Ezra Pound, *Antheil and the Treatise on Harmony* (Chicago: University of Chicago Press, 1927). The quotation continues: "Rhythm is a shape; it exists like the keel line of a yacht, or the lines of an automobile engine, for a definite purpose, and should exist with an efficiency as definite as that which we find in yachts and automobiles." Here shape and rhythm can be united with Ruskin's "leading lines."

31. Kenneth Burke, *Language as Symbolic Action* (Berkeley: University of California Press, 1966), 282.

32. Roland Barthes speaks of both myth and metaphor as a "second-order sign" in "Myth Today," *Mythologies*, trans. Annette Levers (New York: Hill & Wang, 1972), passim. In their new book *Philosophy in the Flesh: The Embodied Mind and Its Challenge to Western Thought* (New York: Basic Books, 1999), George Lakoff and

Mark Johnson assert that the metaphorical activity that constitutes thought is mostly unconscious (ch. 2, "The Cognitive Unconscious"). Any poetic metaphor, however, summons an aesthetic play of thought that knows it seeks an unexpressed pattern between a concept and an image. The idea of "meta-" in metaphysics and in meta-theater means a conscious seeking-after the tacit pattern.

33. Lakoff and Johnson's thesis that cognition itself is embodied by metaphor has been extended to poetry by Raymond W. Gibbs, Jr., *The Poetics of Mind: Figurative Thought, Language, and Understanding* (Cambridge: Cambridge University Press, 1994).

34. See William Wasserstrom's discussion of Willams's earthiness, by way of this passage from *Spring and All*, ch. 9, "The Healing Image," *The Ironies of Progress: Henry Adams and the American Dream* (Carbondale: Southern Illinois Press, 1984), 191. For an excellent discussion of the ways that puns, metaphors, and rhymes "corporealize language," lending a material cause to their agency, see Sigurd Burkhardt, "The Poet and Fool and Priest," *English Literary History* 32 (1956): 279–98.

35. Ludwig Wittgenstein, *Philosophical Investigations*, trans. G.E.M. Anscombe (Oxford: Basil Blackwell, 1967), 230. For introducing me to some of this material about Wittgenstein, especially his meditations about numbers 126–128, I am indebted to Newton Garver, who gave a paper entitled "Why There Can Be No Theses in Philosophy" at SUNY Buffalo (September 1995). See especially Newton Garver, *This Complicated Form of Life: Essays on Wittgenstein* (Chicago: Open Court, 1994).

Conclusion

1. I have written more extensively about conservation laws, symmetry, and equilibrium in "Availing the Physics of Least Action," *New Literary History* 26 (Spring 1995): 419–32. However, I have chosen not to discuss chaos theory, because it has been so extensively studied by postmodern literary theorists and by some creative writers themselves. See, for example, John Barth's fractal applications to his own work and to others in "The Stuttgart Seminars on Postmodernism, Chaos Theory, and the Romantic Arabesque," in *Further Fridays: Essays, Lectures, and Other Nonfiction, 1984–94* (Boston: Little, Brown & Company, 1995), 275–348. For a thorough analysis of the centrality of fractal and chaos theory in recent American narrative, see Joseph Conte, *Design and Debris: A Chaotics of Postmodern Fiction* (University of Alabama Press, forthcoming). See specifically "American Oulipo: Proceduralism in the Novels of Gilbert Sorrentino, Harry Matthews, and John Barth," chapter 4.

2. Albert Einstein, "E = mc²," *Out of My Later Years*, 50–54. Hereafter cited as *E*. I discuss this essay more thoroughly in "The Constancy of an Angular Point of View," in *Fragments of Science: Festschrift for Mendel Sachs* (Singapore: World Scientific Fair, 1999), 9–24. Einstein also predicted that gravity would be eventually understood as a wave function. For a study about researches into the existence of ripples in space-time itself, see David Blair and Geoff McNamara, *Ripples*

on a Cosmic Sea: The Search for Gravitational Waves (Reading, Mass.: Addison-Wesley, 1997).

3. Stephen Hawking, *A Brief History of Time: From the Big Bang to Black Holes* (New York: Bantam, 1988), 65–66.

4. Gene Weltfish, *The Origins of Art* (Indianapolis: Bobbs-Merrill, 1952), 20, 53. Weltfish follows Gottfried Semper in believing that matting and weaving were the source for most protogeometrical designs in art. See Semper, *Der Stil in den technischen und tektonischen Kunsten* (Berlin: 1878–1879). See also a discussion of "The Meander as System," in Mircea Elaide's *A History of Religious Ideas*, vol. 1, trans. Willard Trask (Chicago: University of Chicago Press, 1978), 23.

5. Marcel Griaule, *Conversations with Ogotemmeli*. See his drawings of drums and lashings, 66–67.

6. Washington Matthews, *Navajo Legends* (Boston: American Folklore Society, 1897). Quoted in Alice Kaufman and Christopher Selser's *The Navajo Weaving Tradition 1650 to the Present* (New York: E.P. Dutton, 1985), 130. For a definitive overview, see Marian Rodee, *Southwestern Weaving* (Albuquerque: University of New Mexico Press, 1977).

7. Neil Baldwin, *Legends of the Plumed Serpent*, 52. Miguel Covarrubias, *Mexico South: The Isthmus of Tehuantepec* (New York: Alfred A. Knopf, 1947), 188.

8. Gary Snyder, *Riprap and Cold Mountain Poems* (San Francisco: Four Seasons Foundation, 1969), 30. For an account of how to lay this kind of rock pavement, see Jim Snyder, "Riprap and the Old Ways: Gary Snyder in Yosemite, 1955" in *Gary Snyder: Dimensions of a Life*, ed. Jon Halper (San Francisco: Sierra Club Books, 1991), 35–42.

9. Samuel Taylor Coleridge, *Biographia Literaria*, in *The Collected Works*, vol. 1, 124. For a discussion of this passage as an attack on associationism, as well as its place in Coleridge's own psychological development, see Richard Holmes's admirable biography, *Coleridge: Darker Reflections, 1804–1834* (New York: Pantheon Books, 1998), 396–99.

10. Immanuel Kant, *Critique of Judgement*, trans. J. H. Bernard (New York: Haffner Library, 1951). For Kant's discussion of aesthetic judgment as the "middle term" between understanding and reason, see his Introduction, 13–15. See Terry Eagleton's chapter, "The Kantian Imaginary," in *The Ideology of the Aesthetic* (Oxford: Basil Blackwell, 1990), 70–101, for a contemporary critique of aesthetic judgment.

11. Matila Ghyka, *The Geometry of Art and Life* (New York: Dover, 1977), 85–86.

12. Stewart and Golubitsky, *Fearful Symmetry*, 7–15.

13. For a discussion of the sun and its transformations of helium, see James S. Trefil, *The Unexpected Vista: A Physicist's View of Nature* (New York: Scribner's, 1983), 21–23.

14. Emily Dickinson, "There's a Certain Slant of Light," *The Complete Poems of Emily Dickinson*, ed. Thomas J. Johnson (New York: Little Brown & Co., 1969), No. 258, 118–19.

15. Julia Kristeva, *Desire in Language: A Semiotic Approach to Literature and Art*, ed. Leon S. Roudiez, trans. Thomas Gora, Alice Jardine, and Leon S. Roudiez (New York: Columbia University Press, 1980), 172. Hereafter cited as *DIL*.

16. William Blake, *The Poetry and Prose of William Blake*, ed. David V. Erdman, Commentary by Harold Bloom (New York: Doubleday, 1965), 24. For Blake's use of vortices and curvilinear lines, see W. J. T. Mitchell, "Metamorphoses of the Vortex: Hogarth, Blake, Turner," collected in *Articulate Images: The Sister Arts from Hogarth to Tennyson*, ed. Richard Wendorf (Minneapolis: University of Minnesota Press, 1983). In an earlier study, Mitchell showed how, in his illuminated books, Blake's curved forms serve as dialectical switches from visual to verbal forms. *Blake's Composite Art: A Study of the Illuminated Poetry* (Princeton: Princeton University Press, 1978).

17. Northrop Frye, *Fearful Symmetry* (Princeton: Princeton University Press, 1947). See also A. Zee, *Fearful Symmetry: The Search for Beauty in Modern Physics* (New York: Macmillan, 1986) for a physicist's discussion of physics and beauty by way of symmetry theory. The head of a flaming tiger appears on its cover. In their chapter "Turing's Tiger," Stewart and Golubitsky describe the patterns of symmetry breaking in living things, and they exemplify the thesis in renditions of stripes on a tiger, 149–88. In *Order Out of Chaos*, Prigogine and Stengers discuss "Bifurcations and Symmetry Breaking" in terms of a physical ordering that builds up complexity against the flow of entropy, 160–70. Their discussion is in terms of periodic oscillations. See also a section on chemical waves simulated by a computer, 148–49.

18. The primal sacred fear that a spoken word can summon the presence of the being is discussed as a "momentary deity" by Ernst Cassirer, *Language and Myth*, 44–45. For a discussion of this symbolic phenomenon as being central to Cassirer's relation to semiotics, see Robert E. Innis, *Consciousness and the Play of Signs* (Bloomington: Indiana University Press, 1994), chapter 6. The ability of a poet to compose a series of actions in progressive steps that gradually assemble into a composition, as Homer described Hebe's sequential assemblage of a chariot—wheel, axle, seat, pole, traces, and straps—is celebrated by Gotthold Lessing, *Laocoön: An Essay on the Limits of Painting and Poetry*, trans. Edward Allen McCormick (Baltimore: Johns Hopkins University Press, 1962), 80. Blake's colleague Henry Fuseli translated the work into English, and Blake also engraved a replica of the statuary group, attached with his own visionary graffiti.

19. Charles W. Morris, *Foundations of the Theory of Signs* (Chicago: University of Chicago Press, 1938), 5. Much cognitive science seems to be moving away from its early roots in formal logic. For a brief history of that question, see Lakoff and Johnson, "The First Generation: The Cognitive Science of the Disembodied Mind," *Philosophy in the Flesh*, 75–79. Here is an example of one of those first-generation commitments that they have abandoned. "Symbol manipulation: Cognitive operations, including all forms of thought, are formal operations on symbols without regard to what those symbols mean" (79).

20. Opening stanzas from Walt Whitman's "A Song of the Rolling Earth" in *Walt Whitman*, ed. Floyd Stovall (New York: American Book Co., 1939), 124.

WORKS CITED

Aarsleff, Hans. *From Locke to Saussure: Essays on the Study of Language & Intellectual History*. Minneapolis: University of Minnesota Press, 1982.

Ammons, A. R. *A Coast of Trees: Poems*. New York: W. W. Norton, 1981.

Anbinder, Paul, ed. *Mimbres Pottery of the American Southwest*. New York: Hudson Hills Press, 1983.

Appleman, Philip, ed. *Darwin: A Norton Critical Edition*. New York: W. W. Norton, 1970.

Arendt, Hannah. *Between Past & Future: Six Exercises in Political Thought*. New York: Viking, 1968.

Argyros, Alexander. *A Blessed Rage for Order: Deconstruction, Evolution, & Chaos*. Ann Arbor: University of Michigan Press, 1991.

Aristotle. *Metaphysica*. 2d ed. Edited by W. R. Ross. Oxford: Clarendon Press, 1963.

Armstrong, David, William Stokoe, & Sherman Wilscox. *Gesture & the Nature of Language*. Cambridge: Cambridge University Press, 1995.

Arnheim, Rudolf. *Art & Visual Perception: A Psychology of the Creative Eye*. Berkeley: University of California Press, 1954.

———. *Visual Thinking*. Berkeley: University of California Press, 1969.

Axelos, Kostas. *Alienation, Praxis, & Techne in the Thought of Karl Marx*. Translated by Ronald Bruzina. Austin: University of Texas Press, 1976.

Bachelard, Gaston. *The Poetics of Space*. Translated by Maria Jolas. Boston: Beacon Press, 1969.

Bahktin, Mikhail. "The *Bildungsroman* & Its Significance in the History of Realism (Toward a Historical Typology of the Novel)." *Speech Genres & Other Late Essays*. Edited by Caryl Emerson & Michael Holquist, translated by Vern W. McGee. Austin: University of Texas Press, 1986.

Baldwin, Neil. *Legends of the Plumed Serpent: Biography of a Mexican God.* New York: Public Affairs, 1998.

Baring-Gould, William S., & Cecil Baring-Gould. *The Annotated Mother Goose.* New York: Clarkson N. Potter, 1962.

Barnhardt, Richard. *Peach Blossom Spring.* New York: Metropolitan Museum of Art Catalog, 1983.

Barth, John. *Further Fridays: Essays, Lectures, & other Nonfiction, 1984–94.* Boston: Little, Brown & Company, 1995.

Barthes, Roland. *Mythologies.* Translated by Annette Levers. New York: Hill & Wang, 1972.

Barwise, John, & John Etchemendy. *The Liar: An Essay on Truth & Circularity.* New York: Oxford University Press, 1989.

Bascom, Willard. *Waves & Beaches: The Dynamics of the Ocean Surface.* New York: Anchor, 1960, 1980.

Baudrillard, Jean. *For a Critique of the Political Economy of the Sign.* Translated by Charles Levin. St. Louis: Telos Press, 1981.

Benjamin, Walter. *Benjamin: Philosophy, History, Aesthetics.* Edited by Gary Smith. Chicago: University of Chicago Press, 1989.

Bernstein, Charles, ed. *Close Listening: Poetry & the Performed Word.* New York: Oxford University Press, 1998.

Bernstein, Richard J. *The New Constellation: The Ethical-Political Horizons of Modernity/Postmodernity.* Cambridge: MIT Press, 1992.

Bierhorst, John, ed. & trans. *Four Masterworks of American Literature.* Tucson: University of Arizona Press, 1984.

Black, Max. *Models & Metaphors: Studies in Language & Philosophy.* Ithaca: Cornell University Press, 1962.

Blair, David, & Geoff McNamara. *Ripples on a Cosmic Sea: The Search for Gravitational Waves.* Reading, Mass.: Addison-Wesley, 1997.

Blake, William. *The Poetry & Prose of William Blake.* Edited by David V. Erdman. New York: Doubleday, 1965.

———. William Blake: *The Complete Illuminated Books.* Edited by David Bindman. New York: Thames & Hudson, 2000.

Bloch, Ernst. *Natural Law & Human Dignity.* Translated by Dennis J. Schmidt. Cambridge: MIT Press, 1987.

Bloom, Harold, ed. *A.R. Ammons: Modern Critical Views.* New York: Chelsea House, 1986.

Bloom, Lynn Z., ed. *The Essay Connection: Reading for Writers.* Boston: Houghton Mifflin Co., 1998.

Bloomfield, Morton. "The Syncategorematic in Poetry: From Semantics to Syntactics." *To Honor Roman Jakobson: Essays on the Occasion of His Seventieth Birthday.* Vol 1. The Hague: Mouton, 1967.

Boas, Franz. *Primitive Art.* New York: Dover, 1927, 1955.

Bova, Ben. *The Beauty of Light.* New York: John Wiley & Sons, 1988.

Bowerman, Meslissa. "The Origins of Children's Spatial Semantic Categories: Cognitive vs. Linguistic Categories." In *Rethinking Linguistic Relativity.* Edited by J. J. Gumperz & S. C. Levinson. Cambridge: Cambridge University Press, 1996.

———. "Constructing Spatial Semantic Categories: A Crosslinguistic Perspective." Lecture given at the Center for Cognitive Science, SUNY Buffalo, N.Y.: April, 1998.

Bradstreet, Anne. *The Works of Anne Bradstreet.* Edited by Jeannine Hensley. Cambridge: Harvard University Press, 1967.

Brea, L. Bernabo. *Sicily Before the Greeks.* Rev. ed. New York: Frederick A. Praeger, 1966.

Bredin, Hugh. "Onomatopoeia as a Figure & a Linguistic Principle." *New Literary History* 27 (summer 1996): 555–69.

Bridgman, P. W. *The Nature of Thermodynamics.* Gloucester, Mass.: Peter Smith, 1969.

Brody, J. J. *Mimbres Painted Pottery.* Albuquerque: University of New Mexico Press, 1977.

———. *Mimbres Pottery: Ancient Art of the American Southwest.* New York: Hudson Hills Press, 1983.

Bronowski, Jacob. "Human & Animal Languages." *A Sense of the Future: Essays in Natural Philosophy.* Edited by Pietro E. Ariotti. Cambridge: MIT Press, 1977.

Browning, Robert. *The Complete Poetic & Dramatic Works of Robert Browning.* Boston: Houghton Mifflin Co., 1895.

Brundage, Burr Cartwright. *The Phoenix of the Western World: Quetzalcoatl & the Sky Religion.* Norman: University of Oklahoma Press, 1982.

Bruner, Jerome. *Actual Minds, Possible Worlds.* Cambridge: MIT Press, 1986.

Bruns, Gerald L. *Hermeneutics Ancient & Modern.* New Haven: Yale University Press, 1998.

Bunn, James H. "Circle & Sequence in the Conjectural Lyric," *New Literary History* 3(spring 1972): 511–26.

———. *The Dimensionality of Signs, Tools & Models: An Introduction.* Bloomington: Indiana University Press, 1981.

———. "Shelley's Method in *A Defence of Poetry.*" *English Romanticism: Preludes & Postludes.* Edited by Donald Schoonmaker & John A. Alford. East Lansing: Colleagues Press, 1993.

———. "Availing the Physics of Least Action," *New Literary History* 26 (spring 1995): 419–439.

———. "The Attractive & Repulsive Poles of Coleridge's Science & the Question of a Dialectic in Nature." *Afterimages: A Festschrift in Honor of Irving Massey.* Edited by William Kumbier & Ann Colley. Buffalo: shuffaloff, 1996.

———. "The Artist in the Classroom." *James Kuo: A Retrospective.* Buffalo: Burchfield-Penny Art Center Catalog, 1997.

———. "The Constancy of an Angular Point of View." *Fragments of Science: Festschrift for Mendel Sachs.*" Singapore: World Scientific Fair, 1999.

———. "Universal Grammar or Common Syntax. A Critical Review of Ray Jackendoff's *Patterns in the Mind.*" *Mind & Machines* 10 (February 2000): 119–128, 137–147.

Burger, R. *Phaedo: A Platonic Labyrinth.* New Haven: Yale University Press, 1984.

Burke, Kenneth. *Language as Symbolic Action.* Berkeley: University of California Press, 1966.

Burkhardt, Sigurd. "The Poet & Fool & Priest." *English Literary History* 32 (1956).

Burnett, James, Lord Monboddo. *Of the Origin & Progress of Language.* 6 vols. London: T. Cadell, 1792.

Butlin, Martin. *The Paintings & Drawings of William Blake.* Vol 1. New Haven: Yale University Press, 1981.

Caillois, Roger. "Dynamics of Dissymmetry," *Diogenes* 76 (winter 1971).

Calame-Giraule, Genevieve. *Words & the Dogon World.* Translated by Dierdre LaPin. Philadelphia: Institute for the Study of Human Issues, 1968.

Calvin, William H. *The Cerebral Symphony: Seashore Reflections of the Structure of Consciousness.* New York: Bantam Books, 1990.

Campbell, Jeremy. *Grammatical Man: Information, Entropy, Language, & Life.* New York: Simon & Schuster, 1982.

Carnot, Sadi. *Reflections on the Motive Power of Heat, & on Machines Fitted to Develop that Power.* Translated by R. H. Thurston. New York: American Society of Mechanical Engineers, 1943.

Carrol, Lewis. *Through the Looking Glass.* Baltimore: Puffin Books, 1948, 1963.

Carter, Denny. *Henry Farny.* New York: Watson-Guptill Publications, 1978.

Cassirer, Ernst. *The Myth of the State.* New York: Doubleday Anchor, 1946.

———. *Language & Myth.* Translated by Susanne K. Langer. New York: Harper & Brothers, 1946.

———. *The Problem of Knowledge: Philosophy, Science & History Since Hegel.* Translated by William H. Woglum & Charles W. Hendel. New Haven: Yale University Press, 1950.

———. *The Philosophy of Symbolic Forms.* Translated by Ralph Manheim. Vols. 1–4. New Haven: Yale University Press, 1953.

Chinese Exhibition: The Exhibition of the Archaeological Finds of the People's Republic of China. London: London Times Newspapers Ltd., 1973.

Chomsky, Noam. *Cartesian Linguistics: A Chapter in the History of Rationalist Thought.* New York: Harper & Row, 1966.

———. "The Formal Nature of Language." Appendix to *Biological Foundations of Language* by Eric Lenneberg. New York: John Wiley & Sons, 1967.

———. *Language & the Problem of Knowledge: The Managua Lectures.* Cambridge: MIT Press, 1988.

Chuang-tzu. *Chuang-tzu: Mystic Moralist & Social Reformer.* Translated by Herbert A. Giles. Shanghai: Kelly & Walsh, 1926; New York: AMS Press, 1972.

Clark, Kenneth. *Leonardo da Vinci: An Account of His Development as an Artist.* Hammondsworth, England: Penguin, 1958.

Coleridge, Samuel Taylor. *The Collected Works of Samuel Taylor Coleridge.* Edited by James Engell & W. Jackson Bate. Princeton: Princeton University Press, 1983.

Conte, Joseph. *Design & Debris: A Chaotics of Postmodern Fiction.* Forthcoming.

Cook, Albert. *Myth & Language.* Bloomington: Indiana University Press, 1980.

———. "Metrical Inventions: Zukofsky & Merwin." *College Literature.* (October 1997).

———. "New Music in Poetry: The Senses of Sound." Unpublished manuscript.

Cook, Theodore Andrea. *The Curves of Life: Being an Account of Spiral Formations & Their Application to Growth in Nature, to Science, & to Art: With Special Reference to the Works of Leonardo da Vinci.* 1914; reprint, New York: Dover Publications, 1979.

Corrington, Robert S. *Ecstatic Naturalism: Signs of the World.* Bloomington: Indiana University Press, 1994.

Covarrubias, Miguel. *The Eagle, The Jaguar, The Serpent: Indian Art of the Americas.* New York: Alfred A. Knopf, 1954.

———. *Mexico South: The Isthmus of Tehuantepec.* New York: Alfred A. Knopf, 1947.

Cowen, Painton. *Rose Windows.* London: Thames & Hudson, 1979.

Crease, Robert P., & Charles C. Mann. *The Second Creation: Makers of the Revolution in 20th Century Physics.* New York: Macmillan, 1986.

Creeley, Robert. *Echoes.* New York: New Directions, 1994.

———. *Life & Death.* New York: New Directions, 1998.

Culin, Stewart. *Chess & Playing Cards.* Washington; n.p., 1898.

Culler, Jonathan. *Literary Theory: A Very Short Introduction.* New York: Oxford University Press, 1997.

Cziko, Gary. *Without Miracles: Universal Selection Theory & the Second Darwinian Revolution.* Cambridge: MIT Press, 1995.

Darwin, Charles. *The Descent of Man & Selection in Relation to Sex.* New York: D. Appleton & Co., 1872.

Darwin, Erasmus. *The Temple of Nature, or the Origin of Society, a Poem, with Philosophical Notes.* New York: T. & J. Swords, 1804.

Davie, Donald. *Articulate Energy: An Enquiry into the Syntax of English Poetry.* London: Routledge & Kegan Paul, 1955.

Davies, Paul. *Superforce: The Search for a Grand Unified Theory of Nature.* New York: Simon & Schuster, 1984.

Davies, Paul, & John Cribben. *The Matter Myth.* New York: Simon & Schuster, Touchstone, 1992.

Davis, Philip J., & Ruben Hersh. *The Mathematical Experience.* Boston: Birkhauser, 1981.

Dawkins, Richard. *The Extended Phenotype: The Gene as a Unit of Selection.* San Francisco: W.H. Freeman, 1982.

Deleuze, Gilles, & Félix Guattari. *A Thousand Plateaus: Capitalism & Schizophrenia.* Translated by Brian Massumi. Minneapolis: University of Minnesota Press, 1987.

Dennett, Daniel C. *Darwin's Dangerous Idea: Evolution & the Meanings of Life.* New York: Simon & Schuster, 1995.

Derrida, Jacques. "White Mythology." Translated by F.C.T. Moore. *New Literary History* 6 (autumn 1974).

———. *Margins of Philosophy.* Translated by Alan Bass. Chicago: The University of Chicago Press, 1982.

———. *The Truth in Painting.* Translated by Geoff Bennington & Ian McLeod. Chicago: University of Chicago Press, 1987.

De Voto, Bernard. *The Course of Empire.* Boston: Houghton Mifflin Co., 1952.

Dewey, John. *Art & Experience.* New York: G.P. Putnam's Sons, 1934, 1958.

Dickinson, Emily. *The Complete Poems of Emily Dickinson.* Edited by Thomas J. Johnson. New York: Little Brown & Co., 1969.

Drewes, Charles D., & Charles R. Fourtner. "Helical Swimming in a Freshwater Oligochaete." *Biological Bulletin* 185 (August 1993).

Duchan, Judith, Gail A. Bruder, & Lynne E. Hewitt, eds. *Deixis in Narrative: A Cognitive Science Perspective.* Hillsdale, N.J.: Lawrence Erlbaum Associates, 1995.

Dunbar, Robin. *Grooming, Gossip, & the Evolution of Language.* Cambridge: Harvard University Press, 1996.

Eagleton, Terry, ed. *The Ideology of the Aesthetic.* Oxford: Basil Blackwell, 1990.

———. *Raymond Williams: Critical Perspectives.* London: Polity Press, 1997.

Eddington, A. S. *The Nature of the Physical World.* Cambridge: Cambridge University Press, 1929.

Edelman, Gerald M. *Bright Air, Brilliant Fire: On the Matter of the Mind.* New York: Basic Books, 1992.

Einstein, Albert. *Out of My Later Years.* Totowa, N.J.: Littlefield, Adams & Co., 1967.

Elaide, Mircea. *A History of Religious Ideas.* Vol 1. Translated by Willard Trask. Chicago: University of Chicago Press, 1978.

Emerson, Ralph Waldo. *Selected Essays.* Edited by Larzer Ziff. New York: Penguin, 1982.

Emerton, Norma E. *The Scientific Reinterpretation of Form.* Ithaca: Cornell University Press, 1984.

Engler, H. Rudolf. *Die Sonne als Symbol: Die Schlussel zu den Mysterien.* Kusnacht-Zurich: Helianthus Verlag, 1962.

Fellows, Jay. *Ruskin's Maze: Mystery & Madness in His Art.* Princeton: Princeton University Press, 1981.

Fenollosa, Ernest. *Epochs of Chinese & Japanese Art: An Outline History of East Asiatic Design.* Rev. ed. New York: Frederick A. Stokes Co., 1921.

———. *The Chinese Written Characters as Medium for Poetry.* Edited by Ezra Pound. San Francisco: City Lights Books, 1936.

Ferguson, William M., & Arthur H. Rohn. *Anasazi Ruins in Color.* Albuquerque: University of New Mexico Press, 1987.

Feyerabend, Paul. *Against Method, Outline of an Anarchistic Theory of Knowledge.* London: Verso, 1975.

Feynman, Richard. *The Character of Physical Law.* Cambridge: MIT Press, 1965.

Fleming, Donald. "Charles Darwin: The Anaesthetic Man." *Victorian Studies* 4 (1961).

Fletcher, Angus. "The Image of Lost Direction." *Center & Labyrinth: Essays in Honor of Northrop Frye.* Edited by Elinore Cook et al. Toronto: University of Toronto Press, 1983.

Flew, Anthony. "Rationality & Unnecessitated Choice." *Naturalism & Rationality.* Edited by Newton Garver & Peter H. Hare. Buffalo: Prometheus Books, 1986.

Florescano, Enrique. *The Myth of Quetzalcoatl.* Translated by Raul Velasquez. Baltimore: Johns Hopkins University Press, 1999.

Fodor, Jerry. *The Language of Thought.* Cambridge: Harvard University Press, 1979.

Fong, Wen C. *Beyond Representation: Chinese Painting & Calligraphy 8th–14th Century.* New York: Metropolitan Museum of Art, 1992.

Freud, Sigmund. *A General Selection of the Works of Sigmund Freud.* Edited by John Rickman. New York: Doubleday Anchor, 1957.

———. *Civilization & Its Discontents.* Translated by James Strachey. New York: W. W. Norton, 1962.

Frost, Robert. *The Poetry of Robert Frost: The Collected Poems, Complete & Unabridged.* Edited by Edward Connery Latham. New York: Holt, Rinehart & Winston, 1969.

Frye, Northrop. *Fearful Symmetry.* Princeton: Princeton University Press, 1947.

———. *Anatomy of Criticism.* Princeton: Princeton University Press, 1957.

Gans, Carl. "Locomotion Without Limbs." *Natural History* (February & March 1966).

Gardner, Martin. *The Ambidextrous Universe: Left, Right, & the Fall of Parity.* New York: Mentor, 1969.

Garver, Newton. *This Complicated Form of Life: Essays on Wittgenstein.* Chicago: Open Court, 1994.

———. "Why There Can Be No Theses in Philosophy." Paper presented at SUNY Buffalo, N.Y.: September, 1995.

Gasché, Rodolphe. *The Tane of the Mirror: Derrida & the Philosophy of Reflection.* Cambridge: Harvard University Press, 1986.

Geertz, Armin W. *Hopi Indian Altar Iconography.* Leiden: E. J. Brill, 1987.

Ghyka, Matila. *The Geometry of Art & Life.* New York: Dover, 1977.

Gibbs, Jr., Raymond W. *The Poetics of Mind: Figurative Thought, Language, & Understanding.* Cambridge: Cambridge University Press, 1994.

Gimbutas, Marija. *The Language of the Goddess: Unearthing the Hidden Symbols of Western Civilization.* San Francisco: Harper Collins, 1989.

Godwin, William. "The Voluntary Actions of Men Originate in Their Opinions." *Political Justice.* Edited by Isaac Kramnick. New York: Penguin, 1973.

von Goethe, Johann Wolfgang. *Italian Journey, 1786–1788.* Translated by W.H. Auden & Elizabeth Mayer. New York: Schocken, 1968.

Gombrich, E.H. *The Sense of Order: A Study in the Psychology of Decorative Art.* Ithaca: Cornell University Press, 1979.

Gould, Stephen Jay. "D'Arcy Thompson & the Science of Form." *New Literary History* 2 (winter 1971).

———. *Ontogeny & Phylogeny.* Cambridge: Harvard University Press, 1977.

Graves, Robert. *The Greek Myths.* Vol 1. Baltimore: Pelican Books, 1955.

Griaule, Marcel. *Conversations with Ogotemmeli.* New York: Oxford University Press, 1965.

Grillner, Sten. "Neural Networks for Vertebrate Locomotion." *Scientific American* (January 1996).

Gross, Harvey. *Sound & Form in Modern Poetry: A Study of Prosody from Thomas Harvey to Robert Lowell.* Ann Arbor: University of Michigan Press, 1964.

Grosz, Elizabeth. "Darwin & Feminism: Preliminary Investigations for a Possible Alliance." *Australian Feminist Studies.* Vol. 14, No. 29(1999).

Gummere, Francis Barton. *The Beginnings of English Poetry.* New York: Macmillan, 1908.

Hacking, Ian. *Representing & Intervening: Introductory Topics in the Philosophy of Natural Science.* New York: Cambridge University Press, 1983.

Hammacher, Werner. "Hermeneutic Ellipses." *Transforming the Hermeneutic Context: From Nietzsche to Nancy.* Edited by Gayle L. Ormiston & Alan D. Schrift. Albany: SUNY Press, 1990.

Harrison, Jane Ellen. *Themis: A Study of the Social Origins of Greek Religion.* Cambridge: Cambridge University Press, 1912.

Hartenstein, Julie, ed. *Charles Burchfield: The Charles Rand Penney Collection.* Baltimore: Smithsonian Traveling Exhibition, 1978.

Hartman, Edwin. *Substance, Body, Soul: Aristotelian Investigations.* Princeton: Princeton University Press, 1977.

Hass, Robert. *Twentieth Century Pleasures: Prose on Poetry.* New York: Ecco Press, 1984.

———, ed. *The Essential Haiku: Versions of Basho, Buson & Issa.* New York: Ecco Press, 1994.

Hawking, Stephen. *A Brief History of Time: From the Big Bang to Black Holes.* New York: Bantam, 1988.

Hayles, Katherine. *The Cosmic Web: Scientific Field Models & Literary Strategies in the Twentieth Century.* Ithaca: Cornell University Press, 1984.

Heaney, Seamus. *Selected Poems 1966–1987.* New York: Noonday Press, 1990.

Heidegger, Martin. *The Question Concerning Technology & Other Essays.* Translated by William Lovitt. New York: Harper Colophon Books, 1977.

———. "Being & Time." *The Hermeneutic Tradition: From Ast to Riceour.* Edited by Gayle L. Ormiston & Alan D. Schrift. Albany: SUNY Press, 1990.

von Helmholtz, Hermann. *On the Sensations of Tone as a Physiological Basis for the Theory of Music.* Translated by Alexander J. Ellis. New York: P. Smith, 1948.

———. *Popular Scientific Lectures.* Translated by H.W. Eve. New York: Dover Publications, 1962.

Hendel, Charles W. Introduction to *The Philosophy of Symbolic Forms* by Ernst Cassirer. Translated by Ralph Manheim. New Haven: Yale University Press, 1953.

Henderson, Harold G. *An Introduction to Haiku.* New York: Doubleday, 1958.

Heninger, Jr., S. K., *Touches of Sweet Harmony: Pythagorean Cosmology & Renaissance Poetics.* San Marino: The Huntington Library, 1974.

———. *The Cosmological Glass: Renaissance Diagrams of the Universe.* San Marino: The Huntington Library, 1977.

Herbert, George. *Works of George Herbert.* Edited by E. F. Hutchinson. Oxford: Clarendon Press, 1941.

Herder, Johann Gottfried. *Two Essays on the Origin of Language: Jean-Jacques Rousseau & Johann Gottfried Herder.* Chicago: University of Chicago Press, 1966.

Hippocrates. *Of Airs, Waters, & Places.* Translated by Francis Adams. The Internet Classics Archive: classics.mit.edu.

Hirsh-Pasek, Kathy, & Roberta Michnick Golinkoff. *The Origins of Grammar: Evidence from Early Childhood Comprehension.* Cambridge: MIT Press, 1996.

Hofmann, Hans. *Search for the Real & Other Essays.* Cambridge: MIT Press, 1948.

Hogarth, William. *The Analysis of Beauty.* Edited by Ronald Paulson. New Haven: Yale University Press, 1997.

Hogben, Lancelot. *Mathmatics for the Million.* New York: W. W. Norton, 1968.

Holmes, Richard. *Coleridge: Darker Reflections, 1804–1834.* New York: Pantheon, 1998.

Homer. *The Odyssey of Homer.* Translated by Richard Lattimore. New York: Perennial Library, 1965.

———. *Odyssey, A New Translation.* Translated by Albert Cook. New York: W.W. Norton, 1967.

Horkheimer, Max, & Theodor W. Adorno. "Odysseus or Myth of Enlighten-ment." *Dialectic of Enlightenment.* Translated by John Cumming. New York: Continuum Press, 1972.

Hoskinson, Tom. "Saguaro Wine, Ground Figures, & Power Mountains: Investigations at Sears Point, Arizona." In *Earth & Sky: Visions of the Cosmos in Native American Folklore.* Edited by Ray Williamson & Claire E. Farrer. Albuquerque: The University of New Mexico Press, 1992.

Huizinga, Johann. *Homo Ludens: A Study of the Play Element in Culture.* Boston: Beacon Press, 1955.

Hume, David. *An Enquiry Concerning Human Understanding.* Edited by Eric Steinberg. Indianapolis: Hackett Publishing Co., 1977.

Innis, Robert. E., ed. *Semiotics: An Introductory Anthology.* Bloomington: Indiana University Press, 1985.

———. *Consciousness & the Play of Signs.* Bloomington: Indiana University Press, 1994.

Jackendoff, Ray. *Patterns in the Mind: Language & Human Nature.* New York: Basic Books, 1994.

———. "Response to Bunn." *Mind & Machines* 10 (February, 2000): 129–35.

Jacobs, Carol. *Telling Time: Lévi-Strauss, Ford, Lessing, Benjamin, deMan, Wordsworth, Rilke.* Baltimore: Johns Hopkins University Press, 1993.

Jakobson, Roman. *Main Trends in the Science of Language.* New York: Harper Torchbooks, 1970.

———. *Roman Jakobson: Selected Writings, Major Works, 1976–1980.* Edited by Stephen Rudy. Berlin: Mouton de Gruyter, 1988.

James, William. *Principles of Psychology.* 2 vols. New York: Henry Holt Co., 1890.

Jay, Martin. *Adorno.* Cambridge: Harvard University Press, 1992.

———. *Force Fields: Between Intellectual History & Cultural Critique.* New York: Routledge, 1993.

Jettmar, Frank. *Art of the Steppes.* New York: Crown, 1964.

Johnson, Mark. *The Body in the Mind: The Bodily Basis of Meaning.* Chicago: University of Chicago Press, 1987.

Jones, Owen. *The Grammar of Ornament.* New York: Van Nostrand Reinhold Co., 1982.

Joralemon, Peter David. "In Search of the Olmec Cosmos: Reconstructing the World View of Mexico's First Civilization." *Olmec Art of Ancient Mexico.* Edited by Elizabeth P. Benson & Beatriz de la Fuente. Washington: National Gallery of Art, 1996.

Jusczyk, Peter W. *The Discovery of Spoken Language.* Cambridge: MIT Press, 1997.

Kant, Immanuel. *Critique of Judgement.* Translated by J. H. Bernard. New York: Haffner Library, 1951.

Kauffman, Stuart. *At Home in the Universe: The Search for the Laws of Self-Organization & Complexity.* New York: Oxford University Press, 1995.

Kaufman, Alice, & Christopher Selser. *The Navajo Weaving Tradition 1650 to Present.* New York: E. P. Dutton, 1985.

Kemp-Smith, Norman. *The Philosophy of David Hume.* London: Macmillan, 1949.

Kenner, Hugh. *The Poetry of Ezra Pound.* London: Faber & Faber, 1951.

———. *The Pound Era.* Berkeley: University of California Press, 1971.

Kirk, G. S., J. E. Raven, & M. Schofield, eds. *The Presocratic Philosophers: A Critical History with a Selection of Texts.* 2d ed. London: Cambridge University Press, 1983.

Koch, Kenneth. *Making Your Own Days: The Pleasures of Reading & Writing Poetry.* New York: Scribner's, 1998.

Kristeva, Julia. *Desire in Language: A Semiotic Approach to Literature & Art.* Edited by Leon S. Roudiez. Translated by Thoas Gora, Alice Jardine, & Leon S. Roudiez. New York: Columbia University Press, 1980.

Lakoff, George. *Women, Fire, & Dangerous Things: What Categories Reveal About the Mind.* Chicago: University of Chicago Press, 1987.

Lakoff, George, & Mark Johnson. *Metaphors We Live By.* Chicago: University of Chicago Press, 1980.

————. *Philosophy in the Flesh: The Embodied Mind & Its Challenge to Western Thought.* New York: Basic Books, 1999.

Land, Stephen K. *From Sign to Proposition: The Concept of Form in Eighteenth-Century Semantic Theory.* London: Longman, 1974.

Langer, Susanne K. *Philosophy in a New Key.* New York: Mentor Books, 1948.

Langley, Samuel Pierpoint. *The New Astronomy.* Boston: Houghton Mifflin Co., 1898.

Lawrence, D. H. *The Plumed Serpent.* New York: Knopf, 1926.

————. *Etruscan Places.* 2d ed. New York: Viking Press, 1993.

Lenneberg, Eric. *Biological Foundations of Language.* New York: John Wiley, 1967.

León-Portilla, Miguel. *Mexico-Tenochtitlán: Su espacio y tiempo sagrados.* Instituto Nacional de Anthropologia y Historia, n.d.

Lessing, Gotthold. *Laocoön: An Essay on the Limits of Painting & Poetry.* Translated by Edward Allen McCormick. Baltimore: Johns Hopkins University Press, 1962.

Levertov, Denise. *Poems 1960–1967.* New York: New Directions, 1983.

Lévi-Strauss, Claude. *Structural Anthropology.* Translated by Claire Jacobson & Brooke Grundfest Schoepf. New York: Anchor Books, 1967.

Leyton, Michael. *Symmetry, Causality, Mind.* Cambridge: MIT Press, 1992.

Locke, John. *Essay Concerning Human Understanding.* Oxford: Oxford University Press, 1975.

Lyons, John. *Introduction to Theoretical Linguistics.* Cambridge: Cambridge University Press, 1968.

————. *Language & Linguistics: An Introduction.* Cambridge: Cambridge University Press, 1981.

Macwhinney, Brian, & Elizabeth Bates, eds. *The Crosslinguistic Study of Sentence Processing.* New York: Cambridge University Press, 1989.

Margoliash, D., E. S. Fortune, M. L. Sutter, A. C. Yu, B. D. Hardin, & W. A. Dave. "Distributed Representation in the Song System of Oscines [songbirds]: Evolutionary Implications & Functional Consequences." *Brain, Behavior, & Evolution* 44 (1994).

Marx, Karl. *Capital: A Critique of Political Economy.* Translated by Ernest Untermann. New York: Modern Library, 1936.

Matthews, Washington. *Navajo Legends.* Boston: American Folklore Society, 1897.

Mazzaro, Jerome. "Reconstruction in Art." *A. R. Ammons: Modern Critical Views.* Edited by Harold Bloom. New York: Chelsea House, 1986.

Merrell, Floyd. *Simplicity & Complexity: Pondering Literature, Science & Painting.* Ann Arbor: University of Michigan Press, 1998.

Metropolitan Museum of Art. *From the Land of the Scythians, Ancient Treasures of the U.S.S.R. 3000 B.C.–300 B.C.* New York: Metropolitan Museum of Art, 1975.

Meyer, L. B. *Music, the Arts, Ideas: Patterns & Predictions in Twentieth-Century Culture.* Chicago: University of Chicago Press, 1967.

Mitchell, W. J. T. *Blake's Composite Art: A Study of the Illuminated Poetry.* Princeton: Princeton University Press, 1978.

———. "Metamorphoses of the Vortex: Hogarth, Blake, Turner." *Articulate Images: The Sister Arts from Hogarth to Tennyson.* Edited by Richard Wendorf. Minneapolis: University of Minnesota Press, 1983.

———. *Picture Theory: Essays on Verbal Theory & Visual Representation.* Chicago: University of Chicago Press, 1994.

Monod, Jacques. *Chance & Necessity: An Essay on the Natural Philosophy of Modern Biology.* Translated by Austryn Wainhouse. New York: Alfred Knopf, 1971.

Moran, John H., & Alexander Gode, trans. *Two Essays On the Origin of Language: Jean-Jacques Rousseau & Johann Gottfried Herder.* Chicago: University of Chicago Press, 1966.

Morris, Charles W. *Foundations of the Theory of Signs.* Chicago: University of Chicago Press, 1938.

Muir, John. *The Mountains of California.* New York: Barnes & Noble Books, 1993.

Mundkur, Balaji. *Cult of the Serpent: An Interdisciplinary Survey of Its Manifestations & Origins.* Albany: SUNY Press, 1983.

Napier, A. David. *Masks: Transformation & Paradox.* Berkeley: University of California Press, 1986.

Needham, Joseph. *Physics & Physical Terminology.* Vol. 2 of *Science & Civilization.* Cambridge: Cambridge University Press, 1962.

Neville, Robert Cummings. *Recovery of the Measure, Interpretation & Nature.* Albany: SUNY Press, 1989.

Newman, James R. *The World of Mathematics.* Vol. 1. New York: Simon & Schuster, 1956.

Nussbaum, Allen. *Geometric Optics: An Introduction.* Reading, Mass.: Addison-Wesley, 1968.

Olson, Charles. *Selected Writings.* Edited by Robert Creeley. New York: New Directions, 1966.

Owen, Stephen. *Traditional Chinese Poetry & Poetics: Omen of the World.* Madison: University of Wisconsin Press, 1985.

Panofsky, Erwin. *Perspective as Symbolic Form.* Translated by Christopher S. Wood. New York: Zone Works, 1991.

Paz, Octavio. *El Arco y la Lyra (The Bow & the Lyre: The Poem, the Poetic Revelation, Poetry & History).* Translated by Ruth L. C. Simms. Austin: University of Texas Press, 1973.

Pearson, Karl. *The Grammar of Science.* 1892. Reprint, London: Everyman, 1937.

Peirce, C.S. "Pragmatism in Retrospect: A Last Formulation." In *Philosophical Writings of Peirce.* Edited by Justus Buchler. New York: Dover, 1955.

Pepper, Stephen C. *World Hypotheses: A Study in Evidence.* Berkeley: University of California Press, 1942.

Perloff, Marjorie. *Wittgenstein's Ladder: Poetic Language & the Strangeness of the Ordinary.* Chicago: University of Chicago Press, 1996.

Piaget, Jean. *Structuralism.* Translated by Chaninah Maschler. New York: Basic Books, 1970.

Picasso, Pablo. *Picasso on Art: A Selection of Views.* Edited by Dore Ashton. New York: Viking, 1972.

Pike, Kenneth. *Kenneth L. Pike: Selected Writings.* Edited by Ruth M. Brand. The Hague: Mouton, 1972.

Pinker, Steven. *The Language Instinct: How the Mind Creates Language.* New York: Harper Perennial, 1994.

———. *How the Mind Works.* New York: W. W. Norton, 1997.

Plato. *The Collected Dialogues of Plato.* Edited by Edith Hamilton & Huntington Cairns. New York: Princeton University Press, 1963.

Plotnitsky, Arcady. *Complementarity: Anti-epistemology after Bohr & Derrida.* Durham: Duke University Press, 1994.

Pound, Ezra. *Antheil & the Treatise on Harmony.* Chicago: University of Chicago Press, 1927.

Prigogine, Ilya, & Isabell Stengers. *Order Out of Chaos: Man's New Dialogue with Nature.* New York: Bantam, 1984.

Prout, James H., & Gordon R. Bienvenue. *Acoustics for You.* Malabar, Fla.: Robert E. Krieger Co., 1990.

Pucci, Pietro. "The Poem of the Odyssey." *Arethusa* 15 (1982).

Rapaport, William J. "Syntatic Semantics: Foundations of Computational Natural Language Understanding." *Thinking Computers & Virtual Persons.* Edited by Eric Dietrich. San Diego: Academic Press, 1994.

Read, Herbert. *Phases of English Poetry.* London: L. & V. Woolf at the Hogarth Press, 1928.

———. *English Prose Rhythm.* Boston: Beacon Press, 1952.

———. *The Philosophy of Modern Art.* New York: Horizon Press, 1953.

———. "The Creative Process." *The Forms of Things Unknown: Essays Towards an Aesthetic Philosophy.* New York: Horizon Press, 1960.

———. *The Contrary Experience: Autobiographics.* New York: Horizon Press, 1963.

Ricoeur, Paul. *The Rule of Metaphor: Multidisciplinary Studies of the Creation of Meaning in Language.* Translated by Robert Czerny. Toronto: University of Toronto Press, 1977.

———. *Lectures on Ideology & Utopia.* Edited by George H. Taylor. New York: Columbia University Press, 1986.

————. "Hermeneutics & the Critique of Ideology." Collected in *The Hermeneutic Tradition: From Ast to Ricoeur*. Edited by Gayle L. Ormiston & Alan D. Schrift. Albany: SUNY Press, 1990.

Rodee, Marian. *Southwestern Weaving*. Albuquerque: University of New Mexico Press, 1977.

Rogers, Franklin R. *Painting & Poetry: Form, Metaphor, & the Language of Literature*. Lewisburg: Bucknell University Press, 1985.

Rorty, Richard. *Philosophy & the Mirror of Nature*. Princeton: Princeton University Press, 1979.

Rousseau, Jean-Jacques. *Two Essays on the Origin of Language: Jean-Jacques Rousseau & Johann Gottfried Herder*. Translated by John H. Moran & Alexander Gode. Chicago: University of Chicago Press, 1966.

Rowley, George. *Principles of Chinese Paintings*. 2d ed. Princeton: Princeton University Press, 1959.

Ruskin, John. *The Elements of Drawing in Three Letters to Beginners*. London: Smith, Elder & Co., 1859.

————. *The Queen of the Air: The Complete Works of John Ruskin*. Edited by E. T. Cook & Alexander Wedderburn. London: George Allen, 1907.

Russell, Bertrand. *Philosophy*. New York: W. W. Norton & Company, 1927.

————. *ABC of Relativity*. Rev. ed. New York: Signet, 1958.

Sachs, Curt. *The Wellsprings of Music*. Edited by Jaap Kunst. The Hague: M. Nijhoff, 1962.

Saintsbury, George. *Historical Manual of English Prosody*. London: Macmillan & Co., 1912.

————. *A History of English Prose Rhythm*. London: Macmillan & Co., 1922.

Sambursky, S. *Physics of the Stoics*. London: Routledge & Kegan Paul, 1958.

de Saussure, Ferdinand. *Course in General Linguistics*. Translated by Wade Baskin. New York: McGraw-Hill, 1966.

Schaafsma, Polly. *Indian Rock Art of the Southwest*. Albuquerque: University of New Mexico Press, 1980.

Schneider, Steven B. "The Poet & the Scientist." *A. R. Ammons & the Poetics of Widening Space*. Rutherford, N.J.: Fairleigh Dickinson University Press, 1994.

Schuster, Carl. *Materials for the Study of Social Symbolism in Ancient & Tribal Art: A Record of Tradition & Continuity, Based on the Researches of Carl Schuster*. 3 vols. Edited by Edmund Carpenter. New York: The Rock Foundation, 1986–1988.

Schusterman, Richard. *Pragmatist Aesthetics: Living, Beauty, Rethinking Art*. Oxford: Blackwell, 1992.

Séjourné, Laurette. *Burning Water: Thought & Religion in Ancient Mexico*. New York: Grove Press, 1960.

————. *El universo de Quetzalcoatl*. Mexico City: Fundo de Cultura Económica, 1962.

Semper, Gottfried. *Der Stil in den technischen und tektonischen Kunsten.* Berlin: 1878–1879.

Serres, Michel. "Lucretius: Science & Religion." *Hermes: Literature. Science. Philosophy.* Edited by Josue V. Harari & David Bell. Baltimore: Johns Hopkins University Press, 1982.

———. "The Natural Contract." *Critical Inquiry* 19 (autumn 1992).

———. *The Natural Contract.* Translated by Elizabeth MacArthur & William Paulson. Ann Arbor: University of Michigan Press, 1995.

Shakespeare, William. *William Shakespeare: The Complete Works.* Baltimore: Penguin Books, 1969.

Shell, Marc. *Money, Language & Thought: Literary & Philosophic Economies from the Medieval to the Modern Era.* Berkeley: University of California Press, 1982.

Shelley, Percy Bysshe. *The Complete Works of Percy Bysshe Shelley.* Edited by Roger Ingpen & Walter E. Peck. New York: Gordian Press, 1965.

Shubnikov, A. V. & V. A. Kopsik. *Symmetry in Science & Art.* Translated by G. D. Archard. New York: Plenum, 1974.

Silverman, Kaja. *The Subject of Semiotics.* New York: Oxford University Press, 1983.

Smith, Barbara Herrnstein. "Belief & Resistance: A Symmetrical Account." *Critical Inquiry* 18 (autumn 1991).

Smith, Bradley. *Mexico: A History in Art.* New York: Harper & Row, 1968.

Smith, Cyril Stanley. *The Search for Structure: Selected Essays on Science, Art & History.* Cambridge: MIT Press, 1981.

Snyder, Gary. *Riprap & Cold Mountain Poems.* San Francisco: Four Seasons Foundation, 1969.

Snyder, Jim. "Riprap & the Old Ways: Gary Snyder in Yosemite, 1955." *Gary Snyder: Dimensions of a Life.* Edited by Jon Halper. San Francisco: Sierra Club Books, 1991.

Sobel, Michael L. *Light.* Chicago: University of Chicago Press, 1987.

Spencer, Herbert. "The Origin & Function of Music." In *Essays: Scientific, Political, & Speculative.* London: Williams & Norgate, 1868.

Spivak, Gayatri Chakravorty. Preface to *Of Grammatology* by Jacques Derrida. Baltimore: Johns Hopkins University Press, 1976.

Sproul, Barbara C. *Primal Myths: Creating the World.* San Francisco: Harper & Row, 1979.

Steiner, Wendy. *The Colors of Rhetoric: Problems in the Relation Between Literature & Culture.* Chicago: University of Chicago Press, 1982.

Stevens, Anthony. *Archetype: A Natural History of the Self.* London: Routledge & Kegan Paul, 1982.

Stevens, Peter S. *Patterns in Nature.* Boston: Atlantic Monthly Press, 1974.

Stevenson, Robert Louis. *Essays by Robert Louis Stevenson.* Edited by William Lyon Phelps. New York: Charles Scribner's Sons, 1918.

Stewart, Ian, & Martine Golubitsky. *Fearful Symmetry: Is God a Geometer?* New York: Penguin, 1992.

Stewart, Susan. *Nonsense: Aspects in Folklore & Literature.* Baltimore: Johns Hopkins University Press, 1978.

———. "Letter on Sound." In *Close Listening,* edited by Charles Bernstein. New York: Oxford University Press, 1998.

Stout, Rex. *The Doorbell Rang: A Nero Wolfe Novel.* New York: Viking, 1965.

Strawson P. F. *Subject & Predicate in Logic & Grammar.* London: Methuen, 1974.

Sussman, Henry. *The Hegelian Aftermath: Readings in Hegel, Kierkegaard, Freud, Proust, & James.* Baltimore: Johns Hopkins University Press, 1982.

Sze, Mai-Mai. *The Tao of Painting: A Study of the Ritual Disposition of Chinese Painting.* Vol 1. New York: Bollingen Series, 1956.

———. *The Mustard Seed Garden Manual of Painting.* Vol. 2. New York: Bollingen Series, 1956.

Tedlock, Barbara. "The Beautiful & the Dangerous: Zuni Ritual & Cosmology as an Aesthetic System." *Conjunctions: Bi-annual Volumes of New Writing* 6 (1984).

Tedlock, Dennis. *Popol Vuh: The Definitive Edition of the Mayan Book of the Dawn of Life & the Glories of Gods & Kings.* New York: Simon & Schuster, 1985.

Thistlewood, David. *Herbert Read: Formlessness & Form, an Introduction to His Aesthetics.* London: Routledge & Kegan Paul, 1984.

Thom, René. *Structural Stability & Morphogenesis: An Outline of a General Theory of Models.* Translated by D.H. Fowler. Reading, Mass.: Benjamin/Cummings, 1975.

Thompson, D'Arcy. *On Growth & Form.* Cambridge: Cambridge University Press, 1942.

Todorov, Tzvetan. *Theories of the Symbol.* Translated by Catherine Porter. Ithaca: Cornell University Press, 1982.

Toynbee, Arnold. *A Study of History: A New Edition.* New York: Weathervane Books, 1972.

Trefil, James S. *The Unexpected Vista: A Physicist's View of Nature.* New York: Scribner's, 1983.

Trefil, James S., & Robert M. Hazen. *The Sciences: An Integrated Approach.* New York: John Wiley, 1995.

Turner, Mark. *The Literary Mind.* New York: Oxford University Press, 1996.

Turner, Victor. *The Ritual Process: Structure & Anti-Structure.* New York: Aldine, 1965.

———. *Between Literature & Anthropology: Victor Turner & Cultural Criticism.* Bloomington: Indiana University Press, 1990.

Tzu, Lao. *Tao Te Ching.* Translated by D.C. Lau. Baltimore: Penguin Books, 1963.

Ulam, Stanislaw. *Adventures of a Mathematician.* New York: Scribner's, 1976.

Ulmer, Gregory. *Applied Grammatology: Post(e)-Pedagogy from Jacques Derrida to Joseph Beuys.* Baltimore: Johns Hopkins University Press, 1985.

Verstokt, Mark. *The Genesis of Form: From Chaos to Geometry.* London: Muller, Blond & White, 1987.

Vietor, Karl. *Goethe the Thinker.* Cambridge: Harvard University Press, 1950.

Vitruvius. *Vitruvius on Architecture.* Translated by Frank Granger. 2 vols. New York: G. P. Putnam's Sons, 1931.

———. *The Ten Books on Architecture.* Translated by Morris Hicky Morgan. New York: Dover, 1960.

Vogel, Steven. *Life in Moving Fluids: The Physical Flow of Biology.* 2d ed. Princeton: Princeton University Press, 1994.

Volk, Tyler. *Metapatterns Across Space, Time & Mind.* New York: Columbia University Press, 1995.

Volosinov, V.N. *Marxism & the Philosophy of Language.* Translated by Ladislav Matejka & I.R. Titunik. Cambridge: Harvard University Press, 1986.

Waddington, C.H. *Module, Proportion, Rhythm, Symmetry.* Translated by Gyorgy Kepes. New York: George Braziller Inc., 1966.

———. *Behind Appearance: A Study of the Relations between Painting & the Natural Sciences in this Century.* Cambridge: MIT Press, 1970.

Waley, Arthur. *The Way & Its Power: A Study of the Tao Te Ching & Its Place in Chinese Thought.* London: Allen & Unwin, 1934, 1955.

Wasserstrom, William. *The Ironies of Progress: Henry Adams & the American Dream.* Carbondale: Southern Illinois Press, 1984.

Watson, William. *Style in the Arts of China.* London: Penguin, 1974.

Watts, Alan. *Tao: The Watercourse Way.* London: Jonathan Cape, 1975.

Weil, Simone. *The Iliad or the Poem of Force.* Translated by Mary McCarthy & Dwight Macdonald. Wallingford, Penn.: Pendle Hill, 1956.

Weltfish, Gene. *The Origins of Art.* Indianapolis: Bobbs-Merrill, 1952.

Westerman, Claus. *Genesis 1–11: A Commentary.* Translated by John J. Scullion. Minneapolis: Augsberg Publishing House, 1984.

Weyl, Hermann. *Symmetry.* Princeton: Princeton University Press, 1952.

Wheelwright, Philip. *The Burning Fountain: A Study in the Languages of Symbolism.* Bloomington: Indiana University Press, 1954.

White, Hayden. *Metahistory: The Historical Imagination in Nineteenth-Century Europe.* Baltimore: Johns Hopkins University Press, 1973.

———. *Tropics of Discourse: Essays in Cultural Criticism.* Baltimore: Johns Hopkins University Press, 1978.

Whitehead, Alfred North. *Essays in Science & Philosophy.* London: Rider, 1948.

———. *Process & Reality: An Essay in Cosmology.* New York: Free Press, 1969.

Whitman, Walt. *Walt Whitman.* Edited by Floyd Stovall. New York: American Book Company, 1939.

Whorf, Benjamin Lee. *Language, Thought, Reality: Selected Writings of Benjamin Lee Whorf.* Edited by John B. Carroll. Cambridge: MIT Press, 1956.

Whyte, Lancelot Law. "Towards a Science of Form." *Hudson Review* 23 (winter 1970–71).

Wild, John. *Plato's Modern Enemies & the Theory of Natural Law.* Chicago: University of Chicago Press, 1953.

Wilden, Anthony. *Man & Woman, War & Peace: The Strategist's Companion.* London: Routledge & Kegan Paul, 1987.

———. *The Rules Are No Game: The Strategy of Communication.* London: Routledge & Kegan Paul, 1987.

Williams, Adriana. *Covarrubias.* Edited by Doris Ober. Austin: University of Texas Press, 1994.

Williams, William Carlos. *Selected Poems.* Edited by Randall Jarrell. New York: New Directions, 1963.

Williamson, Ray A. *Living the Sky: The Cosmos of the American Indian.* Boston: Houghton Mifflin, 1984.

Winfree, Arthur T. *When Time Breaks Down: The Three-Dimensional Dynamics of Electrochemical Waves & Cardiac Arrhythmias.* Princeton: Princeton University Press, 1987.

Wittgenstein, Ludwig. *Philosophical Investigations.* Translated by G.E.M. Anscombe. Oxford: Basil Blackwell, 1967.

Wolkstein, Diane, & Samuel Noah Kramer. *Inanna: Queen of Heaven & Earth: Her Stories & Hymns from Sumer.* New York: Harper & Row, 1983.

Wong, Willie, & Horace Barlow. "Pattern Recognition: Tunes & Templates." *Nature* Vol. 404, No. 6781 (April 27, 2000).

Young, J. Z. *An Introduction to the Study of Man.* New York: Oxford University Press, 1971.

Young, M. Jane. "Morning Star, Evening Star: Zuni Traditional Stories." *Earth & Sky Visions of the Cosmos in Native American Folklore.* Edited by Ray Williamson & Claire E. Farrer. Albuquerque: University of New Mexico Press, 1992.

Yutang, Lin. *My Country & My People.* New York: Reynal & Hitchcock, 1935.

———. *The Importance of Living.* New York: The John Day Co., 1937.

Zee, A. *Fearful Symmetry: The Search for Beauty in Modern Physics.* New York: Macmillan, 1986.

Zehou, Li. *The Path of Beauty—A Study of Chinese Aesthetics.* Translated by Gong Lizeng. Beijing: Morning Glory Publishers, 1988.

Zimmer, Heinrich. *Myths & Symbols in Indian Art & Civilization.* Princeton: Princeton University Press, 1946.

Zitkala-Sa. "The School Days of an Indian Girl." In *American Indian Stories.* Lincoln: University of Nebraska Press, 1921.

INDEX

(Italic page numbers indicate an illustration.)